With love

SATELLITE IMAGING INSTRUMENTS
Principles, Technologies and Operational Systems

SATELLITE IMAGING INSTRUMENTS
Principles, Technologies and Operational Systems

C. B. PEASE
Senior Scientist
Royal Aerospace Establishment, Farnborough, UK

ELLIS HORWOOD
NEW YORK   LONDON   TORONTO   SYDNEY   TOKYO   SINGAPORE

First published in 1991
Reprinted in 1994 by
ELLIS HORWOOD LIMITED
Market Cross House, Cooper Street,
Chichester, West Sussex, PO19 1EB, England

A division of
Simon & Schuster International Group
A Paramount Communications Company

Typeset in Times by Ellis Horwood Limited
Printed and bound in Great Britain
by Bookcraft (Bath) Limited, Midsomer Norton, Avon

British Library Cataloguing in Publication Data

Pease, C. B.
Satellite imaging instruments.
1. Artificial satellites. Optical instruments.
I. Title.
629.46
ISBN 0–13–638487–0

Library of Congress Cataloging-in-Publication Data

Pease, C. B. (Bev)
Satellite imaging instruments: principles, technologies, and operational systems /
C. B. Pease.
p. cm. — (Ellis Horwood library of space science and space technology. Series in space technology)
Includes bibliographical references and index.ISBN 0–13–638487–0
1. Scientific satellites — Equipment and supplies. 2. Imaging systems. I. Title. II. Series.
TL798.S3P43   1990
621.36′7–dc20   90–22553
                                                             CIP

Table of contents

Preface 7

Introduction 11

Part I

1 **Orbits** 17

2 **The atmosphere** 30
Consultant: Dr M. D. Steven, University of Nottingham

3 **Telescopes** 41
Consultant: Mr A. Péraldi, Matra

4 **Optical filters** 54

5 **Detectors** 70
Consultant: Mr D. Sloggett, Software Sciences Ltd

6 **The detector electronics** 99
Consultant: Mr S. Braithwaite, University of Southampton

7 **Data quality** 111

Part II

8 **Multispectral Scanner** 125

9 **GOES-E and -W** 134
Consultant: F. Malinowski, SBRC (Hughes A/C Co)

10 **GMS-4** 147

11 **Meteosat** 153
Consultant: Mr A. Péraldi, Matra

12 **AVHRR** 175
Vetted by: staff of ITT, Fort Wayne

13 **The Indian INSAT series** 201
Consultant: Dr S. V. Kibe, D.D., INSAT Project Office

14 **Thematic Mapper (US)** 216
Consultant: Dr Robert Murphy, NASA

15 **IRAS: the Infra-red Astronomical Satellite** 248
Consultant: Dr Helen Walker, RAL, UK

16 **SPOT: the French Commercial Earth-resources satellite (France)** 275
Consultant: Mr A. Péraldi, Matra

17 **MOS-1: the Japanese marine satellite** 299

18 **IRS-1: the Indian Earth-resources satellite** 311
Consultant: Dr George Joseph, ISRO, India

Index 330

Preface

The parents of the Learned Child, his father and his mother,
Were utterly aghast to note the facts he would at random quote,
On sensors curious, rare and wild and, wondering, asked each other,

'An idle little child like this, how is it that he knows
'What years of close analysis are powerless to disclose?
'Our brains are trained, our books are big — and yet we always fail
'To answer why the French SPOT rig was born without a tail†;
'Or why the S.A.R. should cuss in wild unmeaning rhymes,
'Whereas the Indian I.R.uS. will mainly scan the Times'‡

'Perhaps he found a chance to slip, unnoticed, o'er the Sea,
'And gave the NASA boss a tip, or did a Ph.Dee.'
'Oh No' said he in humble tones, with shy but conscious look,
'Such facts I never could have known, but for this little book'.

> With apologies to Hilaire Belloc
> *More Beasts for Worse Children*,

(although the L.C. will scour this particular book in vain for any information on Synthetic Aperture Radar).

I once needed to know what 'momentum thickness' meant. Two phone calls got me to the one man in the company who knew the most about aerodynamics, and he spent a happy half-hour telling me all about boundary-layer theory, laminar and turbulent flow, and why the wind tends to die down at night.

On another occasion, I wanted to know what was so special about an operational amplifier. I quickly found another expert who regaled me with a potted account of control theory, including leads, lags and dead time; also why a politician's cure for a

---

† a mid-I.R. band.
‡ India.

recession inevitably sows the seeds of the next (it has to do with the oscillations that are generated by non-linear feedback loops). By such an informal process, I quickly learned everything that the educated aero-engineer needs to know to communicate with his fellow aero-engineers. By a similar process, I later learned about decibels, quadrupoles, critical bands, and the other tools of the acoustician's trade. Likewise with several other specialities into which I have strayed at various times.

So when I moved into the Space fraternity, I anticipated no difficulty in finding out what a Sun-synchronous orbit is, what IFOV means, or why the Ritchey Chretien telescope is so magic. The shock, when I discovered that none of the sources that I consulted seemed to know, was considerable.

This book is the result.

It is best regarded as a first stab at a compendium of what the well-bred Space buff should know about specialities other than his own. However, as befits a first stab, it covers a strictly limited range of technologies, namely those that are conspicuous in the working of optical imaging remote sensing instruments such as Thematic Mapper and SPOT. Inevitably, it is also selective and incomplete. For want of a better yardstick, I have covered mainly the things that intrigued me. I felt that if I wanted to know then there was a chance that other tyros might too.

The book represents an exercise in communication, rather than in scholarship or academic excellence. It is written for the non-specialist — or the specialist whose field is other than that currently under discussion. Thus the sections on orbital dynamics are intended to be intelligible to the electronics engineer, while the electronics sections are intended to be moderately intelligible to the applications practitioner. Whether the orbital dynamicist will find anything of interest to him in our limited coverage of applications remains to be seen. Hopefully, however, the managers and administrators, whose efforts are so essential to the progress of our activities, will feel that effort devoted to grappling with the book was at least moderately well-spent.

Although every effort has been made to make each chapter authoritative within this limitation, the attempt has certainly been less than 100% successful. I shall of course be delighted to acknowledge constructive criticism, new material or suggestions for improvements. I cherish the hope that a much expanded second edition might eventually materialize, in which many of the holes may be filled.

Finally, an excuse is required for the inclusion of the IRAS telescope in a book otherwise devoted to terrestrial instruments. The reason is simple. A complete report, covering virtually everything I wanted to know, fell into my lap just as I was despairing of ever getting together the material I needed. Its inclusion is therefore a thank-you to the innumerable authors of the IRAS explanatory supplement.

THE CONSULTANTS

It will be clear from the Table of contents how heavily this book is indebted to the efforts put in by a large number of expert consultants. Many more specialists have applied their red pencils to draft chapters, and improved them substantially in the process. The references 'correction in draft' do no more than hint at the magnitude of the task that many of the consultants took on. It is true to say that the concept of the book is viable only to the extent that it has proved possible to attract constructive input from such authoritative sources.

Nevertheless, it is important to point out that responsibility for the scope of the book, for the presentation and for the lucidity or otherwise of the text is entirely mine. Mine also is the responsibility for the errors and omissions. In particular, each consultant has generally only seen the chapter(s) which bears his or her name, and even there, it has not always been possible to avoid amending a chapter after its final vetting.

I record here my gratitude, that so many eminent and busy people took so much time to help an unknown author — to produce a book whose value to the community they could only guess.

ARMAND PÉRALDI

Armand Péraldi was Scientific and Technical Director of Matra until his tragic death in a road accident in the autumn of 1989. He was a communicator par excellence. The book owes a particularly great debt to the pages of red ink that he added, in clear, readable and almost impeccable English, to all the chapters in which he was involved. I would like to dedicate the book, in part, to the memory of the friend and colleague that I never met.

Introduction

Remote sensing means, strictly, measurement (in its broadest sense) from a distance. We are all familiar with it. Our eyes, ears and nose are remote sensing instruments. (The palm of the hand can also be used as a remote sensing instrument, with particular sensitivity to what will later be defined as the 'thermal band'.) Students of these instruments have measured their outputs, both raw and after processing by the brain. Their sensitivities and dynamic ranges are incredible. A good quality ear, in prime condition, can almost pick out the Brownian motion of individual air molecules. Its dynamic range is variously described as 120 to 140 dB, or (in energy terms) times 10^{12} to 10^{14}. Likewise the eye, under optimal conditions, has been shown to stand a reasonable chance of detecting a single photon of light. The author has yet to uncover a typical 'saturation level' for the instrument but its dynamic range is clearly of a similar order, or possibly even greater.

Nevertheless, as instruments, they are both moderately dreadful. They both produce a substantial signal ('dark current') even in the complete absence of stimulation and, when stimulation is present, the true signal is always accompanied by large amounts of spurious nerve firings (noise). It is the task of the brain to separate the wheat from the chaff, and to generate a true and reliable 'image' of the world on the basis of which it can take, possibly life-preserving, action. How it does it is a mystery at present. Nevertheless it shows the way the remote sensing industry must go in the long term. There is almost certainly far more information available in multi-band and multi-temporal imagery than at present ever comes out of it. Much greater use must be made of intelligent interpretation, in particular incorporating prior knowledge obtained from other sources. Fortunately the science of artificial intelligence is developing fast.

One aspect of the brain's behaviour that we might perhaps avoid copying, however, is its tendency to guess. It is sometimes difficult to realize that the crisp image of the world that we see before us is not an instantaneous snapshot as taken by a high quality camera. It is an image of the world, stored in memory, and updated with a very limited amount of just-received information. The quality and the accuracy of the image that we perceive is crucially dependent on the brain's interpretive powers. It clearly has a built-in requirement to come up with an instant and clear cut answer — a crisp image of the world before it — even when it lacks the information to be sure of getting it right. Hence the prevalence of optical illusions. No doubt this behaviour has survival value in the wild; but we must be sure to copy only those aspect's of the brain's methods that suit any particular application.

However, much of that is likely still to be in the future even at the time of reading. At the time of writing, the demand is for noise-free signals and clean imagery which can be interpreted directly with little additional processing. This means that large numbers of photons must be collected, so that their impacts may be averaged out to produce a single reading. By this means it is ensured that each instantaneous reading is a reliable estimate in its own right and, unless the intensity of the source is changing, it should be exactly the same as the next instantaneous reading. This simplifies the processing and interpretation, but imposes heavy burdens on the design of the instrument; which must be made much larger, heavier, more complex and more delicate as a result.

In order to render the creation of this book practicable, its sphere of interest has been strictly confined to a limited class of imaging instruments, as flown on civil-funded satellites. Their function (with one exception) is to observe the ground†. beneath them, and to send back a stream of digital data, which may then be reconstructed into a quasi-natural photograph-like picture. In general, imagery is collected in several narrow wavebands, in the visible and near infra-red (the **reflective** bands), which give different information about the nature of the terrain, and which may be combined into false colour images of varying degrees of naturalism. Many instruments incorporate **thermal-band** sensors, able to estimate the temperature of the ground much in the manner of the outstretched palm. There are two important differences, however. First, the sensors are highly directional, which the palm of the hand is not. So they are able to produce thermal imagery nearly as detailed as the reflective imagery already mentioned. Second, the thermal sensors are highly sensitive, and are able to measure temperatures down to some −100°C. An exception to many of these points is the IRAS instrument, which looks upwards rather than down; which senses radiant energy only, and down to extremely low temperatures (some 30 K); and which is used mainly to produce a map of points rather than imagery. Apart from this IRAS is very similar to the other instruments described.

Instruments identical in principle to those described here are often flown in aircraft; however, the technical and engineering problems are quite different. A satellite environment is totally alien to the normal laboratory environment in which instruments are normally developed; it is beyond the reach of any kind of servicing, replenishment or maintenance; its distance from the ground makes extreme demands on the resolution of the instrument — but as a platform it is highly stable and repeatable. Once the information reaches the ground then it is very easy to use. By contrast, an aircraft is friendly in almost every way, except that it is subject to the vagaries of winds and turbulence. Depending on the application, much work may need to be done to correct for the variations in pointing angle or departures from the correct line as the flight proceeds. While aircraft remotely sensed data are used a

† For current purposes, we take the term to include water and atmospheric manifestations.

great deal, their acquisition tends to be a private affair and for a specific purpose. Only satellite data are widely disseminated on a commercial basis.

As we have already discussed, electromagnetic remote sensing began with the development of the eye, which acquired a sensitivity to the waveband most suitable to the application, the 'visible' waveband. This waveband is suitable because the atmosphere is almost totally transparent to it, and also because the excitation levels happen to be greatest in this region. The Sun is effectively a 'black body' radiating at about 6000 K (Fig. 2.1). Whether it is just a happy accident that the atmosphere is transparent in the waveband of maximum illumination, or whether there is some unsuspected feedback mechanism tending to favour that situation, is an interesting speculation. Not surprisingly perhaps, early remote sensing instruments also concentrated on this waveband. However, it has long been known that active vegetation shows up much more strongly in the near infra-red than does vegetation that is moribund. Branches cut for camouflage, while totally indistinguishable by the naked eye, stand out clearly on infra-red photographs. The reason for this is that chlorophyll reflects strongly in the infra-red — but only while it is actively engaged in photosynthesis. All the instruments covered in this book, therefore, extend their activities into the near infra-red. This has enabled the agricultural community to develop quantitative indexes for estimating the vigour of crops, based on their relative reflectance in the infra-red and the visible bands. Geologists and others have found that different rocks reflect different colours to a different extent, although the effect is much less marked than with chlorophyll. The nature of the underlying rock can also affect the reflectance of the vegetation above it, although this effect is much more difficult to detect. Thus we have a requirement to measure the relative radiance of the ground in a number of wavebands in the visible region, and extending down into the infra-red. There is relatively little interest in the ultraviolet region, both because the atmosphere absorbs strongly in this region, and also because the Sun's radiation falls off strongly at wavelengths shorter than about 0.5 μm. To maximize the sensitivity of the tool, we should have as many different wavebands as possible. However, as will emerge in due course, narrowing the wavebands beyond a certain limit introduces severe practical problems. It will also emerge that, during the period in history that we cover, there was fairly good agreement as to where the optimum compromise lay. Technology is advancing quickly, however, and the new breed of imaging spectrometers will introduce sweeping changes in this area. Whether they will also completely swamp the ground sector with unusable data remains to be seen. It is possibly in this area where advanced automatic processing/interpretation techniques will first find a market.

A totally different area of interest has also been explored in this era of satellite-borne remote sensing, namely that of measuring temperatures and energy fluxes. As Fig. 2.1 shows, the wavelengths emitted by a body at around 300 K is some ten times longer than those emitted by a body at 6000 K. The energies radiated by objects at ground temperature are also a million times weaker than the sunlight that they reflect in the visible band (it being remembered of course that, under cloudy conditions, the satellite cannot see them). On two counts therefore (see Chapter 5) the task of addressing the thermal band is relatively difficult, and not all the major instruments attempt it. Instruments that attempt to cover both bands, as does for example Thematic Mapper, face further problems because of the wide range of frequencies that the telescope must encompass (Chapter 3). For such thermal measurements, it is necessary to select wavebands in which terrestrial objects radiate as strongly as possible, but also to which the atmosphere is transparent. As Fig. 2.2 shows, the region around 10–12 μm is good for this purpose. Another variation between different rock types is their thermal inertia, which leads to differences in day/night temperature change. An instrument that can take thermal band readings both by day and by night is therefore useful to geologists.

Part I

1

Orbits

One of the most important design decisions to be made during the development of an Earth observation satellite system is the choice of orbit that is to be used. Upon this choice hang many other choices, such as the spatial resolution of the imagery, the **swath width** (the width of the strip on the Earth that the instrument observes as it passes overhead), the downlink data rates, the requirements for attitude and position control, and even the value of the data to the end user. These choices then affect the detailed design of the telescope, the focal plane and aspects of detector design and technology — to say nothing of the size and weight of the spacecraft itself. These in turn may have implications for the stability and temperature control requirements. Inevitably, not all these design demands will be realizable, and so the proposed orbit must be modified, and re-modified, until the optimum compromise is achieved. Thus although this chapter may convey the impression that orbit selection is relatively easy, in practice it is anything but.

Two main classes of orbit are currently in use for civil Earth observation satellite systems: the high-altitude geosynchronous orbit, and a range of low-altitude near-polar orbits in the region of 600–1200 km up.

1.1 THE GEOSYNCHRONOUS ORBIT

The reason for using the geosynchronous orbit is straightforward. A satellite in this orbit is synchronized with the Earth's rotation, and therefore appears to be suspended over a particular point on the Equator. From an Earth observation point of view, this means that repeat imagery can be obtained as often as required. On the other hand, at the 'current' state of the art in detector design, spatial resolution is likely to be low. It is no accident that the principal users of imagery from geosynchronous satellites are the weathermen. The other disadvantage of the geosynchronous orbit is that the higher latitudes are viewed at an oblique angle, and are therefore less effectively covered.

Further comments, on the second-order effects that afflict particular satellites, are discussed briefly in Chapters 9 and 11. For completeness, one might mention the **Molniya orbits** although at the time of writing they are mainly used for communications purposes. A Molniya orbit could be described as a 'poor man's geosynchronous orbit', for use at high latitudes. It is a highly eccentric orbit, designed to spend some 8 hours, above a particular high-latitude station, before diving down to a low-altitude perigee at an equally high southern latitude. Three satellites, at different phases of the same orbit, are capable of providing a continuous service. However, controllable ground antennae are required, because the satellites do not remain accurately on station, and because it is necessary to switch satellites before each plunges.

1.2   SUN-SYNCHRONOUS ORBITS

Thus most Earth observation satellites occupy near-polar orbits which give regular coverage of the main inhabited parts of the globe. (The reason why a strictly polar orbit is never chosen will emerge in due course.) The optimum compromise referred to in the first paragraph has so far strongly favoured an orbital altitude of between 700 and 1000 km, which gives an orbital period of about 100 minutes.

We have next to discuss the two main imaging strategies, characterized on the one hand by the Earth resources missions such as Landsat and Spot, and on the other by meteorological series such as AVHRR. Meteorological instruments tend to go for low resolution and wide swaths. To ensure frequent coverage, very large overlap is arranged between the views obtained from different orbits, and therefore the exact path covered is not critical. The Earth resource missions, on the other hand, go for high resolution and minimal overlaps. In addition, it is normally required that scenes cover exactly the same ground at each pass. To meet these requirements, the orbital pattern needs to be closely controlled.

Control is achieved by the imposition of two further constraints. First, the orbital periods are 'quantized', by fitting an exact number of orbits n into an exact number of days d (for example, Landsat 4/5 spacecraft execute precisely 233 orbits in exactly 16 days). The execution of n orbits in d days is called an **Orbital cycle**. This constraint means that the **ground-track** pattern repeats itself exactly once per cycle, and ensures that consistent scene boundaries are maintained. The ground-tracks are not however followed in sequential order. Successive orbits will normally lay down ground-tracks some 26° apart. After one day, approximately 14 of these will have encompassed the globe, but leaving large gaps in the coverage. These gaps will be filled in by the orbits of subsequent days. The order in which the ground tracks are laid down varies from satellite to satellite, as is discussed in more detail later in this chapter.

The second constraint is to arrange that imagery of any given point on the globe is always acquired at approximately the same time of day. For most studies, this is advantageous because it ensures that the Sun elevation angle is roughly the same on each pass. A relatively constant illumination angle eliminates one important dimension of variability. (There exist studies that would benefit from retaining this variability. These have had to be sacrificed.) As a general rule, a local Equator crossing time of 10.30 a.m. emerges as statistically the best from the point of view of obtaining cloud and haze-free coverage. However, different missions will have different priorities in this regard.

1.2.1 Sun-synchronism

The acquisition time is fixed by forcing the orbit to be **Sun-synchronous**, which means that the plane of the orbit maintains a fixed orientation with respect to the sun. The use of a Sun-synchronous orbit also eases two other problem areas, which are possibly even more important. First, it minimizes the period of eclipses, which has a major effect on both power supplies and heat balance. Second it facilitates the control of spacecraft attitude in relation to the Sun, which enables both these factors to be controlled with much improved accuracy. How this is achieved will become clearer after the principles of Sun-synchronism are discussed.

Fig. 1.1 illustrates the problem. Consider a satellite orbiting in a plane which takes in both poles and also the Sun (Fig. 1.1(a)). An external observer would see the satellite passing straight down the middle of the sunlit side of the slowly spinning Earth, and therefore passing over each point in its path at noon. Now consider the orbital plane to be inclined, as shown in Fig. 1.1(b). The satellite will now pass over northern latitudes at some time during the afternoon, will cross the Equator at noon, and will traverse the southern latitudes during the morning. Note, however, that a given point will always be observed at the same local time of day. Fig. 1.1(c) illustrates the actual situation that the orbital dynamicists engineer. The orbital plane is now slewed away from the Sun, but kept constant in relation to it, by a strategy to be described very shortly. The figure shows that most of the populated parts of the globe will be traversed at more or less the same time of day. This time can be controlled by selecting the degree of slew.

1.2.2 Effects of Sun-synchronism

The time of day at which the satellite passes over the Equator is termed the **node**. If the satellite is travelling southwards, during daytime passes, then it is a **descending** node. If it is travelling northwards then it is an **ascending** node. The polar regions are of course the losers in all aspects of this strategy. Meteorological series tend to maintain two satellites, both Sun-synchronous, one with a morning descending node, and the other with an afternoon ascending node.

It should now be clear that the use of a Sun-synchronous orbit eliminates much of the variability in the orientation of the spacecraft to the Sun. The solar paddles can be mounted so that a simple rotation about one axis will keep them always pointing directly towards it. Thermal control is facilitated because there is a face of the vehicle that never sees the Sun. This face may be used for the siting of radiating surfaces. In particular this includes the two-stage radiative cooler which, it will emerge, is featured on all instruments covering the thermal band.

The plane of a Sun-synchronous orbit must follow the Sun in its annual apparent perambulation around the Earth. This means that it must precess at a rate of 360°/365.25, or 0.9856°, per day. It so happens that this can be achieved by taking advantage of the perturbations caused by the slight oblateness (0.3%) that centrifugal force has generated in the Earth. As the following paragraphs show, however,

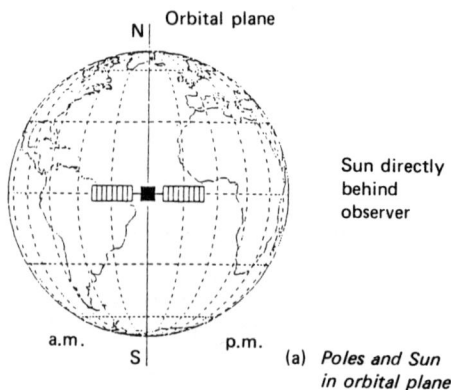

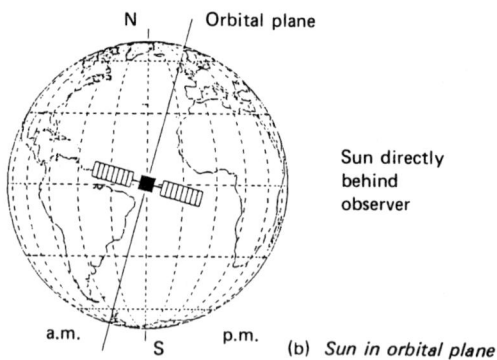

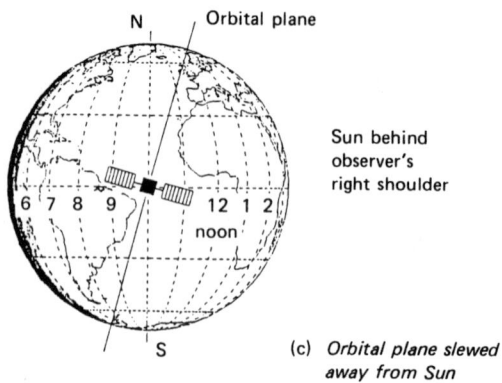

Fig. 1.1 — Sun-synchronous orbits (observer always directly behind satellite).

there is a trade-off between the altitude and inclination required. As the altitude is increased, the inclination must be increased, which leaves an increasing region near the poles uncovered. This forces the compromise towards the lowest possible orbit, consistent with a reasonable life.

In fact the main penalty to be paid in going for a low orbit is not orbital life, but station-keeping. Table 1.1 gives the predicted lives of some of the satellites described

Table 1.1 — Orbital lives

Satellite	Altitude (km)	Orbital life (years)
LANDSAT 4/5	705	80
SPOT	820	300
NOAA 9/10	850	300
LANDSAT 1–3	900	100
IRAS	900	600
GOES/M'SAT	35786	$>10^6$

in this book, taken from RAE (1990). The life, to burn-up, of a satellite is a function of its drag coefficient (which in turn depends mainly on its mass and its shape), and the atmospheric density. Atmospheric density is well known to be a function of altitude. Unfortunately, at orbital altitudes it is also a strong function of solar activity, which is extremely difficult to predict. Thus the figures quoted are highly speculative. Indeed, the difference in predicted life between IRAS and LANDSATS 1-3, both with very similar altitudes, bears testimony to this. However, the predicted lives are all sufficiently above the service lives of all likely missions (at the time of writing at any rate) to be of little significance. Of more significance is the fact that, at 800 km altitude, a satellite is likely to lose upwards of 1 m per day in altitude (CNES, 1988). A lower altitude will mean reduced station-keeping accuracy, or more frequent interruptions of service while the altitude is boosted, or both. Either way, greater fuel capacity will need to be provided — and a greater mass lifted from Earth.

1.2.3 Details of a Sun-synchronous orbit

To provide an accurate long-term prediction of the path of an object in free-fall around the Earth is difficult or impossible. Small second-order perturbations, caused by the axial tilt of the Earth, the gravitational attractions of the Sun and Moon, residual atmospheric drag and solar radiation pressure will all modify the object's orbit. And not all these influences are predictable. However, active artificial satellites are not necessarily in free-fall. Many are flown to a strict timetable, published in advance of launch. (Since the date of launch is unknown at this stage, the timetable is published in terms of 'days from launch'.) To maintain this timetable, regular (but hopefully infrequent) corrections are made, by the expenditure of energy, to maintain the satellites in their chosen nominal orbit. For this reason, the term spacecraft is perhaps more appropriate than satellite, although, in this book, the two terms are used synonymously. The derivation of these nominal orbits, to an accuracy more than adequate for current purposes, proves to be remarkably simple. Thus the time taken for on complete revolution is given by

$$t = \frac{2\pi r^{1.5}}{\sqrt{K}} \text{ s,} \tag{1.1}$$

where K is the geocentric gravitational constant (398601 km³/s²) (Ball & Osborne 1967), and r is the orbital distance from the centre of the Earth (for elliptical orbits, the semi-major axis is used). Fig. 1.2 relates *altitude* to orbital period over the range

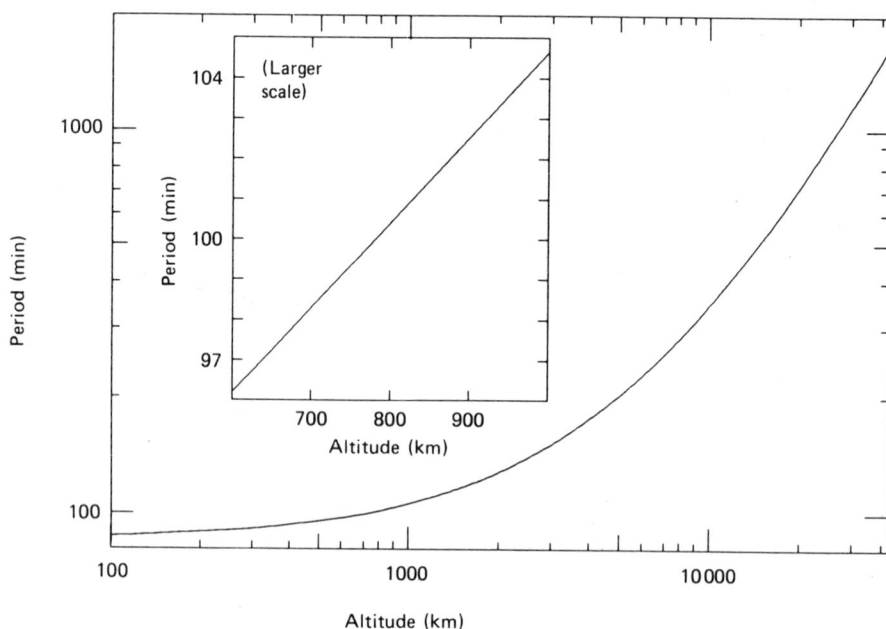

Fig. 1.2 — Period against altitude.

of principal interest. Note that, below about 1000 km, it takes a large relative change in altitude to effect a small change in orbital radius. By substituting, for t, the duration of one **sidereal day** (i.e. 23.93 h, or 1436 min) we can obtain directly the altitude required for a geosynchronous orbit. If R_e, the radius of the Earth at the Equator, is 6378.14 km, then the geosynchronous altitude becomes 35786.5 km.

A **sidereal** day is the time taken for the Earth to make one actual complete revolution, i.e. against the star background. To complete a solar day, the Earth must rotate until the *Sun* is in the same position in the sky. Because, in the interim the Earth has moved nearly 1° further along its own orbit, it must rotate through a total of 361/360° between noon and noon. Therefore a solar day is some 4 min longer than a sidereal day. Sidereal time is the more fundamental but, in most Earthbound applications, it is solar time that applies.

Only geosynchronous orbits are normally Equatorial. To evaluate non-Equatorial orbits, we require an estimate of the mean radius of the Earth in the chosen orbital plane. Ball and Osborne (1967) give the following expression for the radius of the earth, R_ϕ, at latitude ø

$$R_\phi \approx R_e(0.998320047 + 0.001683494\cos 2\phi) \ , \tag{1.2}$$

where R_e is the radius at the Equator. Although not strictly applicable, this expression has been used to estimate the mean radius for all the satellites covered in this book. The orbital periods, and others parameters, calculated on this basis have generally been within 0.3% of the published figures.

The orbital quantization problem can be investigated via a slight modification to Equation (1.1). If n is the number of orbits in a complete cycle, and d is the number of days that the cycle takes then (to a reasonable initial accuracy)

$$1440d = \frac{2\pi n(a + R_e)^{1.5}}{60\sqrt{K}}, \tag{1.1a}$$

where a is the orbital altitude in kilometres (for elliptical orbits, the semi-major axis is used). The value of n dictates the spacing of the ground tracks. The swath width, s, that the instrument must encompass is normally the ground-track spacing at the Equator, plus a margin for overlap. s is given by

$$s = \frac{2\pi R_e}{n} + \text{overlap} \ . \tag{1.2}$$

These expressions may be used, on a trial-and-error basis, to derive an acceptable range of values for a. It might seem reasonable to imagine that, if the number of ground tracks is constrained to fit an exact number of Earth revolutions, then the tracks might well be equally spaced on the ground. In fact it has been shown that it must be so (King, 1976). And if the paths are equally spaced then they must be $360/n°$ apart, or $2\pi R_e/n$ km apart at the Equator. For Landsat, this would be 1.545° or 172.0 km.

The next thing to consider is the ground-track pattern. This is dictated by the fractional part of k, the number of orbits executed in a day. If we write

$$k = \frac{24 \times 60}{\text{period}} = k' + \frac{k''}{d} \ ,$$

then k''/d is the fractional part of k. We can say the following things about suitable values for k'' (Fig. 1.3(a)). It must not be zero, or tomorrow's orbits will coincide exactly with today's and most of the possible tracks will be missed. The fractional part of k must sum to a whole number of orbits after a complete cycle, which involves making k'' an integer also. However, it must not be a factor of d, otherwise not all the possible ground tracks will actually be covered. If k'' is 1, then tomorrow's orbits will occur fractionally *earlier,* as measured at a fixed point on the ground (they will pass a given latitude at the same local time, as we have already discussed). Therefore the Earth will not have rotated quite so far eastwards, and tomorrow's orbit will lie 1 ground-track to the *east* of today's. If k'' is $(d-1)$, as in the earlier Landsats, then tomorrow's orbits will lie 1 ground track to the *west* of today's. If k'' is 9, as in the later Landsats, then tomorrow's orbits will lie 9 tracks to the *east* of today's — or 7 tracks to the *west* which is how it appears on most diagrams. In most traditional applications, the order in which the ground tracks are covered makes little practical difference. However instruments with an off-nadir viewing capability, such as SPOT, have a need to pass within angled-viewing distance of all parts of the globe at frequent intervals. As discussed in Chapter 16, the SPOT ground-track pattern was deliberately tailored with this objective in mind.

Fig. 1.3(b) gives examples of actual ground-track patterns. The diagrams represent, in diagrammatic form, the world map, and show the approximately 14 daytime (southgoing) passes made in 24 hours. The central blow-up shows a hypothetical local area; and depicts the order in which the passes are made. The upper diagram represent Landsats 4/5; and we can see that, if today's pass over this area follows Track 1, then tomorrow's pass will follow Track 8 (9 tracks to the east, or 7 tracks to the west). Track 2 is followed 7 days later (or 6 days earlier) on Day 8. The track separations in kilometres are, of course, as measured at the Equator. The angular distances apply at all latitudes. For SPOT, k'' is 5, and so tomorrow's orbits will be 5 tracks to the east of today's. For IRS-1 (and also for Landsats 1–3) k is $(d-1)$, and so the orbits are sequential. Tomorrow's orbits are $(d-1)$ tracks to the east — or 1 track to the west — of today's.

It is worth remarking at this point that published orbital figures are not always particularly accurate. If accuracy is sought, then it may be obtained from the orbital cycle. The satellite must execute exactly n orbits in precisely d days, and from this fact a precise period may be derived. The remaining parameters may then be evaluated with confidence. Altitude data are particularly uncertain because of the complex shape of the Earth. The orbital radius, however, may be computed with precision.

Having derived the required orbital period, the next step is to select an inclination, i, that will generate Sun-synchronism. The expression

$$\dot{\Omega} = -9.95 \left(\frac{R_c}{r}\right)^{3.5} \cos i \ °/\text{day} \tag{1.3}$$

(Ball & Osborne 1967) gives the precession rate for an orbital radius r and an inclination i. (For elliptical orbits, the semi-major axis is used, and a factor $(1-e^2)^{-2}$

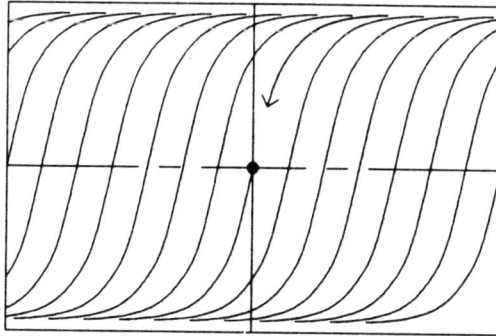

time = 0
or 24 h.

$$k'' = 0$$

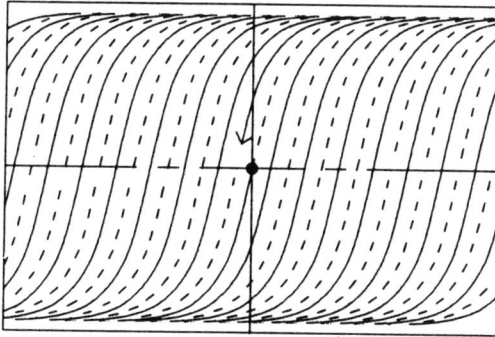

time = 0
or 24 h.

$$\frac{k''}{d} = \frac{1}{2}$$

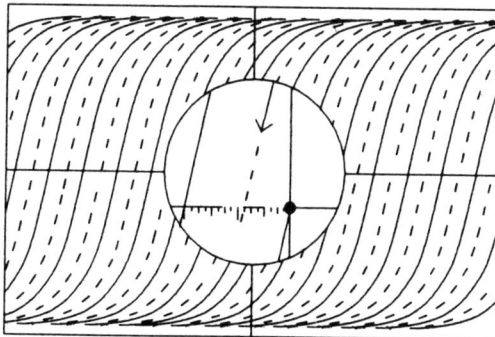

time = 0
or 24 h.

$$\frac{k''}{d} = \frac{9}{16}$$

Fig. 1.3 — (a) Ground track pattern generation (daytime, or southgoing passes only).

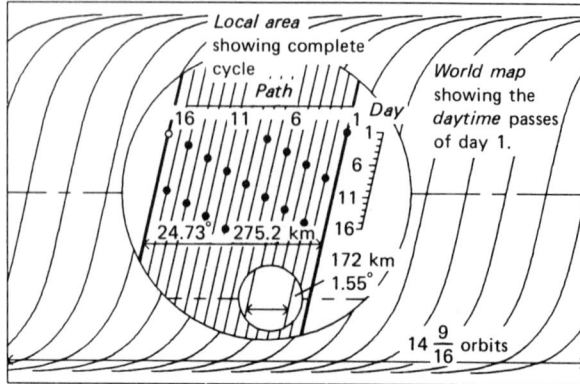

Landsat 4/5

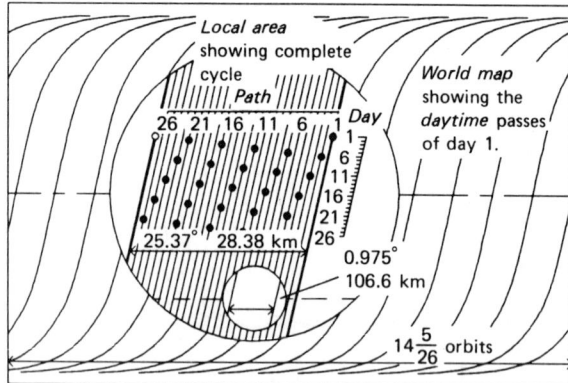

Spot

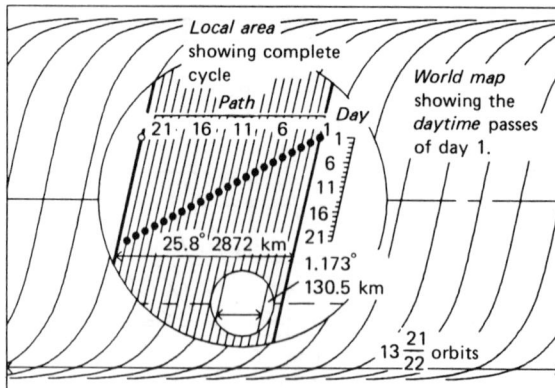

IRS-1

Fig. 1.3 — (b) Examples of ground track patterns.

is inserted before the $\cos i$.) The constant, 9.95, incorporates several fixed quantities. These include K, and the second harmonic oblateness coefficient (J_2) at

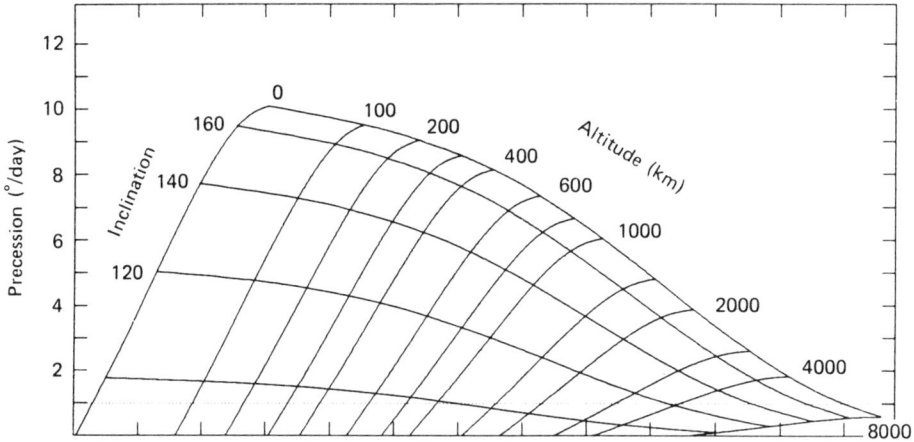

Fig. 1.4 — Precession rate against inclination and altitude.

1082.64×10^{-6}. Fig. 1.4 relates the precession rate to altitude and inclination. Since, for Sun-synchronism, we require a rate of 0.9856°/day, we can see clearly how coverage suffers as altitude is increased.

Table 1.2 shows the orbital parameters for all the satellites covered in this book. The figures in the table have been computed from the orbital cycle parameters, as discussed above. The altitudes quoted are as measured at the Equator. This may explain the systematic descrepancy between these figures and those published in the literature. It must be emphasized that these figures describe the orbit that the satellite is required to follow if it is to keep to the prescribed ground track pattern. Only if the shooting was exceptionally accurate (as was the case, for example, with SPOT 1) will they also describe the orbit into which the satellite was initially injected. Thus there are also discrepancies between the figures presented here and those contained in the Earth Satellites 'Bible' (RAE, 1987).

The entries for AVHRR (AVH-1 and AVH-2) require some explanation. NOAA (1987) gives two orbital altitudes for the TIROS-N series of spacecraft, namely 833 and 870 km. The two orbital cycles used for illustration generate the closest approximation to these two altitudes that could be found. It has also been suggested that the orbits approximate to a 5-day and a 10-day cycle respectively. If these cycles are used, then slightly different figures are obtained. This uncertainty demonstrates the relative unimportance of the matter to the meteorological community.

Table 1.2 — Computed orbital details

Satellite	Cycle		Orbits per day	k''	Period (min)	Radius (km)	Altitude (km)	Incl. (°)	Shift/orbit		G/T spacing		Ground speed (km/s)
	orbits	days							(°)	(km)	(°)	(km)	
MSS	251	18	13.944	17.00	103.27	7291.24	934.10	99.10	25.82	2873.91	1.43	159.66	6.45
AVH-1	341	24	14.208	5.00	101.35	7200.68	842.67	98.71	25.34	2820.53	1.06	117.52	6.57
AVH-2	1397	99	14.111	11.00	102.05	7233.71	876.77	98.85	25.51	2839.97	0.26	28.69	6.52
TM	233	16	14.563	9.00	98.88	7083.45	717.00	98.22	24.72	2751.94	1.55	172.00	6.73
IRAS	14	1	14.000	0.00	102.86	7271.93	915.16	99.02	25.71	2862.51	25.71	2862.51	6.47
SPOT	369	26	14.192	5.00	101.46	7206.10	848.32	98.73	25.37	2823.72	0.98	108.60	6.56
MOS-1	237	17	13.941	16.00	103.29	7292.38	935.21	99.11	25.82	2874.58	1.52	169.09	6.44
IRS-1	307	22	13.955	21.00	103.19	7287.72	930.67	99.09	25.80	2871.83	1.17	130.54	6.45
GEOS	1	1	1.003	0.00	1436.07	42164.21	35786.06	0.00	359.02	39965.71	360.00	40075.10	0.46

REFERENCES

Ball, K. F. & Osborne, G. F. (1967) *Space Vehicle Dynamics* Ball & Osborne, Oxford 1967.

CNES (1988) *Spot User's Handbook* English Edition.

King, J. C. (1976) 'Quantization and symmetry in periodic coverage patterns with applications to earth observation' *J. Aerospace Sciences* **19** 4.

NOAA (1987) 'The TIROS-N/NOAA A-G Satellite Series', NOAA NESS 95, reprinted May 1987.

RAE (1990) *The R.A.E. Table of Earth Satellites 1957–1989* Royal Aerospace Establishment.

2

The atmosphere

Consultant: **Dr M. D. Steven**
Department of Geography, University of Nottingham†

Passive remote sensing, such as we are interested in here, can only use wavelengths in which either the source is bathed in a reasonable level of illumination or it radiates at adequate intensity in its own right. In the former case, we are measuring the reflective properties of the target in one or more of the wavelengths available to us (see Introduction) and inferring what we can about its nature. In the latter case we are more concerned with its temperature. Meteorologists are extremely interested in cloud-top heights, which can be inferred from this information. They are equally interested in the much smaller differences in sea surface temperature, because this has a considerable effect on the down-wind weather prospects. Geologists are also able to infer a good deal about the nature of surface rocks from small local temperature differences measured by day and by night.

In either case a useful proportion of the radiation leaving the target has successfully to run the gauntlet, at least once, of a nominal 100 km of atmospheric soup. Unfortunately, the optical properties of this soup are highly variable. Experimentalists in most walks of life have to be prepared to calibrate their systems before they can take accurate measurements. Most can assume however that, once calibrated, their system will remain tolerably stable for a reasonable length of time. It falls to relatively few to have to deal with a system whose properties vary widely and unpredictably from day to day — and which is also difficult or impossible, to calibrate. Such, however, is the position concerning the optical properties of the atmosphere.

By far the most active ingredient in the atmosphere is water, in both vapour and droplet form, and yet it is also the most variable. Water droplets (cloud) can generate complete opacity to all the optical wavelengths, and under such conditions ground observation is impossible. Statistically, the extent of coverage by cloud increases during the day, as heat from the Sun stirs up the atmosphere's various layers.

† Department of Geography, University of Nottingham.

However, early morning mist and fog is often burned off by that same Sun, and the optimum time for the passage of Earth resource satellites turns out to be about 10:30, local time. Synoptic meteorological measurements, on the other hand, are likely to be required at other local times. They are intended to define the state of the entire atmosphere at particular instants, and are made simultaneously throughout the world at fixed times GMT. At the same time, they are concerned as much with clouds as with the ground (or, more usually, the sea); hence the emphasis on geosynchronous instruments in meteorological remote sensing.

Reflective remote sensing relies, as Fig. 2.1 illustrates, on the measurement of radiation that has made two passages through the atmosphere. A proportion of the sunlight will be absorbed or scattered, and never reach the target area. Some of the missing light will find its way back into the telescope as, effectively, an additional source of white noise. It is actually a deterministic signal, which can be used to determine the aerosol properties of the atmosphere (see below). However, in the present context, the term 'noise' seems a reasonable description. In fact an image, acquired on a hazy day, gives a very good visual impression of the effect of noise contamination on a signal. Contrast is lost, and low-energy components of the signal are swamped. Some of the scattered light will reach the target — but from many different directions — thus causing confusion to any experiment that is incidence-angle-sensitive. The reflected radiation of course has to run a gauntlet of similar hazards.

2.1 ATMOSPHERIC ATTENUATORS

A ray of light, travelling through the atmosphere, can be scattered or absorbed by attenuators of all sizes, from molecules to raindrops (free atoms are virtually non-existent in the atmosphere). The most obvious of these are water droplets, of radii around $10\,\mu$m, which, in the mass, constitute clouds. They attenuate a direct beam mainly by scattering. The vast majority are large compared with the wavelength, and therefore exhibit **Mie scattering**. This is a complex phenomenon with a strong — but oscillating — dependence on wavelength. However in practice Mie scatterers come in a wide range of sizes; and so the effect of wavelength is almost totally smoothed out (Steven, 1990).

The next most effective attenuators are particles of dust and smoke, or salt, of radii between 0.1 and $10\,\mu$m. A dispersion of such **particulates** is normally called an **aerosol**. Again, the particle sizes are mostly greater than the wavelengths of interest to us, and an aerosol also acts mainly as a very efficient Mie scatterer. However, it may also absorb. The effectiveness of an aerosol is increased under humid conditions, when the particles form nuclei for the formation of small water droplets. Particulates are to be found mainly in the Earth's 'boundary layer' and, because of their small size, are likely to remain suspended for days or weeks unless dispersed by winds. The Earth's **boundary layer** is the lowest 1 to 2 km of the atmosphere which are stirred up by the passage of wind over the ground. Where the effect of convection is not significant, (e.g. under windy conditions) the Earth's boundary layer is very similar to that generated over the sails of a yacht or the surfaces of an aircraft. The scattering properties of an aerosol vary with its major constituents, and four major classes have been identified. These are **continental**, comprising mainly dust picked up from dry ground; **maritime**, made up of salt from sea spray; **urban** comprising smoke

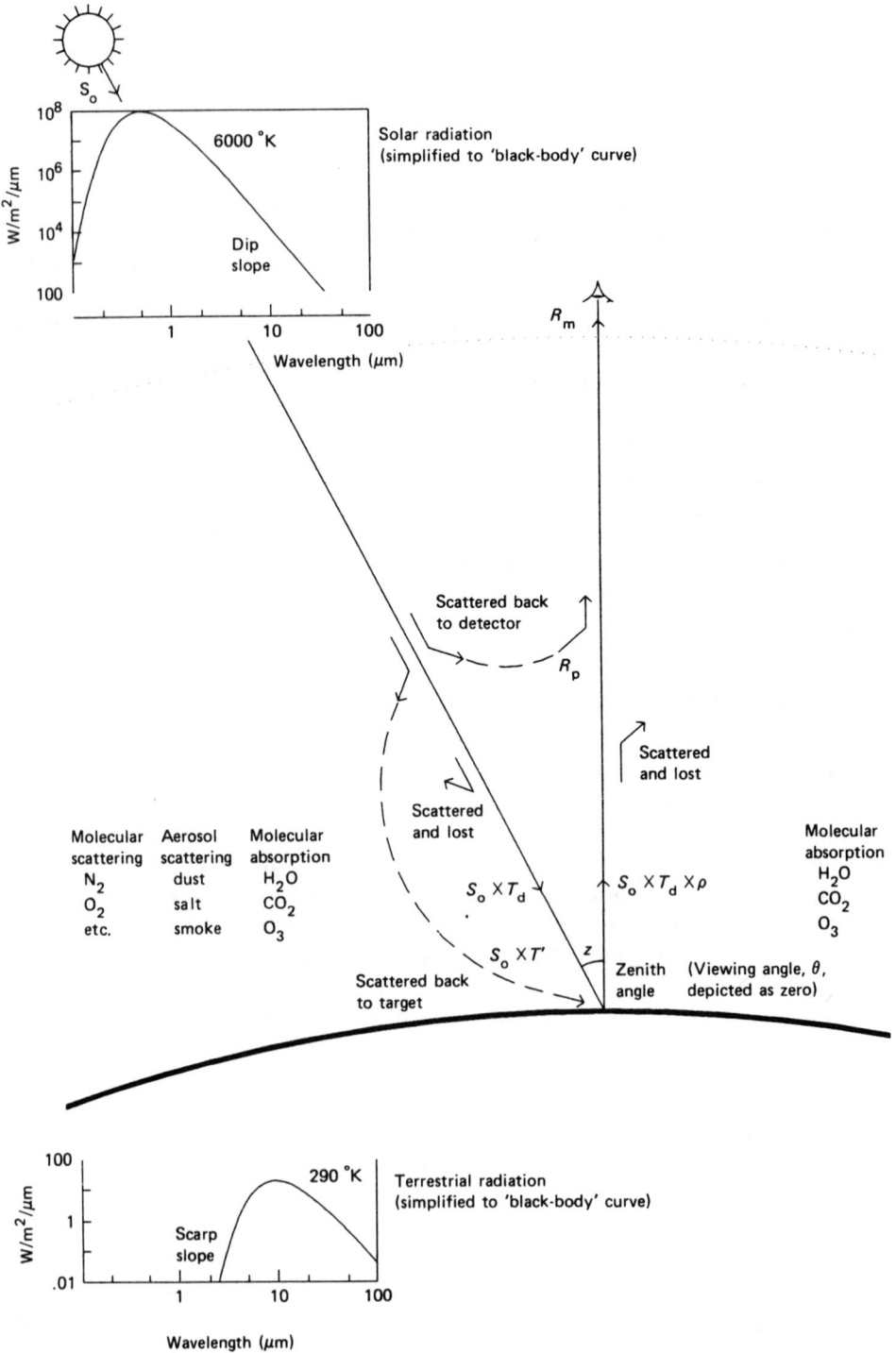

S_o

10^8

10^6

10^4

100

W/m^2/µm

6000 °K

Solar radiation
(simplified to 'black-body' curve)

Dip
slope

1   10   100

Wavelength (µm)

R_m

Scattered back
to detector

R_p

Scattered
and lost

Molecular scattering	Aerosol scattering	Molecular absorption
N_2	dust	H_2O
O_2	salt	CO_2
etc.	smoke	O_3

Molecular
absorption
H_2O
CO_2
O_3

Scattered
and lost

$S_o \times T_d$

$S_o \times T_d \times \rho$

$S_o \times T'$

z

Scattered back
to target

Zenith
angle

(Viewing angle, θ,
depicted as zero)

100

1

$.01$

W/m^2/µm

290 °K

Terrestrial radiation
(simplified to 'black-body' curve)

Scarp
slope

1   10   100

Wavelength (µm)

Fig. 2.1 — The effect of the atmosphere.

particles; and finally **rural**, which contains relatively few particulates of any kind. In the absence of any better measure, the type is normally assumed from the locality, while the amount present is estimated from measurements of meteorological 'extinction'. Extinction is normally measured in kilometres, it is almost, though not quite, synonymous with horizontal visibility. At the shorter wavelengths, the effect of aerosols can be severe. Fortunately, however, the broad-band nature of the effect means that they tend to serve each of the reflective bands more or less equally, while their effects on the thermal bands is altogether less marked. Thus a well-designed experiment should be able to handle the uncertainties in aerosol absorptivities.

Active volcanism can in fact throw aerosols of small particulates up as far as the stratosphere, where they can persist for months or even years. However, only if the contamination is particularly strong is there likely to be any noticeable effect on vertical or near-vertical viewing. The effect, when it occurs, is fairly broad-band, though it appears to be greatest in the visible region (see Table 2.1).

Finally we have the molecules, mainly water, carbon dioxide and ozone. Molecules both scatter and absorb. It is molecular scattering which creates the effect of a blue sky, as a result of the strong wavelength-dependence of the mechanism. Direct sunlight is thus yellower than it would be in space. (Red sunsets are generated under the extreme conditions of grazing incidence; and do not affect the present argument.) A simple model to explain atmospheric scattering was proposed by Lord Rayleigh (1842–1919). It was based on the assumption that molecules were somewhat akin to billiard balls, and were large enough to be significant in comparison to optical wavelengths. It was Einstein (*c.* 1905) who pointed out that this assumption was invalid. The modern Rayleigh model, as applied to molecules, assumes that the refraction is caused by the cumulative effect of a large number of molecules, and that their Brownian (thermal) motion causes small random fluctuations in density, and hence in refractive index. The effect is greatest where the number of molecules per unit volume is small. Thus molecular scattering occurs mainly at the higher altitudes. The intensity of pure Rayleigh scattering is inversely proportional to the fourth power of the wavelength. However it is now known that, for even the clearest atmosphere, the total variation is much more nearly inversely proportional to wavelength (Slater, 1980). This implies that real atmospheres always contain a proportion of particles which are not small in relation to the range of wavelengths of interest. Molecular scattering normally becomes negligible by about $3\,\mu$m.

Molecular absorption is a resonance effect between electron orbits, and occurs at specific wavelengths only. An atom or molecule absorbs a photon by using its energy to raise an electron to a more excited state. According to quantum mechanics, the likelihood of absorption of a photon is high only when its energy closely matches that required to achieve a specific transition. An equivalent phenomenon, occurring in metals and semiconductors, is the basis of most modern detection mechanisms. They are discussed further in Chapter 5. Atomic absorption (and emission) lines are typically extremely narrow — to the extent that they are used for length and time standards. However, in a molecule, the number of possible transitions may well be large, and so the number of different photon energies that may be absorbed is sometimes also large. In particular there is a tendency for the lines to cluster, and to produce a narrow absorption *band*. The water vapour molecule, for example, is extremely complex notwithstanding the simple chemical formula normally attri-

Table 2.1 — Atmospheric transmission coefficients (Source: LOWTRAN)

Effect of surface visibility (U.S. Standard Atmosphere)

Thematic Mapper band							Code
1	2	3	4	5	7	6	
0.485	0.56	0.66	0.83	1.65	2.215	11.45 μm	

Navy B/L model — Extinction (vis) = 39.5 km

Spring/summer							
0.66	0.71	0.76	0.76	0.86	0.84	0.80	UNs2
Fall/winter							
0.68	0.73	0.78	0.77	0.86	0.84	0.80	UNw2

Rural or maritime B/L model — Extinction (vis) = 23 km

Spring/summer							
0.56	0.63	0.70	0.73	0.91	0.90	0.84	URs2
0.58	0.63	0.69	0.70	0.84	0.84	0.84	UMs2
Fall/winter							
0.58	0.65	0.72	0.75	0.91	0.91	0.84	URw2
0.60	0.65	0.70	0.71	0.84	0.84	0.84	UMw2

Rural or urban B/L model — Extinction (vis) = 5 km

Spring/summer							
0.21	0.27	0.35	0.43	0.76	0.81	0.79	USs2
0.21	0.27	0.35	0.42	0.71	0.76	0.78	UUs2
Fall/winter							
0.21	0.28	0.36	0.44	0.76	0.81	0.79	USw2
0.22	0.28	0.36	0.42	0.71	0.76	0.78	UUw2

Codes

Column: 1	2	3	4
Atmos. model	Boundary Layer model (vis. in km)	Season	Volcanism (stratosphere)
A=Sb.Arc smr	M=M'time (23)	s=sprg/smr	2=aged/aged moderate
B=Sb.Arc wtr	N=Navy (40)	w=fall/wtr	5=fresh/fresh moderate
M=Midlat smr	R=Rural (23)		4=aged/aged high
N=Midlat wtr	S=Rural (5)		3=fresh/fresh high
T=Tropical	U=Urban (5)		
U=U.S. std			

Continued next page

Table 2.1 (*Contd.*) — Atmospheric transmission coefficients (Source: LOWTRAN)

Effect of latitude (various atmospheric models)

			Thematic Mapper band				Code
1	2	3	4	5	7	6	
0.485	0.56	0.66	0.83	1.65	2.215	11.45 μm	

Rural or maritime B/L model — Extinction (vis) = 23 km

Spring/summer (arctic/midlatitudes/tropics)

0.56	0.63	0.70	0.72	0.90	0.88	0.74	ARs
0.56	0.63	0.70	0.70	0.89	0.86	0.62	MRs2
0.56	0.63	0.70	0.68	0.88	0.84	0.46	TRs2

Fall/winter (arctic/midlatitudes/tropics)

0.58	0.64	0.72	0.78	0.92	0.94	0.95	BRw2
0.58	0.64	0.72	0.76	0.92	0.92	0.90	NRw2
0.58	0.65	0.72	0.69	0.89	0.84	0.46	TRw2

Rural or urban B/L model — Extinction (vis) = 5 km

Spring/summer (arctic/midlatitudes/tropics)

0.21	0.27	0.35	0.42	0.75	0.79	0.70	ASs2
0.21	0.27	0.35	0.42	0.74	0.77	0.59	MSs2
0.21	0.28	0.35	0.40	0.73	0.74	0.43	TSs2

Fall/winter (arctic/midlatitudes/tropics)

0.21	0.28	0.36	0.46	0.76	0.83	0.89	BSw2
0.21	0.28	0.36	0.45	0.76	0.82	0.85	NSw2
0.21	0.28	0.36	0.41	0.73	0.74	0.43	TSw2

Effect of volcanism (U.S. standard atmosphere)

Rural B/L: *Spring/summer* (see code)

0.56	0.63	0.70	0.73	0.91	0.90	0.84	URs2
0.56	0.63	0.70	0.73	0.89	0.89	0.84	URs5
0.52	0.59	0.66	0.70	0.90	0.90	0.84	URs4
0.53	0.59	0.66	0.69	0.84	0.84	0.82	URs3

Codes

Column: 1	2	3	4
Atmos. model	Boundary Layer model (vis. in km)	Season	Volcanism (stratosphere)
A=Sb.Arc smr	M=M'time (23)	s=sprg/smr	2=aged/aged moderate
B=Sb.Arc wtr	N=Navy (40)	w=fall/wtr	5=fresh/fresh moderate
M=Midlat smr	R=Rural (23)		4=aged/aged high
N=Midlat wtr	S=Rural (5)		3=fresh/fresh high
T=Tropical	U=Urban (5)		
U=U.S. std			

buted to it, and it absorbs over a large number of bands over the range of current interest, as Fig. 2.2 shows. The concentrations of carbon dioxide and ozone vary relatively little over the seasons and around the globe. That of water vapour varies considerably, however, even from hour to hour.

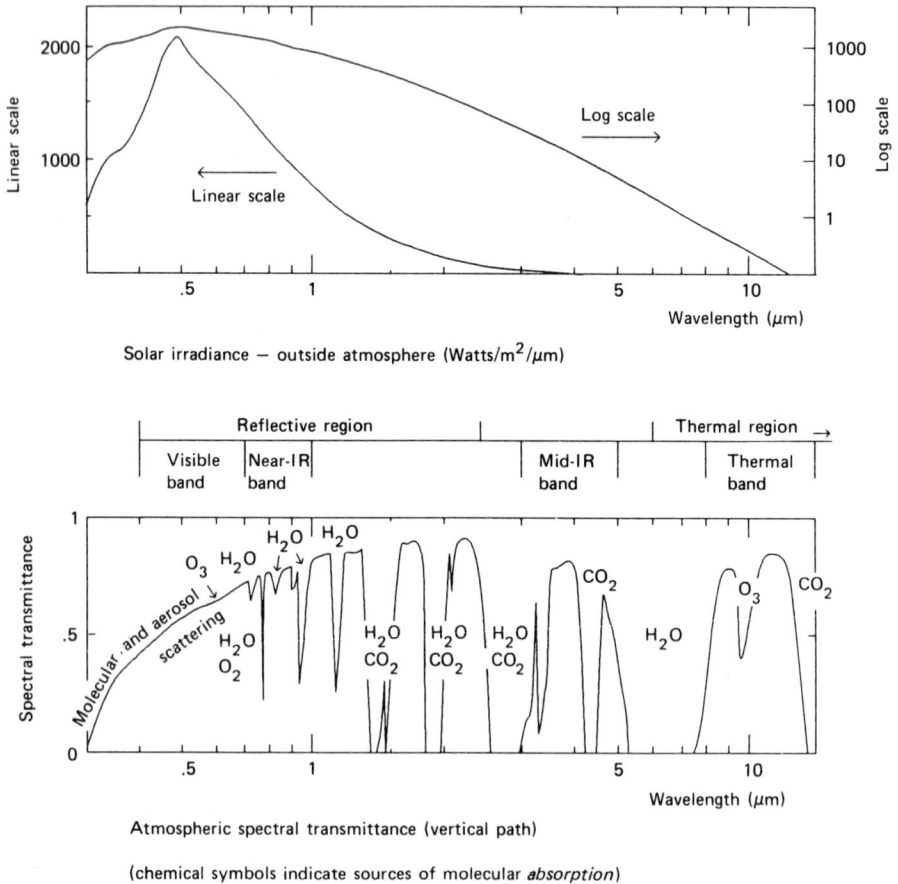

Solar irradiance — outside atmosphere (Watts/m^2/μm)

Atmospheric spectral transmittance (vertical path)

(chemical symbols indicate sources of molecular *absorption*)

U.S. Standard Atmosphere 1962

Fig. 2.2 — Predicted atmospheric properties. (Source: LOWTRAN model.)

2.2 ESTIMATING ATTENUATION

To obtain a reliable estimate of atmospheric attenuation at the time that any particular image was acquired requires two steps to be taken. First the state of the atmosphere at the time must be defined. This is difficult. Indeed, for many experimenters, it is next to impossible. Then this information must be plugged into a suitable atmospheric model. After many years of continuous development, some of these are now extremely good. Model atmosphere data are normally available for plugging gaps in the known data. These may incorporate seasonal and latitude variations, as well as assumed aerosol types. But, whether the resulting hybrid data set will lead to answers that truly reflect the conditions that existed on the day of the experiment, is entirely up to the experimenter to guess. It is not surprising that many experimenters design their measurements to be effectively self-calibrating. They calibrate partly by the obsessive taking of ground-truth data, and partly by relying on inter-band comparisons; a process which cancels out much of the variability. If it can be arranged that there is deep water in the locality then its property, of being virtually totally absorbing in the near infra-red, can be used to assist in the calibration of this band. Other related bands of interest can then be calibrated indirectly (Aranuvachapun, 1986).

Atmospheric absorption curves are available from a variety of sources, and covering a variety of different 'standard' atmospheres. The curves presented in Fig. 2.2 may look somewhat unfamiliar to experienced readers. They are derived from the **LOWTRAN** atmospheric model (Kneizys *et al.*, 1983), and depict its representation of the 1962 U.S. standard atmosphere. The use of numerical data from a model facilitates the matching of the presentation to the particular conventions of this book, and also enables the variability to be illustrated. The figure illustrates the effect of the atmosphere on measurements of ground irradiances when measured from space.

A word of warning is appropriate concerning the derivation of spectral energy or other 'dimensioned' data. Such data has to be defined in terms of energy (or other quantity) per unit bandwidth. Unfortunately there are two candidate measures of bandwidth. These are unit wavelength, much used by optical engineers and physicists, and unit frequency, favoured by electronic engineers. Both are equally valid, but give markedly differently shaped curves on account of one being inversely proportional to the other. Energy per unit frequency might be favoured logically, because increasing frequency goes with increasing energy and thus graphs and tables automatically fall out in a way that appears natural. However the remote sensing community has tended to follow the physicists, for the zones that interest us, and the electrical engineers for the microwave region. These conventions are perfectly acceptable as long as everybody sticks to them. It is undesirable for a particular author to devise his own convention, and it is reprehensible for any work to switch between the two unless there is a very good reason. (This problem arises again in Chapter 4.)

The traditional unit of frequency is the cycle per second or **hertz**. However, the frequencies of visible and thermal radiation are very large, and the **wavenumber** is usually preferred because of the more tractable numbers that it produces. The wavenumber is the reciprocal of the wavelength, and is equal to the frequency divided by the speed of, in this case, light — at 3.0×10^{14} μm/s. (Another measure, much used in control theory and mathematical modelling, is the **angular frequency**. It is measured in radians per second, and is equal to frequency multiplied by 2π.)

Fig. 2.2 gives the solar intensity (in linear and logarithmic units) at an altitude outside the atmosphere. The Sun is generally represented as a black body, with a temperature of around 6000 K. If this were strictly true then its radiation profile would be that of the simple black body curve depicted (for simplicity) in Fig. 2.1. In fact of course the structure of the Sun is anything but simple; and the energy radiated at different wavelengths originates at different depths, and from material at different temperatures. The curve in Fig. 2.2 is taken from the LOWTRAN model, and represents something of a simplification of the actual measured curve. Even over the short span of visible wavelengths, the apparent black body temperature of the Sun varies between about 5700 and nearly 6000 K. At shorter wavelengths, although the fact is of little interest to us, the effective temperature reduces, falling to 5000 K at $0.2\,\mu$m. At longer wavelengths, the apparent temperature increases. By the time we again lose interest, at say 3–5 μm, the apparent temperature has reached 6000 K. At some of the longer wavelengths the apparent temperature of the Sun rises to between 10^6 and 10^{10} K.

The figure also gives a 'typical' transmission coefficient for the atmosphere, namely that of the 1962 U.S. standard as predicted by the LOWTRAN model. The spectral intensity reaching a *horizontal* ground surface will be roughly the product of the two curves (very roughly, under wintertime conditions). For the higher-frequency reflective bands, the spectral intensity reaching the detector is the all-important reflectivity of the target times the product of solar intensity and (again, roughly) the *square* of the atmospheric attenuation.

More accurately, solar irradiance on a horizontal ground surface is given by:

$$S_g = S_0(T_d + T') \ ,$$

where S_0 is the solar irradiance on a horizontal area just outside the atmosphere. T_d is the downward transmittance, at a zenith angle z (see Fig. 2.1), for the radiation following a direct path. T' is the downward transmittance for the additional radiation which reaches the target after scattering. This contribution is not included in Fig. 2.2, but can be estimated by a good atmospheric model.

The radiance that reaches the detector is given by:

$$R_m = S_g \rho T_u + R_p$$

where ρ is the all-important reflectivity of the surface. T_u is the upward transmittance. R_p is the 'path radiance' — the total radiance scattered back into the upward path, from a range of levels of the atmosphere, and without reaching the ground.

T_u and T_d can be estimated from an estimate of the vertical transmittance using the expression:

$$T_{angle} = T_{vertical} \sec(\text{angle})$$

where angle is the zenith angle z, or the viewing angle Θ (Steven & Rollin, 1986).

The figure shows clearly the separation of the reflective and the thermal regimes.

Reflective measurements demand that there be good solar illumination at the ground at the wavelength of interest, whereas thermal measurements require there to be none. Only at wavelengths greater than 5 μm, therefore, may thermal measurements be made during daytime passes. The shortest wavelengths available for such measurements are therefore those in the 8–13-μm window, and these are the ones normally used.

2.3 MODEL PREDICTIONS

However, LOWTRAN also includes data for a number of other model atmospheres, covering different parts of the globe and different seasons, and for some of these the figures are markedly different. Table 2.1 shows some of the atmospheric transmittance predictions at the centre wavelengths of the seven Thematic Mapper bands. These figures are given purely to illustrate the range of variability in atmospheric attenuation that exists in the various bands of interest to us. The author takes no responsibility for the consequences of using them in any experiment. For example, the variation with latitude that the model gives is generated entirely by the temperatures assumed, and hence on the amount of moisture that the atmosphere is assumed to have taken up. On any particular day, the actual moisture content could well be very different.

The basic atmospheric model defines the temperature and humidity profiles, as well as the 'upper atmosphere' ozone concentration. To this must be added an 'aerosol' model, defined in terms of 4 altitude bands; namely the 'boundary layer' (below 2 km), the 'troposphere' (2–10 km), the 'stratosphere' (10–30 km) and the 'upper atmosphere' (30–100 km). In fact, however, only the boundary layer and the stratosphere normally contain aerosols.

As the table shows clearly, the boundary layer region has by far the greatest effect on simple near-vertical viewing such as interests us. This region is also much the most variable and unpredictable. Boundary layer attenuation is a broad-band effect, but one which tapers off towards the long wavelengths, becoming almost negligible by 12 μm. In the reflective wavebands, however, the intensity can easily be halved after a single passage. The boundary-layer attenuation factor is closely correlated with horizontal visibility. It is probable that a well-designed experiment would reveal data of doubtful quality early enough to avoid risk of erroneous interpretations.

Latitude, on the other hand, is shown to have a much greater effect at the longer wavelengths. This is due entirely to assumed probable moisture content. The predicted effect is entirely negligible in the reflective bands down to, say, 0.7 μm. It is noticeable from 0.7 to around 2 μm, and at 12 μm there can be up to a factor of 2 between the 'subarctic' region and the tropics (with the tropics coming off worst). A slight effect is noticeable between spring/summer and autumn/winter, with the air being clearer in the autumn/winter. This result, however, assumes that the surface visibility is the same in both cases, which it probably seldom is. Finally, volcanic eruptions appear to have little effect on the kinds of measurements that we are considering here, except perhaps in the immediate aftermath. More exotic measurements, in which the path lies exclusively in the higher altitude regions, will of course be affected much more.

Chapter 16 includes a review of the position, from a SPOT-oriented viewpoint, derived from the SPOT handbook.

REFERENCES

Aranuvachapun, S. (1986) *Int. J. Remote Sensing* **7**, 499–514.

Kneizys, F. X., Shettle, E. P., Gallery, W. D., Chetwynd, J. H. Jr., Abreu, L. W., Selby, J. E. A., Clough, S. A. & Fenn, R. W. (1983) *Atmospheric transmittance/radiance computer code LOWTRAN 6*, Air Force Cambridge Research Labs, Bedford, MA, USA, 1983.

Slater, P. N. (1980) *Remote Sensing: Optics and Optical Systems*, Addison Wesley.

Steven, M. D. & Rollin, E. M. (1986) *Int. J. Remote Sensing* **7**, 481–498.

Steven, M. D. (1990) Correction in draft.

3

Telescopes

Consultants: **Mr A. Péraldi, Mr C. Chipaux†**
Sources: **Péraldi, (1989), Chipaux (1990)**

3.1 BASIC OPTICS

The essence of the working of a telescope, as applied to remote sensing instruments, is extremely simple. Much of it can be discussed with the aid of geometric optics and a simple 'thin' lens (Fig. 3.1). The results derived apply equally to reflecting systems.

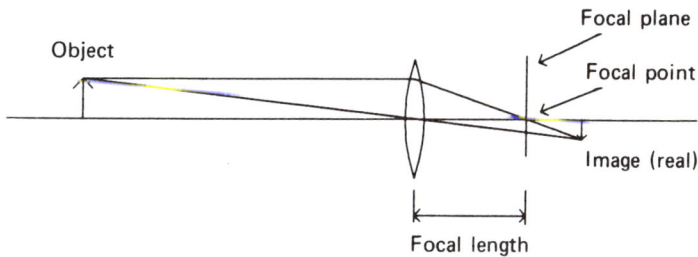

Fig. 3.1 — Geometric optics.

Two rules of geometric optics are particularly important. The first is that any ray of light entering a lens parallel to its axis will be deflected to pass through a single point, also on the axis. The point is called the **focal point** of the lens, and its distance from the lens centreline is its **focal length**. The **focal plane** of the lens is the plane, perpendicular to the axis, that passes through the focal point. The second rule is that a ray passing through the centre point of the lens will continue on its way undeflected. A parallel beam of light entering the lens off-axis will be focussed at a point on the

† Mr Armand Péraldi was Scientific and Technical Director of Matra until his tragic death in late 1989. Mr Claude Chipaux is a senior engineer, also at Matra.

focal plane. Thus multiple beams, emanating from a distant extended object, will form a 'real' image at the focal plane. The image plane will recede as the object is brought closer, as shown in the figure. The term 'real' is used because, unlike a 'virtual' image, it actually exists. A piece of card, placed at the focal plane, will be illuminated by an inverted image of the object.

The image produced by a single lens or mirror is small, and of little use for viewing by eye. Therefore, in normal telescopes, the real image is viewed through an eyepiece lens so positioned as to generate a magnified virtual image a convenient distance from the eye. However the eyepiece lens does nothing to enhance the optical performance of the telescope, and is required purely to accommodate the limited resolving power of the human eye. For remote sensing applications it may be dispensed with, as long as the detector array can be made small enough to resolve the real image adequately. Paradoxically, the size of the real image is inversely proportional to the power of the lens. A long focal length (weak) lens is required if the detector aperture is not small. This is unfortunate because it makes the telescope large and heavy.

Some consultants have criticised the use of the word 'aperture' in the context of the detector, preferring to use it only to mean the 'light gathering area of the telescope as a whole'. In plain English, however, an aperture is an opening, and this is exactly what we mean here. No alternative descriptor has been offered which is equally lucid. However, the reader is warned of the dangers of using plain English in a scientific context.

The **instantaneous field of view** (IFOV) of a remote sensing system is defined as the geometric image of the detector aperture on the ground. In other words, it is defined in terms of the simple geometric optics that we have just discussed, and takes no account of the second order optical effects that we are about to consider, or problems associated with sampling rate, or anything else. Alternatively, it may be thought of as the size of the ground patch that will image onto the aperture of a single detector (Fig. 3.2). (The situation is complicated in the case of the SPOT multispec-

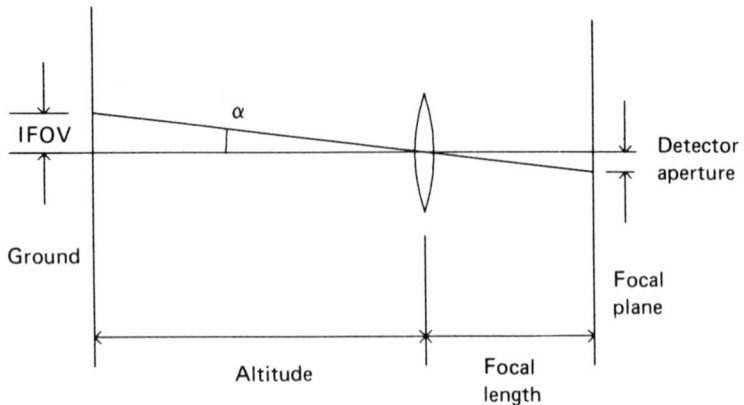

Fig. 3.2 — Instantaneous field of view (IFOV).

tral bands, where the IFOV of a single detector is 10 m; but where the output of two detectors is summed, and then averaged over two sample lines, to produce effective 20-m pixels (Chapter 16). In discussing the SPOT XS bands, therefore, it is necessary to be clear whether one is speaking of the combined IFOV of a single detector, or the *effective* IFOV of a transmitted pixel.)

From the figure, it can be seen that:

$$\text{IFOV} = \frac{\text{Detector aperture}}{\text{Focal length}} \text{ radians ,}$$

$$= \frac{\text{Detector aperture}}{\text{Focal length}} \times \text{altitude m.}$$

The performance of a remote sensing instrument is governed by two major parameters: its resolution and its sensitivity. We have already introduced the question of resolution, although more follows. The sensitivity of an instrument is its ability to respond to small signals, in this case to low light levels. An important contribution to sensitivity is made by the light gathering power of the telescope. Apart from losses, this is proportional simply to the *area* of the 'entrance pupil' or aperture of the telescope. Doubling the area of the telescope aperture will halve the illumination required to obtain a given level of performance from any particular detector.

3.1.1   Wave effects

The size of the aperture has an even larger effect on the resolution of the telescope, although, to discuss the effect, it is necessary to resort to physical or wave optics. Rays, emanating from the same point on the object, but taking different paths through the telescope, will take slightly different times to reach the same point on the image (Fig. 3.1). (If the lens is 'thick' then there may also be a significant time penalty for the path lengths of the rays through the denser medium of the glass, but this is only ever a second order effect.) Where the time differences are small in relation to the wavelength of the radiation, the rays will reinforce one another, and the effect will be the same as predicted by geometric optics. However, rays whose path difference is equal to half a wavelength will cancel each other out. A parallel bundle of rays emanating from a distant point, and captured by the full area of the telescope's entrance pupil, will generate the classic Airy diffraction pattern shown in Fig. 3.3. Readers from a number of backgrounds will recognize the form of this curve. It is the Bessel function $(J_1(x)/x)^2$. Bessel functions of this type can be regarded as the two-dimensional equivalent of the sinusoidal function, where circular symmetry is involved. Thus the curve is similar in form to the $(\sin x/x)^2$ curve (although the side lobes of the latter are about twice as high as those shown). For the same reason, it is also similar in form to the autocorrelation function for band-limited white noise (expressed in energy terms). In both one- and two-dimensional situations, the curve is known as the **point spread function**. The form of the pattern does not change appreciably if the object is not distant. The bright central spot is called the **Airy disc**, and it contains 84% of the total energy, within the first 'zero' ring. Thus the

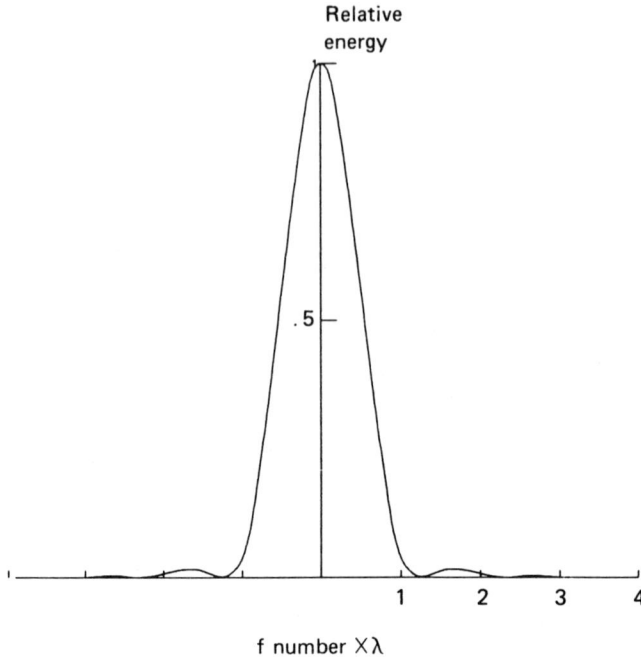

Fig. 3.3 — The Airy diffraction pattern.

Airy disc is very similar to the **80% blur circle**, which is defined as the circle encompassing exactly 80% of the energy. Other blur circles are also met occasionally. As the diagram shows, the disc will be surrounded by a series of circular fringes of rapidly diminishing intensity.

3.1.2 Resolution

The **resolution** of a telescope is defined as the separation of two point sources that can just be resolved. It is dictated largely by the size of the disc, but it is also affected by the intensity of the fringes — particularly if the two objects have low contrast with the background. The size of the disc is *inversely* proportional to the diameter of the telescope aperture. This is because the angles involved are greater and so the effect works itself out over a smaller region of the focal plane. Many practical telescopes, as we shall shortly see, feature a central blockage, normally a small mirror. This blockage clearly reduces the light gathered somewhat, although the effect is not normally significant. What it does do, however, is reduce the resolution. It does this mainly by increasing the proportion of the energy that goes into the fringes. This is logical because it is the interference between rays of maximally differing path lengths that generate the fringes. Light passing through the centre of the lens will go mainly into the disc.

The effect is analogous to the process of 'optical spatial filtering'. The reader must accept (or not) that the focussing of a beam of light is a two-dimensional Fourier transformation process, and that, by blocking off different parts of the incoming beam, both high and low pass filtering can be carried out. The central part of the beam holds the low frequency information. Introducing a central blockage therefore represents an unintentional high pass filtering operation.

The size of the central disc will also be proportional to wavelength. This too is logical since, if the wavelength is doubled, then the path difference needed to produce a given interference effect is also doubled. Therefore the resolving power of a given telescope is inversely proportional to wavelength. For a wide spectral range instrument, such as TM or IRAS, this becomes a significant effect (Chapters 14 and 15).

3.1.3 Aberrations

There are two other major sources of deficiency in practical telescopes, namely geometric and chromatic aberrations. **Chromatic aberration** does not occur in mirror-only systems. It arises from the variation of refractive index with wavelength, and will be discussed in a later paragraph.

Geometric aberrations arise from deficiencies in the lens (or mirror) profile. They are likely to be generated by any realizable surface, but they are at their worst when a spherical profile is used. In practice, only rays travelling close to the axis of a lens will follow the simple rules of geometric optics. (A mirror is considerably less likely to feature a spherical profile.) Other rays will come to a focus elsewhere on the focal plane, or even on an adjacent plane. The equation describing where all possible rays will come to a focus is complicated, and beyond the scope of this book. Mathematically the problem (for spherical profiles at least) is associated with the sine expansion

$$\sin \Theta = \Theta - \frac{\Theta^3}{3!} + \frac{\Theta^5}{5!} - \frac{\Theta^7}{7!} + \dots .$$

The convenient results of elementary geometric optics are based on the assumption that $\sin \Theta = \Theta$. When the Θ^3 term is added to the equations, then a series of higher order terms appear, each of which describes a different symptom of the same basic weakness. The effects all vary non-linearly with lens size and radius of curvature. It is therefore normally possible to devise a combination of lens surfaces that will reduce or eliminate any particular aberration. (Although it must not be forgotten that if these terms are corrected out too well, then the higher order terms will begin to intrude.)

The following descriptions of the various symptoms are included for the benefit of those, such as the author, who are not expecting to go deeply into the subject. Readers are warned that they present an over-simplified view of the nature of these aberrations (Chipaux, 1990).

Distortion is a rubber sheeting effect giving rise to such phenomena as 'pincushion' or 'barrel' distortion. Pincushion distortion is often seen in wide-angle photographs, and can result in people near the edges of pictures appearing to lean outwards. The use of a parabolic profile eliminates distortion, at least for distant objects.

Astigmatism, in this context, is different from the eye deficiency of the same name. The eye deficiency arises when the lens is slightly cylindrical, and therefore focusses lines at different distances depending on their orientation. A manufactured lens should never suffer from this problem. However it will produce a similar effect when the object is off-axis. The use of a parabolic profile does nothing to reduce astigmatism or coma (which follows).

Coma is also associated with non-axial objects. Peripheral rays will strike the focal plane at different points along a radial line passing through the principle image. This leads to an effect not dissimilar to a straightened-out comma. The tail of the comma may point inwards or outwards or may vanish, depending on the precise shape of the lens. Coma is sometimes regarded as the worst of the symptoms of geometric aberration, and therefore the most important to correct for.

Spherical aberration is simplest to describe in relation to a distant axial object. It arises, as with the other effects, because a spherical (convex) profile deflects rays distant from the axis, as it were, too much. This means that the rays at the edge of the beam will strike the axis in front of the focal plane, and will have diverged again by the time the focal plane itself is reached. The consequences are therefore not dissimilar to those of the interference effects discussed earlier, except that in this case the effect gets more severe as the aperture increases. Spherical aberration is a function of the relative shapes of the front and back faces of the lens. That produced by a plano-convex lens, for example, will differ greatly depending on which way round the lens is mounted. Spherical aberration can be eliminated by the use of a parabolic profile.

Field curvature, is a tendency, over and above the effects already described, for the focal surface to be saucer-shaped. The effect is a function of the power of the lens. However, it is also non-linear, so that the 'positive' field curvature produced by a convex lens can be negated by a suitable concave lens placed immediately behind it.

3.2 TELESCOPES

Galileo's original telescope used a convex lens to focus the incoming light rays onto the focal plane, and most terrestrial telescopes do the same. However large lenses are heavy, and difficult to make. A more serious problem concerns the profile that the lens faces should have. The ideal profile for a lens profile is a curve of infinite order, incorporating all the terms of the sine expansion already mentioned. However the fourth order curve

$$z = ar^2 + br^4 \ ,$$

is an adequate approximation for all practical purposes, where z is the distance along the axis, and r is the radius. For a lens, a and b depend on the refractive index of the glass. Where b emerges as small, this profile approximates to a parabola. A well-designed lens to such a profile would be capable of providing 'perfect' focussing of an axial object out to an arbitrarily large diameter. However it is impracticable, even in the early 1990s, to generate such 'advanced' profiles on a large lens. Indeed, as we shall be discussing later, lens profiles are always spherical except in the most exceptional circumstances.

3.2.1   Reflecting telescopes

The situation for a mirror, on the other hand, is slightly different. Conic section theory dictates that, for a distant object on the axis, the ideal profile is a parabola. Fortunately, it is considerably easier to form a large mirror into a reasonable approximation to a paraboloid form, and this fact led Newton to invent the reflecting telescope. It has also enabled countless amateur astronomers to grind and figure their own mirrors in their own homes. A **Newtonian** telescope comprises a single large parabolic mirror which forms a real image on the axis of the telescope in the path of the incoming light (see Fig. 3.4). In large astronomical telescopes, the observer often used to sit at this point together with his paraphernalia. In smaller designs, a plane mirror was used to deflect the converging beam out to the side, or back through a hole in the centre of the main mirror. Unfortunately the plane mirror, when used, has to be rather large, so that it generates a not inconsiderable blockage. This, as we have seen, has a degrading effect on performance. Newtonian telescopes are still used by amateur astronomers because of their simplicity. For professional work, however, the design has been entirely superseded by those about to be described.

In the Newtonian design, the problem of off-axis objects remains unsolved, and with it the field-of-view problem. Indeed the main reason for the decline of the Newtonian telescope is its severely limited viewing angle. Widening this angle is a major thrust area in modern telescope design. However, there is no theoretical solution to the problem. It has to be tackled by the application of fine tuning and compromise. What is required is to provide as many degrees of freedom as possible in the design, and then to juggle them to provide the best possible practical solution for the requirement in hand.

In reflecting telescope design, these degrees of freedom are obtained by using aspherical profiles, and by using more than one of them. The profiles get progressively more difficult to design and generate as one moves from spherical through parabolic to hyperbolic.

The first stage of improvement is the **Cassegrain** telescope. The Cassegrain instrument retains the parabolic main mirror of the original Newtonian telescope, but it introduces a secondary mirror which, being much smaller, may reasonably easily be made to a hyperbolic profile. Part of the effect of this mirror arises purely from its being convex. It increases the focal length, much in the manner of the bifocal spectacle lens (where a small convex lens is let into the large concave main lens). The main mirror may therefore be made to a shorter focal length than the final design requires, which is much easier to do. In addition, the secondary mirror can be placed closer to the focus of the main mirror, where it can be much smaller, and still divert the overall focal plane to a plane behind it. This reduces the blockage substantially.

Another benefit of the two-mirror design is that it can make the telescope more compact. If high-curvature mirrors are used then the length of the telescope can be shortened very considerably, without sacrificing focal length.

However the 'advanced profile' effect of the two mirror system may be thought of

	Typical aperture dia. (m)	Max. field angle (total)	Spectral band-width	Compactness ratio
Newtonian (Alternative) Parabolic 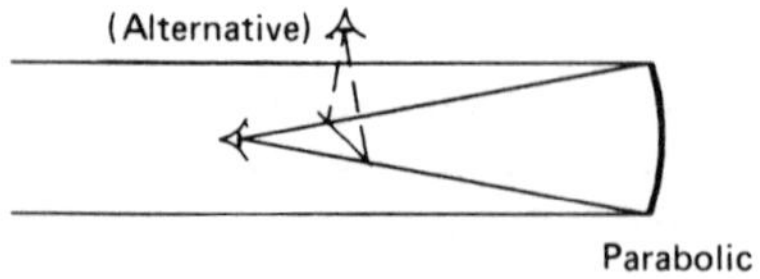	1	0.2°	Wide	1
Cassegrain 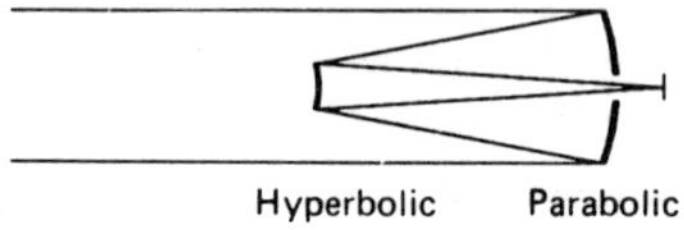 Hyperbolic Parabolic	1	0.2°	Wide	Up to 20
Ritchey Chretien 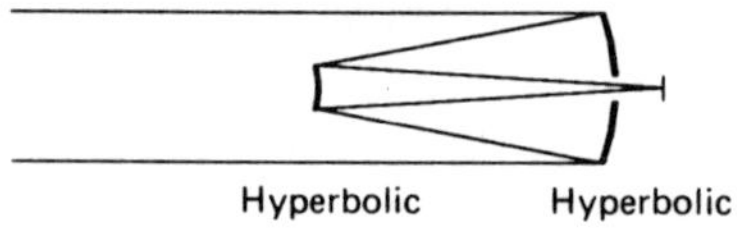 Hyperbolic Hyperbolic	5 for ground use	0.8°	Wide	Up to 20
AVHRR 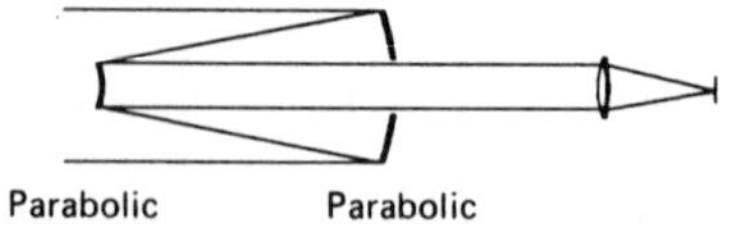 Parabolic Parabolic	See Chapter 12			
Schmidt 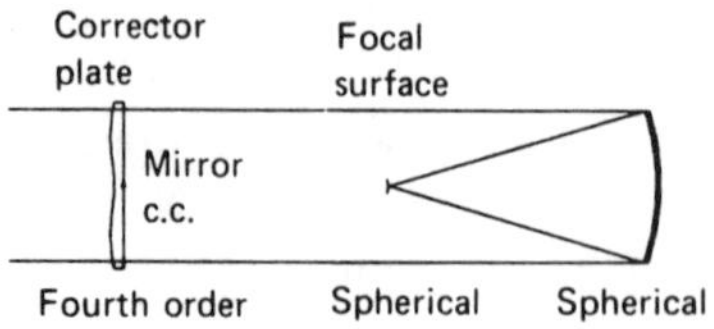 Corrector plate Focal surface Mirror c.c. Fourth order Spherical Spherical	Up to 1	5°	Limited	0.5
Refracting 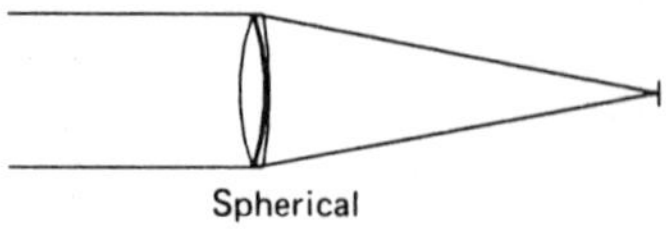 Spherical	0.1 (max. 0.3)	Up to 100°	Limited	1

Fig. 3.4 — Telescope designs. (Source: Matra).

thus. The second mirror, in partly negating the first, also enhances the effect of the second-order differences between them. The Cassegrain design provides one extra degree of freedom which may be juggled to provide improved performance. The benefit that is obtained is limited, however. AVHRR, where off-axis performance is not required, uses a design which is sometimes called a modified Cassegrain. In fact however both AVHRR's mirrors are parabolic.

The development of the Cassegrain, which is used in many of the most advanced instruments, is the **Ritchey Chretien**. This is currently the ultimate in two-mirror telescope design, although it must be assumed that fourth order profiles will eventually be introduced. In the Ritchey Chretien, the bullet is bitten, and the main mirror is manufactured to a hyperbolic profile. The secondary mirror is also hyperbolic. As the figures show, the Ritchey Chretien is visually indistinguishable from the Cassegrain, and of course it features the same advantages. However, the use of two hyperbolic profiles provides a second additional degree of freedom in the design. It is possible to generate the effect of a significantly more advanced profile than before, and to achieve a fourfold increase in view field. Most modern astronomical telescopes are Ritchey Chretien, as are those used in MSS, TM, IRAS, GOES and Meteosat. In general, the Ritchey Chretien is not suitable for pushbroom systems where an even wider angle of view is normally needed.

The variation on the Cassegrain telescope employed by AVHRR is also used in x-ray telescopes. These, like AVHRR, do not require a wide field of view. It is described as an 'afocal' design, because the optical power of the secondary (convex) mirror exactly matches that of the primary, and therefore exactly counteracts its convergence. The primary optics therefore recreate the incoming parallel beam, but concentrated in proportion to the ratio of diameters of the two mirrors (in fact, 8 : 1). The reason that AVHRR uses this arrangement is that it allows plenty of room for the installation of beam splitters and filters before refractive optics take over. The design is discussed further in Chapter 12, as are the reasons why the philosophy is practicable in the case of AVHRR.

A number of other designs have been created, using three or even four mirrors, in an attempt to increase the angle of view still further. In this they are successful, and fields of view up to 5° may 'easily' be obtained (Murphy, 1989). (Note that the field of view is always quoted in terms of the total included angle, except in this book, where the '±' is sometimes used to eliminate, neatly, any possible ambiguity.) However, they all suffer from severe deficiencies. Either there is excessive blockage by one of the mirrors, or the focal plane is generated in an inaccessible position. They are inevitably extremely expensive, although this could be acceptable in a space application. Such designs were, for example, considered and rejected for the SPOT system.

By far the most popular wide-angle telescope has long been the **Schmidt**. It is a hybrid design, and was developed as part of an astronomical camera for recording large areas of the night sky. The Schmidt telescope uses the simplest possible spherical mirror. However a 'corrector plate' is mounted in front of the mirror, located at its centre of curvature, to provide two extra degrees of freedom. The corrector plate is profiled to provide the second order correction mentioned earlier, and illustrated in the figure. On its own, this profile is cheap to form. It is adding it to a spherical or parabolic profile that is difficult. The correction is not perfect, but it enables the field of view to be extended to about 5°. The main remaining weakness of this arrangement is that it produces an abnormally curved focal 'plane'. In some astronomical applications, saucer-shaped photographic plates are used to cope with the effect.

The SPOT telescope is basically to the Schmidt design, with modifications to render it more robust, and to remove the focal plane curvature. These are discussed in more detail in Chapter 16. The modifications all involve refractive elements which, as we shall shortly see, introduce the spectre of chromatic aberration. A great deal of additional work was necessary to bring the chromatic aberration back to negligible proportions. Even so, it was only possible to meet the very tight specification of the SPOT instrument within a limited spectral range. To achieve the increased spectral range of SPOT 4 (to 1.6 μm), the use of exotic optical glasses has proved necessary.

The task of designing a reflecting telescope is far from easy, and later chapters record very substantial problems encountered by many of the instrument builders. Conceptually, however, it is relatively simple. The task is basically to select a profile for two mirrors; and then to construct them to that profile, to the necessary accuracy, out of materials that will maintain their dimensions over the full range of environments to be expected. The mirrors then have to be mounted in their correct positions, by means of supports that will hold them there under all circumstances.

3.2.2 Refracting telescopes
By contrast, the design of a refracting instrument is conceptually intractable though, mechanically, considerably simpler. Refracting telescopes are becoming increasingly popular, as design techniques improve and also as the 'bottom line' performance requirements become clearer. In particular, it is often possible to arrive at specifications that only require moderate apertures (less than, say, some 0.3 m). This is achieved by going for moderate resolutions but also, usually, by combining more than one telescope into a single instrument. The use of multiple telescopes can eliminate the light loss associated with beam splitters, and also greatly reduce the spectral range that each must handle. As will emerge shortly, this last is of vital importance in allowing refracting techniques to give good results.

However the design of the single lens of a refracting telescope is still, in the early 1990s, something of a black art. The basic reason for this is that it is still not economic to use aspheric profiles, except where no solution is otherwise possible. **Geometric aberrations** must therefore be countered by the use of two or more individual elements, each of which has strictly spherical profiles. For moderate apertures, this proves to be an economic exercise because lens faces can be polished to highly accurate spherical profiles very cheaply. It also proves to be remarkably effective. The best refracting telescopes match the best reflecting instruments in every respect — apart from a serious problem over spectral bandwidth as we shall shortly see. This performance is achieved over at least twice the field of view of the best reflecting instrument. And, as long as the aperture is not too large, refracting telescopes are extremely cheap. So cheap (and light) are they that the spectral bandwidth problem may be overcome by having a separate telescope for each waveband. This is done in several modern instruments, including IRS-1. The number of lens elements needed is, of course, a function of the degree of fidelity required.

However, the fact that the radiation has to pass through the diverting medium (which of course, with a mirror, it does not) introduces the additional complication of **chromatic aberration**. Because the refractive index of all materials is wavelength-dependent, the position of the focal plane will vary with wavelength. Unfortunately, the effect is large enough to affect even single bands of the average remote sensing instrument. It is countered by replacing each element of the geometrically corrected lens with two of different materials. The materials are chosen to have different 'dispersivities' or variation of refractive index with wavelength. By this means one element can, to some extent, counteract the aberration generated by the other. By normal photographic standards, it is possible to achieve a truly excellent performance, but this is (as we have already discussed) because of the limited spatial resolution of the human eye. A modern remote sensing instrument requires much better performance; and it is not possible, at the present state of the art, to produce a chromatically corrected lens covering more than a limited waveband — at least when other requirements, such as wide field of view and low distortion, are incorporated into the specification.

The net result of these considerations is what might be described as a surfeit of degrees of freedom, each of which is moderately inflexible in use. We have generated a design problem with far too many variables to be amenable to an analytical solution. For example, for the IRS-1 telescope, the total number of elements finally arrived at is eight. For each of these elements, we have to choose two radii of curvature, one lens thickness and one refractive index. To complete the problem, we must add seven inter-lens spacings, giving a grand total of 39 degrees of freedom to be optimized. Until computers and optimization algorithms have grown to cope with such a complex task, a more pragmatic approach has to be adopted.

Each specialist company will have its own library of tried and tested basic designs suitable for a range of possible duties. Two examples are shown in Fig. 3.5. Also shown is a 'hypothetical aspheric' lens whose performance could match that of one of them. The aspheric profiles depicted are, however, purely illustrative, and are not based on any possible design. The refracting portion of the SPOT telescope is based on the Petzval design, whereas the double Gauss is the basis for the IRS-1 telescopes. For a given application, one such design is selected whose basic properties appear suitable. This choice fixes most of the basic parameters, including the number of elements and their basic shapes. The number of variables left to be optimized is then sufficiently small to be within the compass of practical least squares optimization routines.

However, as we have already suggested, this performance is only achievable over a strictly limited range of wavelengths. It is usual, therefore, for a refracting remote sensing instrument to incorporate multiple telescopes: one (or sometimes more) for each waveband. For example IRS-1 (Chapter 18) carries a total of 12 telescopes. Eight of them are identical, and four are of identical design but to half scale. The four small telescopes provide low resolution (80 μrad pixel size) coverage, over four reflecting bands; using a single 2048-element charge-coupled device (CCD) chip each. The eight large telescopes provide doubled resolution coverage over the same four bands. Doubling up each telescope enables identical CCD chips to be used, but without the need for the complex butting arrangements that the SPOT instrument

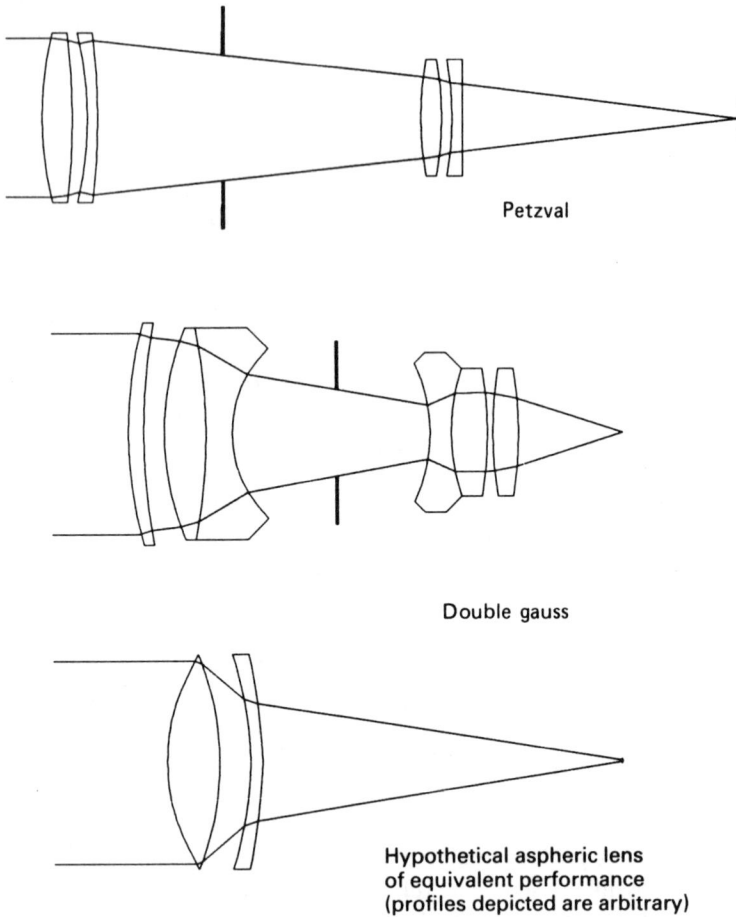

Petzval

Double gauss

Hypothetical aspheric lens
of equivalent performance
(profiles depicted are arbitrary)

Fig. 3.5 — Standard lens profiles.

requires (Chapter 16). All the telescopes are designed — and aligned — to provide inter-band registration to within a quarter of a pixel throughout the image.

All 12 telescopes are to an identical design. This cuts the design costs, but it also means that the small telescopes had to be designed to a somewhat higher specification for distortion than might otherwise have been necessary. It also means that the large telescopes had to be designed for a wider field of view than might appear to be required. However, in practice the specified linearity could only be achieved by pointing all telescopes straight downwards. Therefore all telescopes are designed to cover the complete swath, and the CCD array is mounted on the left- or the right-hand side of the focal plane as appropriate. Thus all the telescopes are designed to provide adequate performance over a field of view of 9.4°.

In some respects the MOMS system provides an even more striking example of the use of multiple refractive optics, although it is not covered elsewhere in this book for lack of information. MOMS is a pushbroom CCD system, that flies on Shuttle and incorporates two bands only. As with IRS-1, each band is served by its own telescope array. MOMS uses four CCD chips per band to obtain high-resolution coverage over a 26° swath. However the CCD chips are not mounted each on a separate telescope as with IRS-1. Instead they are interleaved on two telescopes. This saves telescopes, and overcomes the problems, experienced with SPOT (Chapter 16), of mounting several CCD chips end-to-end without gaps in coverage. Thus one telescope carries, say, chips 1 and 3 while the other carries chips 2 and 4. However, it demands a high performance, over a very wide angle of view, if a realistic specification is to be met.

3.2.3 The systems compared

The main criteria for comparisons between different possible designs are aperture diameter, field angle, spectral range, compactness ratio (i.e. focal-length/telescope-length) and cost.

Broadly, reflecting systems are able to generate large apertures and a long focal length without becoming unwieldy, and to operate over a wide spectral range. But their fields of view are strictly limited and they tend to be expensive or very expensive.

Refracting systems can provide a wide field of view and can also be extremely light and cheap. However they are only practical for limited apertures, and over limited or very limited spectral ranges.

The hybrid Schmidt system combines, paradoxically, both the best and the worst of both types. It is cheap, and provides a wide field of view and reasonably wide apertures. But its spectral range is limited and it becomes extremely cumbersome at the longer focal lengths. It also produces a curved focal surface which may, or may not, be a serious drawback depending on the application.

The relative performance of the various types are indicated in Fig. 3.4 (Chipaux, 1990).

REFERENCES

Chipaux, C. (1990) Major corrections in draft. Mr Claude Chipaux, MATRA, Vélizy-Villacoublay.

Murphy, R. (1989) Dr. Robert Murphy, NASA. Correction, in draft, to Chapter 14.

Péraldi, A (1989) Informal source material and major corrections in draft. Mr Armand Péraldi, Technical Director, MATRA, Toulouse.

4

Optical filters

Sources: **Holah (1982), Rancourt (1987)**

Remote sensing, of the kind that concerns us here, is based on the use of filters with a narrow pass band. Unlike some other applications (the isolation of a single wavelength from a laser, for example) it is important that the filter profile should have steep sides and a flat top. This however is easier said than done, as many of the diagrams in Part II show. The steepness of the sides in particular will become increasingly important as the fashion for image spectroscopy gets under way.

In this book, we are concerned with the visible and near infra-red wavelengths, up to about $12\,\mu$m; with the sole exception of the IRAS mission, whose range of interest extends to about $120\,\mu$m. Unlike electronic filters, the mechanisms used to generate optical filters cannot be relied upon to provide massive attenuation at all wavelengths outside their designed pass band. One or more additional filters, formulated for their opacity in these regions rather than for the sharpness of their cut-off, may have to be added. Black polythene is particularly useful for eliminating, totally, all wavelengths shorter than $4\,\mu$m.

There are basically two ways of limiting the spectral range of light passing through a medium. The first is the **bulk absorption** filter. It uses the bulk properties of the medium, as in stained glass or the 'gelatins' used in theatre lighting. If a medium can be engineered to have a pass-band which meets the requirement, then this is much the best kind of filter to use. It is simple, robust, easy to make, and insensitive to the angle of the incident light (although both the shape and the position of the pass-band may be affected by temperature).

Unfortunately, however, the range of options open to the designer of a bulk absorption filter is strictly limited. He has to rely on finding materials, more or less 'off the shelf', which can be combined together to give him the transmission characteristics that he requires. In the early days, users had little choice but to take what they could get. But now they are able to be a great deal more particular, and are exercising that privilege. As a result, these techniques are getting rapidly less attractive as a prime solution. They retain an important place in the filter designer's armoury, as will emerge in due course.

The reason for this change is the development of the **interference** filter. Using interference techniques, it is possible to fabricate a filter to almost any reasonable specification. An example of an interference filter is the well-known case of the oil film on water. Paradoxically, because the boundaries of the oil film are almost transparent, it only produces its rainbow effect when observed by the reflected rays. A fish sees nothing to marvel at in an oil film. As will emerge in due course, an effective **transmission** filter tends to require highly reflecting interfaces between its various media. Interference filters are sensitive to the angle of incidence, and are therefore only suitable for applications involving a relatively narrow viewing angle. Fortunately, even the $\pm 5°$ of IRS-1, may be defined as narrow for these purposes. In such an application, however, it is desirable to site interference filters in front of the telescope or other focussing device. This is because, after it, the rays are converging, which imposes an additional variation in incidence angle which might be quite considerable.

A well-tailored transmission filter is a complex structure involving multiple layers of precisely defined thicknesses of precisely formulated materials. They are difficult to design and far from easy to make. Under the stimulus of popular demand, however, techniques for the design and production of interference filters are advancing rapidly. Indeed the SPOT filters, designed in the late 1970s, employ a hybrid design, using bulk absorption techniques for the long-wavelength cutoffs and 'resorting to' interference techniques only for the short-wavelength cutoffs. By contrast the IRS-1 filters, built more recently, employ interference techniques exclusively — and, as Figs 15.6 and 16.6 show, they are much better filters. (As will emerge in Chapter 16, the design of the SPOT HRV instruments precluded the siting of the filters in front of the telescope, which may have been part of the reason for preferring to avoid interference techniques.)

A word of comment is required on terminology. Because of the possible confusion between the two spectral units (wavelength and frequency or wave number), there is a conflict between precision and lucidity when it comes to the discussion of spectral details. This book cheats. It takes the view that 'high' naturally implies high energies, even among the 'wavelength' brigade. However, the alternative is generally supplied in parentheses just to make sure.

Although two highly informative books have now been published on optical filter design (Rancourt, 1987 and Thelen, 1989), the author was unable to find an authority willing to act as consultant for the topic. In view of its importance this reticence seems unfortunate, and the current chapter is offered on a best-endeavours basis.

4.1 BULK-ABSORPTION FILTERS

There are several mechanisms available for generating selective absorption. Most of them depend on the absorption of photons by energy-state transitions of atoms or molecules within the material.

The classical mechanism is simply the vibration of the atoms in a crystal lattice. Filters using this mechanism are called **Reststrahlen** filters. The word derives from the German word for 'remainder', on account of the fact that, after several reflections, only radiation within the Reststrahlen band remains (Rancourt, 1987). The possible number of resonance modes in a lattice is, of course, large, but it may be discussed by the invocation of Hooke's law and assumed interatomic forces and masses. The effect tends to produce band-pass filters, but they are not very good, and most of the available crystalline compounds work at wavelengths greater than 24 μm. Amorphous substances exhibit a similar effect, although their behaviour is broader band than with crystalline materials. They have uses as blocking materials; and also in the coating of mirrors (see Section 4.2.5).

Holah (1982) discusses a second mechanism, namely **lattice vibrational absorption,** as being different from the Reststrahlen effect. In one variant of the mechanism, the vibrating molecules are incorporated into long polymer chains, such as polythene or PTFE (polytetrafluoroethylene). Black polythene cuts out all energies above (shorter wavelength than) 4 μm, with a tolerably sharp cut-on, being virtually transparent below (longer wavelength than) 60 μm. PTFE has a useful absorption edge at around 50 μm, but transmits again at shorter wavelengths around the visible region. Quartz exhibits an edge near 40 μm. These three materials tend to be used mainly for blocking out large regions of the spectrum, where some other mechanism is used to provide the sharp edge. However, none of them is perfect, even for this lower grade duty, and it is generally necessary to use all three to ensure complete opacity at all longer wavelengths.

More commonly, however, it is molecular resonance within ionic crystals that is invoked. The mechanism is useful, as a low (long-wavelength) pass filter, mainly at cryogenic temperatures (which in some applications is no impediment). This is because the thermal vibration of the molecules interacts with the radiation-induced vibration, and smears the absorption edge. Since, apart from IRAS, remote sensing instruments seldom operate below 70–80 K, this phenomenon is of little use to us. However, the mechanism also has some use as a high(energy)-pass filter. Materials are available covering certain energies below (longer wavelength than) about 5 μm, although the cut-off wavelength can be adjusted somewhat by varying the material thickness.

Another mechanism is **semiconductor interband absorption.** We will be discussing this mechanism in more detail in the next chapter, in relation to detector design. Clearly, however, if a photon passing through a layer of semiconductor material fails to be absorbed by whatever mechanism is generating detection, then it may pass on through the layer and out at the other side. Thus a detector may make a band stop filter at the wavelengths to which it is sensitive. Band stop filters are not, of course, used with any great regularity in remote sensing activities (but see Section 4.2.5). However, the cut-off at the low (long wavelength) end of the sensitivity band is normally extremely sharp, and so we have the means of generating the *high* (short-wavelength) cut-off of a band-pass filter. The transmission of such semiconductor materials gradually falls off again at the longer wavelengths, but too slowly to form a complete band-pass filter, even if the specification were fairly relaxed. The mechanism is available at wavelengths between about 1 and 7 μm (Holah, 1982). Doping can be used to adjust the threshold, and to control the cut-off within reasonably fine limits. This is a topic that is of supreme importance in detector design, and it is discussed in greater detail in the next chapter.

The semiconductor materials that have proved suitable for duty as low (long-wavelength) pass filters are in fact virtually the same as those commonly used as detectors, namely silicon, germanium, indium arsenide and indium antimonide. Cadmium–mercury–telluride (CdHgTe) is of vital importance as a detector material, because its response range can be tailored by adjusting the proportions of cadmium and mercury, and because the resulting material is sensitive to longer wavelengths than any other usable material. However its use as a filter material is severely limited by absorption from extraneous electrons (and 'holes') which the doping process tends to generate.

Semiconductors are also used as substrates for interference filters, as a method of providing the additional low pass filtering mechanism which it will emerge that they require.

All bulk absorption filters should, of course, be made as thin as possible, to minimize absorption in the pass band, and it would be normal to apply an anti-reflective coating to each surface. The techniques are discussed in Section 4.2.2. It takes the form of a quarter-wavelength layer of a material of intermediate refractive index. A single anti-reflecting layer can be designed to cancel out reflections at some fundamental wavelength plus harmonics. At intermediate wavelengths they will provide no benefit.

4.2   INTERFERENCE FILTERS

Interference filters are similar to a wide range of systems, including electronic filters, vibration isolators and car exhaust systems in exploiting resonance phenomena to achieve their effects. There is one important difference, however. In most other applications, **damping** is employed as one of the means of obtaining the desired performance. In car suspensions, for example, the damping is massive. In electronic filters damping is employed, in the quest for a response, that is both sharp and free from ripple and other spurious defects. An electronic filter is, however, normally built into the feedback circuit of an amplifier (see Chapter 6), which conceals the attenuation introduced. Unfortunately, a noise-free optical amplifier is not generally available to us (except for the photomultiplier tube, which is currently out of fashion, see Chapter 5), and, since detectors usually need all the illumination that they can get, damping (absorptivity) must always be kept to the minimum technically possible. In fact the best analogy is probably the travelling wave oscillator, such as an organ pipe or guitar string, because they can also respond to harmonics. Musical instruments also tend to avoid unnecessary damping.

This means that a good optical filter is much more difficult to design than a good electronic filter. The design of refractive telescopes was described in the previous chapter (with the permission of one of the key players in the field) as 'something of a black art'. Although the 'grapevine' failed to throw up an equivalent player in the current field, it will become clear that identical considerations apply.

4.2.1   The quarter-wave cell

While an electronic filter is built up from elementary RC circuits, and a vibration damper is a mass/spring system, an interference filter is an optical delay line (Fig. 4.1(a)).

The spectral unit used mainly in this chapter is wavelength (λ) for consistency with the rest of the book and with remote sensing usage. This is, unfortunately, at variance with practice in most other aspects of technology — and in particular with that of optical filter design. In these areas, frequency or **wave number** ($1/\lambda$) is more common. Frequency produces scales which fit natural expectation, in that large numbers go with large energies. In optical filter design, they also produce response curves which are clear and symmetrical. The response peaks which rapidly become excessively cramped when plotted against λ, maintain their shape and remain equally spaced when plotted against frequency.

Any wave motion, when faced with an abrupt change in impedance/refractive index/dielectric constant, always reflects some of its energy back into the incident medium. This can be seen clearly at the seaside; where waves reflect back strongly from the sea wall, but dissipate their energies totally upon the sloping beach. The ability of a gradual change in impedance to suppress reflections is used frequently in high-frequency electronics and in anechoic chamber design. We will be discussing, in the next section, how it is occasionally possible to use it directly in optics.

In general, however, optical interfaces are assumed to generate step changes in properties. The relative intensities of the transmitted and reflected rays are then dictated totally by the change in refractive index. The intensity of the reflected ray is given by the **Fresnel** formula:

$$r = \frac{(n_0 - n_0)}{(n_0 + n_1)} , \tag{4.1}$$

where n is refractive index. If the medium is absorbing then n is complex. However, we are confining our discussions to *non-absorbing* media. In some areas, the **dielectric constant**, ε, is preferred to refractive index. For a non-absorbing medium, ε is equal to n^2.

The relative amplitude of the transmitted ray is given by $1 - r$.

An increase in refractive index generates a reflected ray with a 180° phase change. To return to our oceanographic analogy, a sea wave peak is reflected back as a peak and not as a trough. However, a wave peak that meets a sudden decrease in impedance reflects a trough back. To visualize this, we can imagine a pressure peak escaping from the end of an organ pipe (or a sea wave generated in some mysterious fashion inside a long narrow harbour and propagating out towards the sea). On losing the constraint of the walls, the wave is suddenly able to flow out to the sides as well as to the fore. The combination of the potential energy pushing the fluid out to the sides and the established axial momentum pushing it forwards leads to an excessive displacement, and to a trough being reflected back up the pipe. Mathematically, the two negatives generate a positive, and lead to zero change in phase angle.

The transmitted ray is propagated through an interface between two non-absorbing media without phase change.

An absorbing medium has a complex refractive index leading, sometimes, to

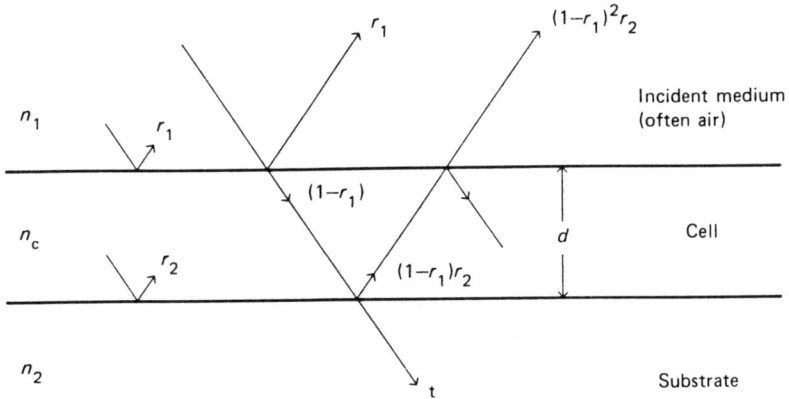

Notes

1. Refractive effects are omitted, because the deflection could go in either direction.
2. Absorption effects are also neglected.
3. If $d = \dfrac{\lambda_{important}}{4}$ then this becomes a quarter-wave cell.

(a)

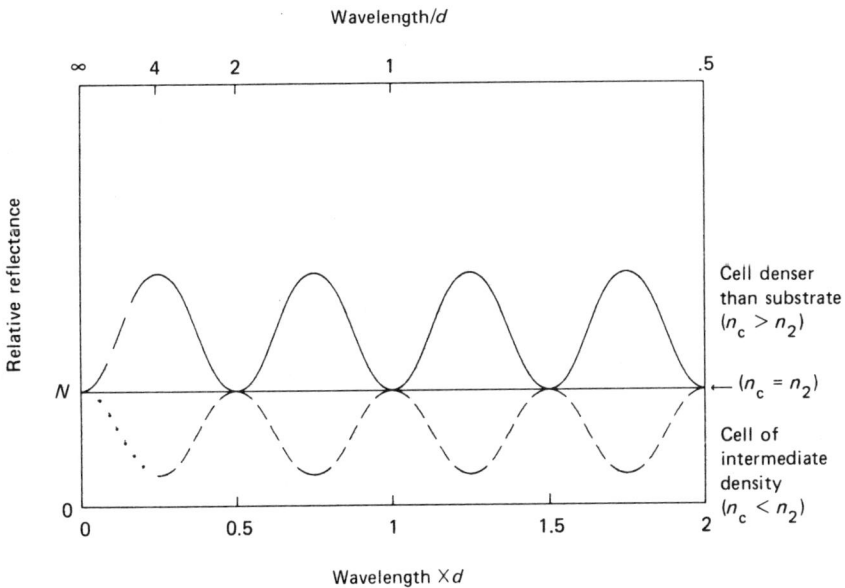

Notes

1. Only one curve applies in any one case.
2. The plot is against *wavenumber*.
3. N is the relectance of the naked substrate.

(b)

Fig. 4.1 — A simple interference filter. (Acknowledgement: Rancourt, 1987).

dramatic departures from the above simple concepts. However, mirrors (and bulk absorption filters) are the only optical elements that willingly use absorptive media, and so we will not consider this problem further.

The basic building block of the interference filter is the **quarter-wave cell.** This is effectively that depicted in Fig. 4.1, except that the cell optical thickness, d, is constrained to be a quarter of the principle wavelength of interest. (The term **cavity** is often used, where we have used 'cell' for reasons that will become clear in Section 4.2.4.) It comprises a layer of (normally) non-absorbing material sandwiched between two other media of different refractive indices. Multiple internal reflections within the cell will generate a family of rays reflected back towards the source. Another family will be transmitted on without deflection, though with a slight lateral displacement. The net effect of the cell will be the summation of both families. Rancourt, 1987 quotes the following expression, which is derived directly from Equation (4.1):

$$r_{\text{nett}} = \frac{r_1 + r_2 e^{-2i\beta}}{1 + r_1 r_2 e^{-2i\beta}} , \tag{4.2}$$

where r_1 and r_2 are the relative intensities of the first reflections, as given by Equation (4.1), i is the complex operator and β is the phase change introduced by a single passage across the cell. β is given by:

$$\beta = \frac{2\pi}{\lambda} n \times (\text{physical thickness}) .$$

As before, the net relative amplitude of the transmitted ray is given by $1 - r_{\text{nett}}$.

The complete behaviour of a single cell may be evaluated by an analysis of these expressions, as illustrated in Fig. 4.1(b). It emerges that the total energy reflected hits an extremity at wavelengths of four times the optical thickness, and at all odd harmonics thereof, hence the name. (Musicians will recognize this behaviour as analogous to that of the clarinet.) At the even harmonics, the cell has no effect, and the behaviour of the system will be dictated by the relative refractive indices of the two media on either side.

If the refractive index of the cell is greater than those of the media on either side, then the reflectivity will peak (at perhaps 35%) at the odd harmonics, and the transmission will be at a minimum. We therefore have the makings of a band-selective filter. In fact the same applies if the cell is less dense than both the adjacent media. If, however, the index of the cell is intermediate between those of the adjacent media then the transmission will peak and the reflectivity will trough. This gives us quite a useful anti-reflection coating in its own right. Indeed, if the index of the cell is exactly equal to the geometric mean of those on either side then the reflection troughs will drop to zero, and transmission will be 100%.

As is clear from Equation (4.1), it is advantageous to have as wide a range of refractive indices to play with as possible.

The degree of modification to the incident ray that is achieved by a single cell is relatively modest, and up to a couple of dozen cells are normally stacked to provide enhanced performance. Both Holah (1982) and Rancourt (1987) state that the behaviour of a multiple stack may be described in terms of a hypothetical single cell, although neither describes how the assumed properties of this supercell are derived. Nor is it explained how untoward interactions between individual cells are avoided. Readers familiar with complex vibration or control-system problems will be particularly concerned about the latter point. Such readers will have to be content with the assurance that the mathematical development takes all possible rays into account. Rancourt does, however, discuss the problem of interference between complete filters. Where these are widely separated (as in AVHRR) then the problem vanishes because the 'coherence length' of normal light is only a few wavelengths. If two filters are to be placed close together, however, then interaction may be avoided by ensuring that they are not quite parallel.

4.2.2 Anti-reflection coatings

As we have seen, the degree of reflection at the interface between media of differing refractive indices is a direct function of the change in refractive index. Thus ordinary soda glass ($n \approx 1.5$) in air reflects about 4% of the incident light, while dense flint glass (n up to 1.8) may reflect up to 8%. In a multi-element refractive telescope, such as we discussed in the previous chapter, the total loss could be very considerable. Worse, all this stray light, issuing from all the internal surfaces, can generate considerable veiling glare. The semiconductor substrates used in infra-red systems can have much higher refractive indices still, and can generate reflectances of 30% or more.

As we have seen, the ideal anti-reflection coating would generate a gradual change in refractive index, just as the wedges on the sides of an anechoic chamber generate a gradual increase in acoustic impedance, or a beach absorbs the energy of ocean rollers. Such optical coatings have in fact been generated. The trick is worked by selecting a glass with at least two phases. One phase is progressively etched away to a depth of at least a quarter of the longest wavelength of interest. The objective is to lose a decreasing amount of material as the depth is increased, leaving a microstructure of the remaining phases with air replacing the lost glass. Unfortunately the many fine capillaries left will pick up liquid contaminants, which would be very difficult to remove. In theory, an alternative would be to generate many thin layers of media of gradually reducing index. In practice, however, nature provides an insufficiency of materials to do better than a crude approximation. However Rancourt describes an effective coating for a germanium substrate employing just three layers. Its performance is shown in Fig. 4.2. Germanium, in common with many other infra-red materials, has a particularly high refractive index. This makes the provision of a good anti-reflective coating especially important. Fortunately, however, it also makes such relatively easy to design, because the availability of materials with intermediate indices is increased. As the figure indicates, coatings that span several octaves are not uncommon. (Note that the comparison of Figs 4.1(b) and 4.2 illustrates well the problems of plotting filter responses against wavelength.)

The simplest practical anti-reflection coating was in fact described in the previous section. It comprises a thin layer of a material whose index is (ideally) the geometric mean of the media on either side. The thickness should be approximately $\frac{1}{4}\lambda$ at the principle wavelength of interest. Such a coating will generate zero reflectance at this wavelength, and at every odd harmonic thereof. At the even harmonics, the coating

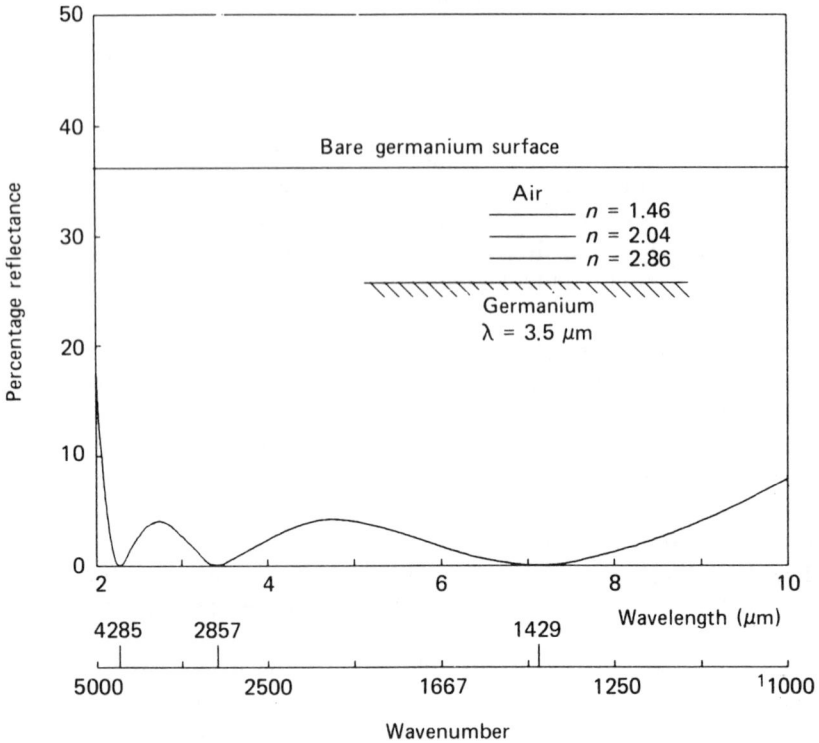

Fig. 4.2 — A practical AR coating. (Acknowledgement: Rancourt, 1987).

Note:
1. The price paid, in terms of symmetry, for the use of a wavelength scale.
2. In terms of wavenumber, the troughs are equally spaced.

will have no effect. Depending on the application, this behaviour could be highly beneficial or it could be of little benefit. By the addition of further layers, tuned to slightly different wavelengths, it is possible to broaden the reflectance trough, but at the expense of its maximum depth.

4.2.3 Edge filters
The edge filter normally provides the steep cut-off of a high-pass or low-pass filter. Unlike the equivalent electronic filter, and for reasons already discussed, an edge filter is unlikely to provide the full effect required. A high(short-wavelength)-pass filter will normally require a long-wavelength blocking filter to kill its longer-wavelength pass bands. It may also require considerable design skill to stretch the pass band upwards (towards the shorter wavelengths) as far as is required. A low(long-wavelength)-pass filter should not have spurious pass bands, but at very

long wavelengths, other properties of the materials used could generate undesired blockage.

Edge filters are generally made from all dielectric materials, because attenuation in the pass band is not normally acceptable. They therefore exhibit the variation in cut-off with angle of incidence described in Section 4.2.6.

4.2.4   Band-pass filters

These are the filters of greatest interest to us. The simplest way of producing a band-pass filter is in fact to place two edge filters in series, and where it will meet the case, this is very often done. The strategy means that the two edges may be designed separately. This simplifies the procedure considerably. Also, because there are fewer design constraints on each filter, the remaining constraints can normally be satisfied better. In particular, it is possible to produce particularly good transmission, because each filter may be optimized with this in mind (Rancourt, 1987). (It is of course necessary to prevent interaction, although in practice this does not appear to be a problem.) The principle limitation on this design is the narrowness of the pass band. According to Rancourt, the minimum practicable bandwidth, using two edge filters in series, would appear to be around 2%, in which case the strategy should be eminently suitable for many remote sensing applications.

There are a number of different ways of mounting the filters. The cheapest method could be to fabricate one filter directly on top of the other, because this way, only one surface of the substrate needs to be cleaned. However, if the substrate is thin then stresses might be a problem. To avoid warping, it may be prudent to mount one filter on each side. If the filters are especially difficult to make, then it may be wise to fabricate each on a separate substrate to maximize yields.

In applications where two edge filters cannot be used, then it is necessary to resort to a specifically designed band-pass filter. The classic band-pass filter design is the **Fabry–Pérot** filter, or **etalon**. The device is similar to the simple cell that we have already described, except for two important differences. Traditionally, a Fabry–Pérot filter comprises two plane half-silvered mirrors separated by a *half-wave* air gap.† (This is of course the explanation of the term 'cavity' referred to earlier.) Theoretically, a Fabry–Pérot filter is adjustable. However, since extreme accuracy is required, the air gap is often replaced by a transparent medium, which destroys this feature.

The half-silvered surfaces of the Fabry–Pérot filter generate far more internal reflections than obtain in the cells that we have considered so far, and so the net transmitted and reflected rays are the sum of many more individual rays. A Fabry–Pérot cell is therefore more highly tuned, and is capable of producing much sharper response peaks. However, the shape of the peak can only be controlled by varying the reflectivity which, as Fig. 4.3 shows, also greatly affects the out-of-band rejection. Furthermore, the response will always have a peaked top.

Apart from specific applications, therefore, band-pass filters are generally created by fabricating a number of Fabry–Pérot cavities in series. The cavities are separated by quarter-wave spacers of higher refractive index. In fact, to minimize the effect of incidence angle, the highest index possible is used.

---

† Readers are warned that, at one point, Rancourt describes an Etalon as being a *full-wave* cavity. However Holah (1982) and Rancourt, on another page, refer to it being a half-wave structure.

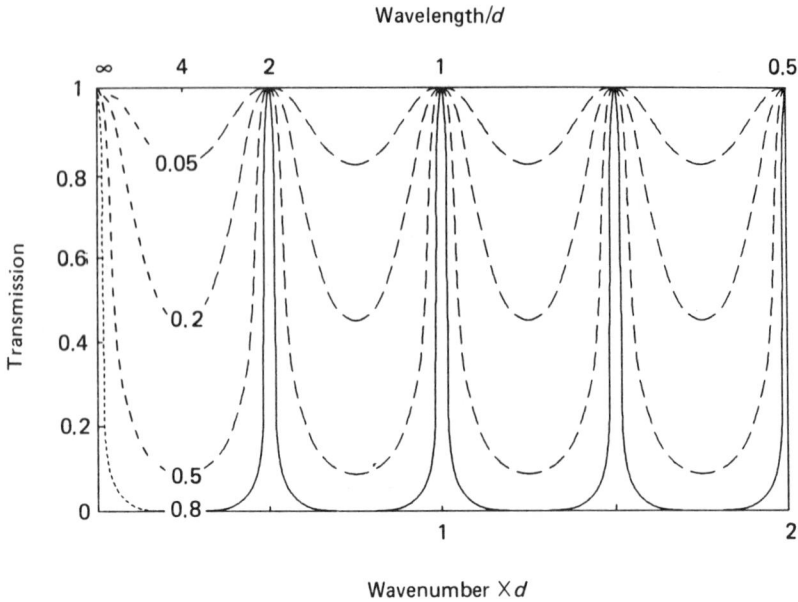

Fig. 4.3 — Transmission of a Fabry–Pérot cell. (Acknowledgement: Holah, 1982).

Because they are highly tuned, the degree of interaction between Fabry–Pérot units is much increased. Within a single filter the control of all the rays generated is part of the design. But if two separate filters are placed close together then there is a danger of a third, uncontrolled, filter being generated. Any chance of this must be eliminated, perhaps by the strategy (mentioned earlier) of placing the substrates at a very slight angle.

Because of the need for silvering, Fabry–Pérot filters can never generate more than, perhaps, 50% transmission, and the filters will always have a 'wavy' top, as the practical figures in Part II show. However the strategic introduction of a third refractive index enables much improved 'super square' profiles to be created, in which these effects are reduced. Fig. 4.4 shows a comparison between a Fabry–Pérot filter and a comparable 'square' filter. The Fabry–Pérot filter is described as '$(HL)^2L^2(LH)^2$', and the square filter as '$(LH)^2L^2(HL)^4(LH)^4L^2(HL)^2$', where the indices of the H and L layers are 3.5 and 2.

4.2.5 Reflectors

The traditional reflecting surface is, of course, polished metal, which was common at the time of the siege of Troy. In later years, a thin film of silver was evaporated on to

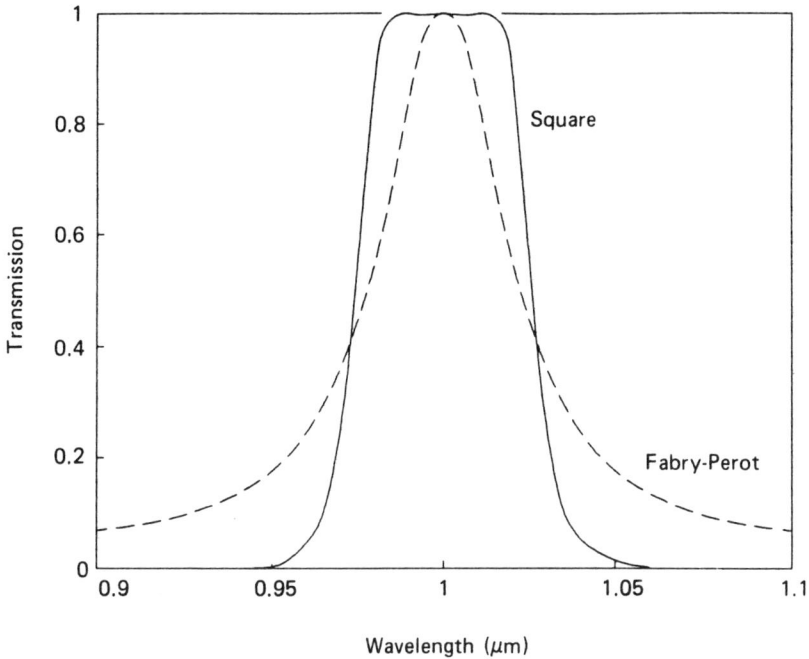

Fig. 4.4 — Comparison between a Fabry–Pérot and a square band-pass filter. (Acknowledgement: Rancourt, 1987).

the back of a sheet of glass. This provides a particularly flat face, and protects the reflecting surface. For precision mirrors, the silver (or other metal) is generally deposited onto the front face of the glass to eliminate double reflections. Other methods of protection then have to be considered. A metal that provides optimum reflectivity over part of the desired spectral range may require assistance over other parts. Occasionally, it might be necessary to generate supremely high reflectivities over a severely limited spectral range. These are the reasons for the inclusion of this section.

Glass is generally used for the substrate of precision mirrors because it is hard, stable and easy to grind and polish to extremely accurate profiles. However, some of the mirrors covered in Part II are made from a metal such as beryllium.

Metals (silver, gold or aluminium) provide the best general purpose reflecting surfaces, covering as they do a wide range of wavelengths and angles of incidence with minimum untoward effects. Silver has the highest reflectance of any metal in the visible and infra-red. In the visible, it averages out at around 98%. However it tends to tarnish. Astronomical silver mirrors need to be recoated regularly to retain optimum performance. In Space, of course, that is not possible, and protection is required. Silver is not a good reflector in the ultraviolet. Gold is good in the infra-red, and does not tarnish. However, it does not adhere well to glass, and requires an intermediate layer of a material such as chrome which adheres to both. Pure gold is also extremely soft, and so a gold mirror is easily damaged.

Aluminium is less good than silver or gold, with an average reflectance in the visible of around 92%. However, it remains a popular coating material because it is inexpensive and easy to apply. Aluminium starts to oxidize at once, but a thin layer of the oxide is transparent over a wide spectral range and so the effect is not excessively serious. However the deposition processes tend to generate roughness when the thickness approaches $0.1\,\mu$m or more, and so the layer must be kept thin. A coating of optimum thickness will render a strong incandescent bulb placed behind the mirror just barely visible. Aluminium is good in the ultraviolet. However, it has a dip, to about 86%, at around $0.825\,\mu$m. Aluminium would be quite unsuitable for a multi-mirror telescope such as we discussed briefly in the previous chapter, because four or five surfaces would generate a severe loss. The same number of silver surfaces would still be quite acceptable.

The mirrors in the IRAS telescope (Chapter 15) were fashioned from beryllium for its stability over a wide temperature range. The secondary mirror only was coated with aluminium to improve its performance in the visible region.

Both silver and gold mirrors can be overcoated, for protection, with materials such as silicon dioxide. However, the overcoat materials tend to be transparent only over a limited spectral range which, for wideband instruments, can be a serious problem. The thickness of the overcoat has to be carefully considered. For durability it has to have an optical thickness of at least quarter of a wavelength, in the visible. But it will reduce reflectivity in the spectral region around its quarter-wavelength point, so it may well have to be thicker than that. A possible strategy is to make its optical thickness half a wavelength at the centre of the region of main interest.

It is possible to use a stack of coatings to enhance the reflectivity of a metal mirror. The effect works over a limited spectral range only, but it is possible to increase the reflectance of an aluminium mirror to about 99% over the visible range. The price, however, will be a deep trough, down to perhaps 75%, at the low (long-wavelength) end of the enhancement band.

Over relatively narrow spectral bands, even higher reflectivities can be obtained by the use of dielectric stacks on their own. Effectively they are band-stop transmission filters. Dielectric materials are durable and so no protection problems occur. The performance of a stack with a given number of layers is critically dependent on the range of refractive indices available. For example, if an index ratio of 3 were available, a 99.9% reflectance could be obtained with the use of 8 or 10 layers. However if the available ratio was 1.5 then some 20 layers would be needed. Therefore these mirrors are much easier to make for the infra-red.

Fig. 4.5 illustrates the performance of a narrow-band high-reflectance filter, derived from Rancourt (1987). It is described in the reference as '$(\frac{L}{2}H\frac{L}{2})^5$'. The two refractive indices are 1.35 and 2.50. For comparison, the performance of a single '$\frac{L}{2}H\frac{L}{2}$' stack is also shown.

4.2.6 Beam splitters

Beam splitters come basically in two types. **Neutral-density** beam splitters function like the traditional half-silvered mirror. Something under half the energy is transmitted, something under half is reflected, and a not insignificant proportion is absorbed. By contrast, **dichroic** beam splitters transmit most of the energy in one waveband, and reflect most of the energy in another. It is of course necessary that the

Wavelength (μm)

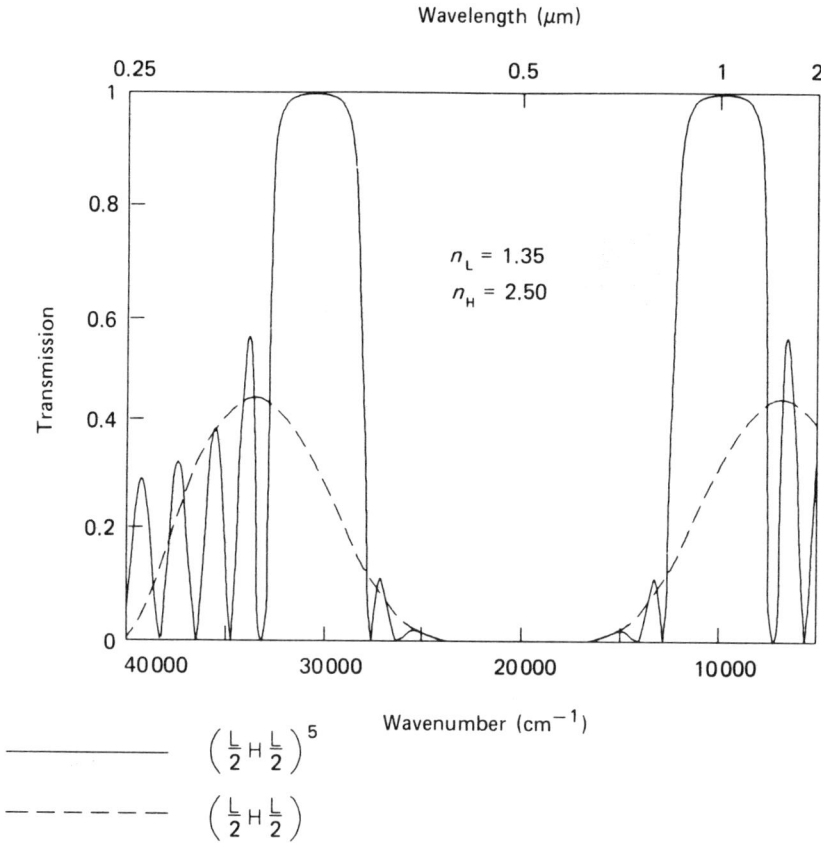

Fig. 4.5 — A narrow-band high-reflectance mirror. (Acknowledgement: Rancourt, 1987).

splitter should have a high rejection of the unwanted radiation in each case. Where there is a sufficient difference in the wavebands to be separated, dichroic beam splitters are clearly much more efficient. Edge filter technology is normally used and, by careful design, they can also double as part or all of the band-pass filter. AVHRR (Chapter 12) provides an excellent example of the use of both types of splitter.

However, beam splitters suffer from a problem that has been ignored up until now, namely polarization. In general the filters and mirrors described in this book are used at very nearly normal incidence, and so the problem is not normally significant. However, beam splitters tend to be used at 45° incidence, and to generate a perpendicular reflected ray while leaving the transmitted ray undeflected. Having said that, however, AVHRR incorporates one splitter which returns the reflected ray as closely as practicable whence it came. This is done specifically to minimize the polarization effect. An additional mirror is incorporated into one of the beams, specifically to ensure that all the beams are afflicted equally.

The source material does not cover polarization effects in relation to dichroic filters; however Fig. 4.6 illustrates the problem as generated by a neutral-density

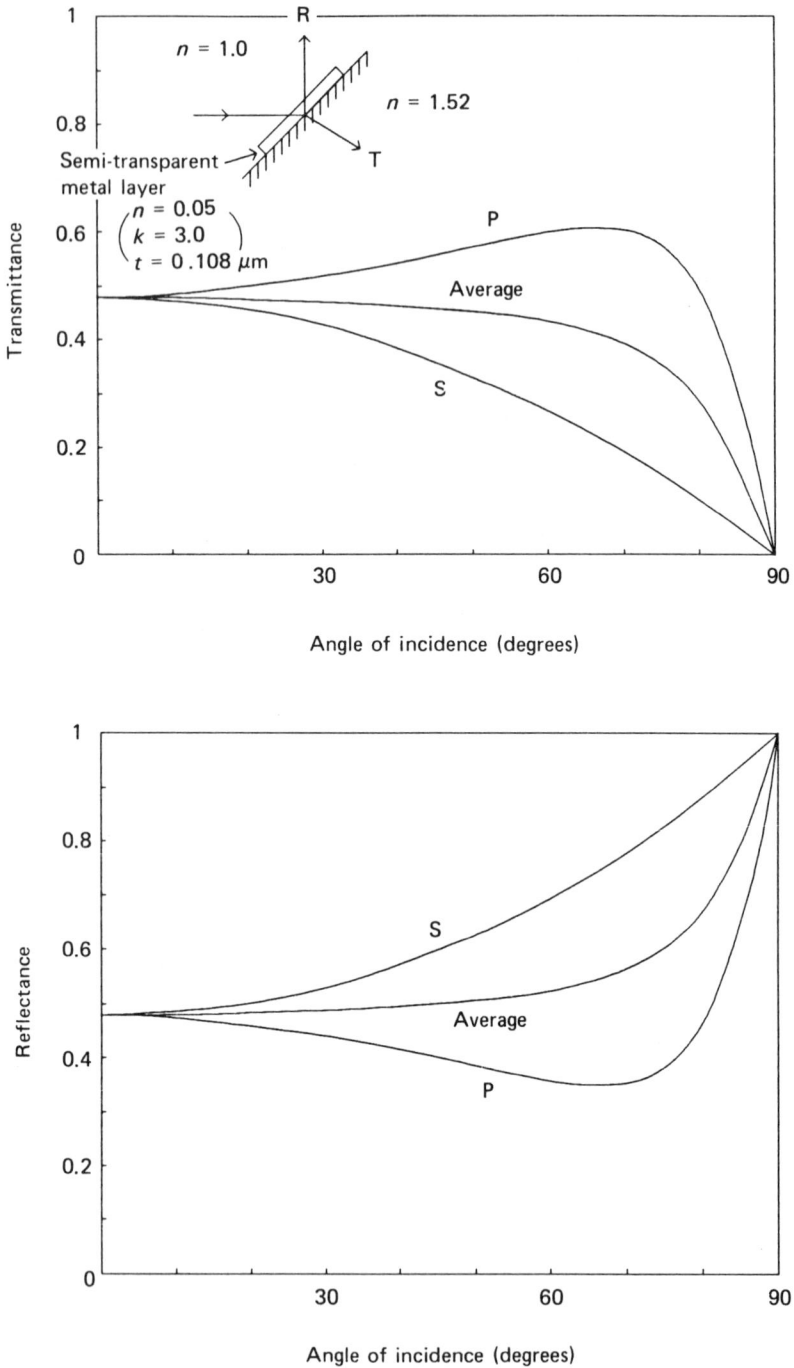

Fig. 4.6 — The effect of incidence angle and polarization on reflectance and transmittance.
(Acknowledgement: Rancourt, 1987).

splitter. The terms 's' and 'p' describe the plane of polarization as being perpendicular or parallel to the plane of incidence. (The term 's' derives from the German word for perpendicular, namely 'senkrecht'.) The splitter illustrated is made from a 0.0168-μm layer of silver, whose complex refractive index is $(0.05 + 3.0i)$. The incident medium is air and the index of the substrate is 1.52. It is an extremely efficient splitter, in that the maximum absorptance is about $2\frac{1}{4}$,%. A similar splitter, using chromium (index $3 + 5i$), has a maximum absorptance of nearly 50%. (This absorptance is more or less constant, out to an incidence of some 50°, whence it drops progressively to reach zero at 90°.

4.2.7 Emissivity control
This topic is really beyond the scope of this book; however, a few comments are probably justified. The simplest emissivity-reducing coating is familiar to all who have looked inside a dewar flask. It comprises an opaque coating of a high infra-red reflectivity material such as silver or gold. If the coating must be transparent to visible radiation then 'ITO' (an oxide of indium and tin) can be used. The coating needs to be on the 'vacuum' side of the cavity.

Emissivity-enhancing coatings are also available. These may be used to improve the thermal coupling between a heat source and its environment. At the same time, it may be necessary to reflect external radiation. A suitable thermal control surface for a spacecraft is given by Rancourt, 1987. It is an emissive layer of fused silica sheet, coated on the inside with a reflecting layer. The reflecting layer is made from a dielectric reflectance-enhancing stack over a layer of silver. This arrangement provides a surface which can generate a useful outflow of heat, at a realistic environmental temperature, even in direct sunlight.

REFERENCES

Holah, G. D. (1982) 'Far-infrared and submillimeter wavelengths filters', *Infrared and Millimeter Waves* Vol 6, Academic Press.
Rancourt, J. D. (1987), *Optical Thin Films User's Handbook* Macmillan.
Thelen, A. (1989) *Design of Optical Interference Coatings*, Macmillan.

5

Detectors

Consultant: **D. R. Sloggett**†

The detectors are, in some respects, the key to the design of the whole instrument. The factors that govern their performance are very similar to the requirements of a photographic film. The speed rating of a film dictates the exposure levels that a camera must provide, and both the aperture of the lens (in 'f' numbers, or fractions of its focal length) and the exposure time (in fractions of a second) can be controlled to ensure that the correct amount of light reaches the film. Similar considerations apply to the detectors of an Earth observation instrument.

5.1 NOISE

Noise is generated in all electronic systems, and the principle mechanisms of interest to us are discussed briefly below. Most metering systems start with a very small incoming signal, which must then be greatly amplified before being presented to the user as a usable output. Noise which is generated by the early links in this chain receives exactly the same amplification as the signal. Therefore it is to the early links that one must look for trouble. Noise generated by later links, such as the A/D converter, must be at a very much higher level before they begin to contaminate the data stream.

5.1.1 Units

The non-specialist reader may well find the situation regarding units confusing. It will emerge that a wide range of detectors produce an output *signal*, in amps or volts, that is proportional to the *power* of the incident radiation. The reason for this is that a quantum of incident radiation, which represents a packet of energy, releases an electron within the detector, which represents a packet of charge. Roughly, it could be argued that the energy of a photon is expended in releasing a single electron from the crystal structure, thus making it available to contribute to the detector current. A

† Principle Consultant, Software Sciences Ltd, England.

flow of energy constitutes an incident power, and it generates a flow of electrons, ideally at a rate of 1 : 1, which generates a current. Thus it is correct to define a 'noise equivalent power', in watts, in terms of an rms *voltage* out of the detector.

In fact it is the output current that most directly characterizes the response of a detector, although most electronic circuits are designed to be sensitive to voltage. The two are, of course, related by $I = V/R$. This means that the voltage is also a function of the detector resistance, which may vary slightly with operating or other conditions. (The problem may be overcome by backing the detector by a 'current-driven' amplifier; see Chapter 6).

5.1.2 Detector noise
There is a fundamental limit to the sensitivity that can be obtained from any detector, a limit which comes straight from quantum mechanics. Electromagnetic radiation is propagated, not as a continuous flood, but as a stream of packets or photons, and at low illumination levels a single photon can carry a significant proportion of the total energy flux.

The effect of this is to superimpose a random fluctuation, or **noise**, onto the steady signal that we might otherwise expect (Fig. 5.1). This noise is called, variously, **photon, quantum** or **radiation** noise. The statistics of the incoming flux mean that the photon noise power varies as the square root of the wanted flux, so that increasing the illumination reduces the uncertainty in any particular reading. In practice, photon noise is not easy to separate from the 'shot' noise generated by the detector itself (see Section 5.1.5). For reasons that will become clear, the statistics of shot noise are identical with those of photon noise, and the usual expression for the total detector noise signal is

$$N i_d = \sqrt{(2ei\Delta f)} \text{ amps}$$
or   (5.1)
$$N v_d = \sqrt{(2evR\Delta f)} \text{ volts },$$

where e is the charge on the electron, at 1.6×10^{-19} Coulomb, i is the current passing through the detector at the time, and Δf is the electronic bandwidth. v is the voltage generated across the detector, and R is its resistance. If i is generated entirely by the photon flux then $N i_d$ (or $N v_d$) is defined by Norwood & Lansing (1983) as being the photon noise itself. On the other hand ITT†, in their analyses for AVHRR and related instruments, show no sign of acknowledging the existence of photon noise as an entity separate from the detector shot noise.

5.1.3 Detector noise control
The implications of Equation (5.1) are that the signal-dependent noise signal (photon, shot or both) increases as the square root of the incoming illumination, which confirms that increasing the light reaching the detector reduces the *relative* noise level of the signal. The noise also increases as the square root of the electronic bandwidth. Reducing the electronic bandwidth has the effect of averaging the

† (see, for example, ITT 1982).

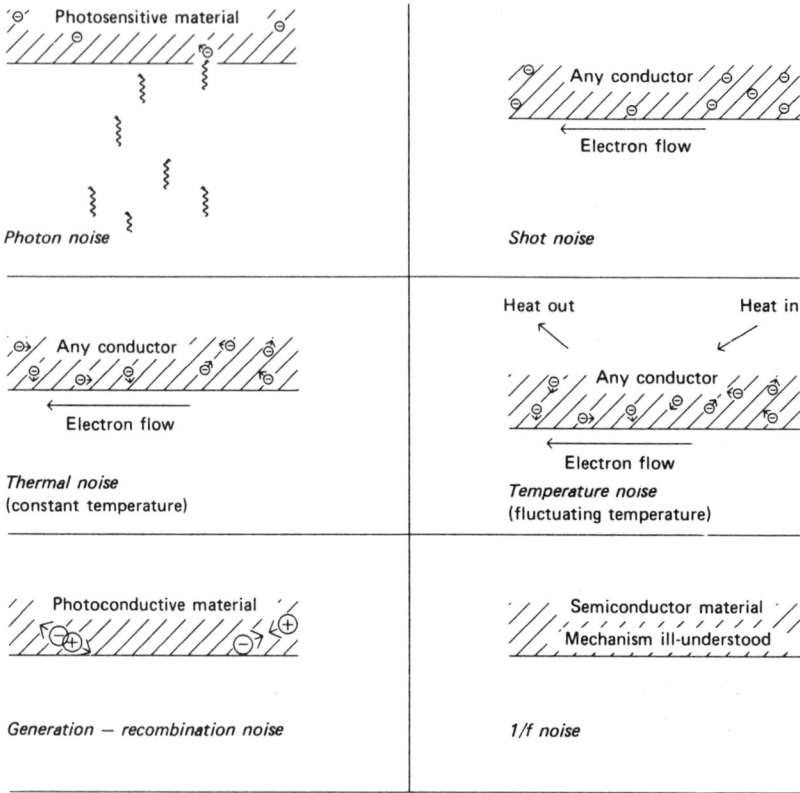

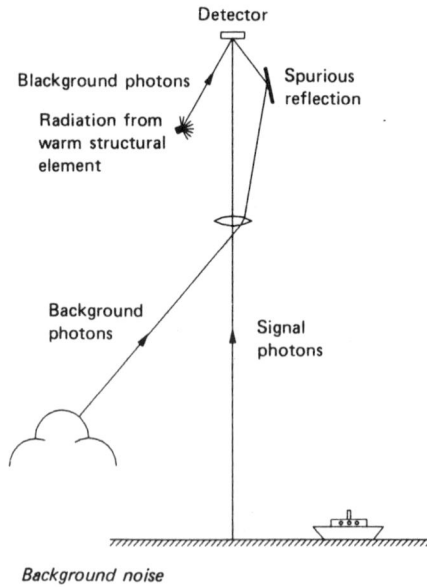

Fig. 5.1 — Noise sources.

incoming signal over a longer time. Note that the optical bandwidth does not appear as an explicit factor. It is equally important, however, as we shall see in Section 5.1.6, and its effect is included in the value obtained for i or v.

Increasing the exposure, averages out the noise fluctuations and enables a more accurate reading to be obtained. However, the exposure can only be increased at a price, either in terms of cost or of other aspects of device performance. The design study must ensure an adequate performance for the purpose in hand, while at the same time minimizing cost, size, weight, power requirement, downlink capacity, and so on. For example, most terrestrial applications require results to within, say, 10%, and this forces the designers of Earth observation instruments to design for moderately large exposures. The IRAS mission, on the other hand, was mainly interested in identifying sources, not in measuring their precise strengths. Results were quoted in the logarithmic magnitude scale. The designers were therefore able to turn up the electronic gains, and obtain sensitivities several orders of magnitude higher than would have been possible if the same instrument had been intended to be turned onto the Earth. (The detectors were also cooled with liquid helium; the significance of which will become apparent.)

The exposure is controlled by four separate design features. These are illustrated in Fig. 5.2. First, increasing the telescope **aperture** increases the number of photons per unit time that reach the detectors. However there is a limit to how large the telescope can be made without a dramatic increase in cost and weight penalties. Second, increasing the 'transmission' (transparency) of the light path minimizes the demand for a large telescope. Unfortunately design problems with the spectral filters and separation of the different bands are eased considerably if light loss can be tolerated. Third, each spectral 'slice' can be made as wide as the requirements will permit. Doubling the range of wavelengths over which photons are collected doubles the exposure obtained. Fourth, the signal can be averaged over a longer time, to smooth out the fluctuations. Owing to the satellite motion, however, this may limit the spatial resolution that can be obtained, or demand a bulky array of paralleled detectors, which will in turn feed back to the telescope design.

It will emerge in Part II that photon noise is by no means always a significant design problem. The visible bands are fairly well provided with incoming irradiance, and ample photon flux can easily fall out as a by-product of other considerations. On perusal of Chapter 14, the reader might expect photon noise to be a difficulty for Thematic Mapper, but no reference to it was found in the source material. In fact, the only reference found, to photon noise as a controlling factor was in relation to the mid-infra-red band of AVHRR (Chapter 12).

5.1.4 Background noise

Background irradiance is energy that reaches the detector by extraneous routes or from extraneous sources. It could originate from the wanted source (or from another bright external source), but arrive via spurious reflections from the structure etc. For the IRAS mission (Chapter 15) this was a real danger. Or it could originate from local sources such as the spacecraft structure. This mechanism is most likely to be troublesome on high-gain thermal instruments, such as AVHRR Band 3, because the associated structure is probably relatively warm. The minimization of background irradiance was in fact a serious consideration in the design of AVHRR

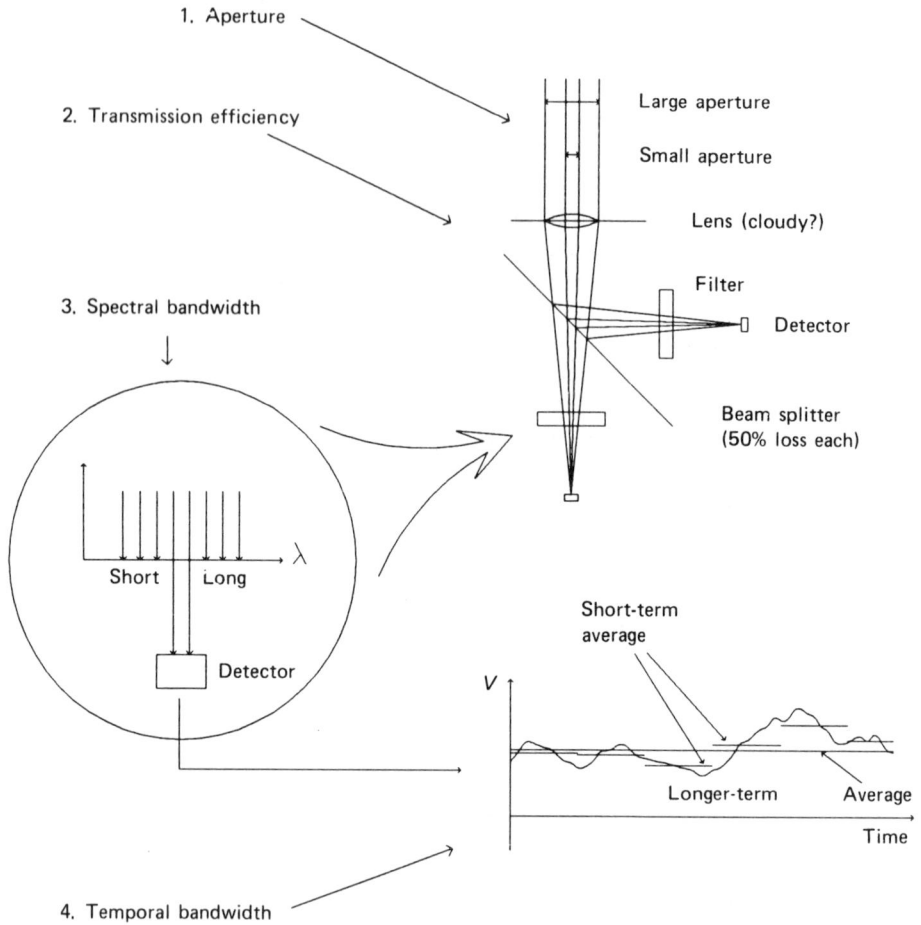

Fig. 5.2 — The factors contributing to detector exposure.

(Chapter 12). **Background noise** is the additional noise generated by the background irradiance. (The constant background signal, if not too large, can be treated like 'dark current'; see Section 5.4.)

5.1.5 System noise

However, even if the photon flux is not low enough to generate significant photon noise, it may well excite but a small response from the detector. In this situation, noise generated by the detector itself — or by the high-gain pre-amplifier that must follow it — is likely to become the limiting factor.

Electronic circuits generate a noise very similar to photon noise, on account of the quantized nature of electricity. The flow of electrons past any point in the circuit will fluctuate slightly, in the manner of traffic on a motorway. This will lead to fluctuations in the instantaneous measured current, which will manifest itself as noise. The same statistics apply as for photon noise, and the expression for **shot**, or **Schottky noise** is identical to Equation (5.1). Shot noise is particularly severe in semiconductor devices, where it will emerge that electrons have to cross boundaries between differently doped regions of semiconductor material. Shot noise constitutes that part of Ni_d, (Equation 5.1), that is not generated by the incoming photons.

Because it is proportional to the square root of the current (or voltage), shot noise is at its worst in low impedance circuits, where currents tend to be considerably greater. In Hughes (1984), it is described how the design of the thermal-band detectors for Thematic Mapper was in part dictated by the need to maximize their impedance.

An additional source of noise is generated, in resistive elements, by *thermal* agitation. Fluctuations in the energy of the electrons cause further instantaneous fluctuations in current. Fluctuations in the vibrational energy of the conducting material cause instantaneous fluctuations in the effective resistance. These effects are proportional to the square root of absolute temperature. They combine to generate **thermal, Nyquist** or **Johnson noise**. It is given by:

$$Ni_t = \sqrt{(4kT\Delta f/R)} \text{ A },\qquad\qquad (5.2)$$

or

$$Nv_t = \sqrt{(4kT\Delta fR)} \text{ V },\qquad\qquad (5.2a)$$

where k is Boltzmann's constant (1.38×10^{-23} J/K), T the absolute temperature in kelvins and R the resistance in ohms. If R is 1 kΩ, and the bandwidth is 1 MHz, then (at ambient) the thermal noise is 1 μV.

An additional source of thermal noise, in semiconductor elements, is fluctuations in the generation and/or recombination of electron/hole pairs. (As we shall be discussing in Section 5.4.1, a hole is a gap where an electron should be.) This is called **generation** or **generation/recombination** noise.

The thermal noise voltage is proportional to the square root of the resistance of the element, and is therefore at its worst in *high*-impedance circuits. It will emerge in the next chapter that high-gain amplifiers, such as are required by detectors producing small outputs, normally involve large 'feedback' resistors. These generate substantial quantities of thermal noise, and this may in fact become the factor limiting the sensitivity of the channel. An example is the thermal channel of INSAT (Chapter 13). In Part II, we will be discussing a number of circuits in which the preamplifier's feedback resistor is mounted in the cool zone beside the detector. The dependence of thermal noise on temperature is the reason why some of the detectors used in remote sensing require to be heavily cooled. Cooling adds weight, design complication, and possibly the need for cryogenics, and is therefore always kept to a minimum consistent with acceptable results. The effect of thermal noise is also minimized by keeping the light levels high and by averaging over a longer time, as already discussed.

Thermal noise is not the same as **temperature noise**, which is due to small fluctuations in the actual temperature of the detector, caused by fluctuations in heat loads etc. Temperature noise is mainly a problem with thermal detectors (Section 5.4.2), but the response of 'quantum' detectors is also affected to some degree.

Nowadays, signals are virtually always digitized, for transmission and/or storage, and this modifies somewhat the problem of defining the maximum acceptable noise level. The continuous grey scale that represents each band within the analogue parts of the circuitry will be split into 64, or 256, or some other number of discrete levels by the digitization process. What is required is that the superimposed noise should not push more than a small percentage of the pixels into the wrong level.

The noises that we have discussed turn out to have similar statistical properties. They are all essentially 'random', in that the amplitude of the noise component at any given time is completely unpredictable. On the other hand, if a sufficiently long sample of the noise is analysed, then its mean value will average out to zero. The fluctuations will be found to contain all possible frequencies all at equal amplitude. In other words, it is 'white' noise (as distinct from $1/f$ or 'pink' noise, in which the lower frequencies are stronger). All noise is in fact invariably 'band limited' by practical considerations. It will not normally go down to zero frequency, and something in the system will certainly filter out the very highest frequencies.

The types of noise just discussed are the most fundamental, and normally the most significant, contaminants of optical signals, but there are many others which depend more on the techniques adopted and the components used in the system. For example, capacitors introduce 'dielectric loss noise', semiconductor detectors generate **flicker**, **excess low frequency** or **1/f** ('one over f') **noise**, in a manner not fully understood (Wilson & Hawkes, 1983). Individual detectors, built into large arrays, will all differ slightly in their gains, offsets and noise levels. This leads to **pattern noise**, usually manifested in imagery as 'striping'. Sampled data systems (i.e. virtually all imagery) may well suffer from **aliasing noise**, as discussed further in Section 5.3.1. CCDs are subject to a number of specialized noise problems as a result of the complicated way in which they operate, as is discussed in more detail later in Section 5.5.5.

The total noise power generated by a circuit is the sum of the individual noise powers, so the total noise signal (voltage or current) is given by:

$$N_T = \sqrt{(N_d^2 + N_t^2 + \text{etc})} \ . \tag{5.3}$$

5.1.6 Noise bandwidth

When bandwidths are being discussed it is important, at all times, to be aware whether it is the optical or the electronic bandwidths that are meant. The optical bandwidth of a channel is dictated by the response of the optical filters and/or the detectors. The electronic bandwidth is normally governed by the IFOV and the scanning rate (except for CCD systems which are somewhat different).

In general, the electronic bandwidth should be adjusted to the rate at which the incoming signal may be expected to change. A rapid response system for measuring the speed of a ship, for example, would not represent good design. The effect of noise is to introduce uncertainty into individual readings. If all else fails, the effective bandwidth of a measurement can be reduced by means of pencil and paper. If pairs of individual readings are averaged, then the uncertainty will be halved. However, individual readings must be independent estimates, so it will not necessarily be possible simply to take the average of twice as many readings. Therefore the cost is a halving of the rate of change in the signal that can be picked up.

Many noise sources, theoretically at least, generate over an infinite range of frequencies. This leads to the paradox of sources apparently able to generate infinite noise levels, which is of course nonsense. Fortunately, however, as we have already mentioned, any practical instrument will respond only to energy within a limited range of frequencies (or wavelengths). If noise is a problem, then filters can be incorporated to reject all energy outside a specific range of interest. The narrower the instrument's 'pass band' the smaller the amount of noise that gets through to contaminate the wanted signal. Thus the question of bandwidth is fundamental to any discussion of noise levels.

Many parameters are defined as **spectral** quantities; i.e. they are defined 'per unit bandwidth'. In such quantities, wavelength (or frequency) will appear specifically in the dimensions. Others are defined in terms of the pass band of the instrument, or individual channel thereof, or in some other relatively loose way. Here, frequency/ wavelength will not necessarily appear in the dimensions, although the bandwidth assumed will still control the quoted figures. It is important to be clear about this, because not all sources are as careful over their definitions as might be desirable. We are about to define some units as spectral quantities, according to the above definition, and others not. However we cannot guarantee that the reader will not meet different variations on the theme as he peruses the extensive literature.

5.1.7 Measures of noise performance

The noise source(s) that predominate in any particular channel will depend on the frequency band, on the performance requirements, and on the main design parameters that are chosen to meet the requirements.

What is common to all designs, however, is that there will be a noise problem, that it will limit the sensitivity that can be obtained from the instrument, and that its magnitude requires to be described.

In the sources consulted, the noise descriptor tends to be defined in terms of the incoming noise power in watts, rather than in terms of the outgoing noise signal in volts (or amps), though whether this convention is universal cannot be guaranteed.

Possibly the most universal noise level descriptor is:

• **SNR (signal to noise ratio)**. This is a measure that will be familiar to many readers. When SNR is unity then the statistical uncertainty in any one reading is equal to its value. However, it needs to be related to signal level (e.g. full scale or 'minimum signal'). If none is given then it becomes a little difficult to interpret. It is sometimes quoted at a particular **albedo** (ground/atmospheric reflectivity); in which case a solar irradiance must also be quoted or assumed.

  In quoting SNRs, it is generally important to be clear whether the units are power or amplitude related, since power (in watts) is proportional to the square of the amplitude (in volts, amps etc.). (This problem vanishes, of course, where the figure is converted to decibels, because the difference is taken into account in the conversion.) In our application, it is necessary to remember that the detector output voltage ratio is nominally *equal* to the incident power ratio.

Many of the noise level descriptors used in the current context are based on equating the basic minimum 'floor' noise level to a useful parameter. We have:

- **NEI (noise equivalent irradiance)**, in watts/λ. This is the amount of (spectral) 'radiant power' incident on the detector that would give the same output signal as the total (spectral) noise signal, including all noise sources. It also represents the level of wanted signal that would produce an SNR of unity. NEI is an absolute measure in that, unlike some other measures, it requires no assumptions concerning the state of the atmosphere etc. It is therefore of particular use to instrument designers because it gives them numbers that they can relate directly to their design constraints.
- **NEP (noise equivalent power)**, also in watts/λ. This is the power that a noise-free detector, which was otherwise identical to that under discussion, would generate when illuminated by the NEI. It must be remembered, of course, that by 'power' we actually mean the detector output in volts or amps. The reader is warned that the two definitions are sometimes used interchangeably.
- **NEFD (noise equivalent flux density)**, in watts per square metre. This is the flux density incident upon the detector that would give the same output power as the total spectral noise signal. We have not defined it as a spectral quantity, although it is undoubtedly sometimes used as such. If it is, then NEFD becomes equivalent to NEI per unit area.

It is argued that a good 'figure of merit' should rise with good performance. Therefore a **detectivity (D)** has been defined which is the reciprocal of NEP (or possibly of NEI). When discussing detector materials, it is desirable to generate a figure of merit which is independent of detector design, and in particular of detector size. Thus we have:

- **D* (normalized or specific detectivity)**. This is the detectivity per unit detector area and per unit bandwidth. It is given by:

$$D^* = \frac{\sqrt{(\text{detector area} \times \Delta f)}}{\text{NEP}}$$

- **D** (D double star)** takes into account the detector field of view. It is only of significance in particularly harsh photon-noise limited regimes.
- **D^*_{blip} (D star blip)** is the theoretical maximum possible detectivity. It occurs at the point where the background noise is equal to the incoming signal.

Finally we have two measures of real interest to the user:

- **NE$\Delta\rho$ (noise equivalent ground reflectance)**. This is a measure of significance for the reflective bands only. It gives the change in ground reflectance that would be masked by the total noise signal. It is what the user really wants to know, but it is affected by atmospheric conditions and is therefore of little use for specification purposes.
- **NEΔT (noise equivalent temperature difference)**. This is a measure for thermal

channels. It gives the temperature difference that would be masked by the total noise signal. The atmosphere is relatively transparent at thermal wavelengths, and therefore NEΔT is reasonably useful as a performance specifier.

5.1.8 Worked examples (AVHRR)

It has to be said that the sources used vary considerably in their approach to the noise problem, and in the assiduity with which they attempt to acquaint their readers with their reasoning. AVHRR is used for the following examples, partly because ITT (1982) and ITT (1977) between them produce one of the better accounts, and partly because the AVHRR instrument covers three very different regimes. ITT do not discuss photon noise as a separate entity, and their convention is followed here. The instrument itself is described in more detail in Chapter 12. Briefly, AVHRR has two visible channels (1 and 2), two thermal channels (4 and 5) and one channel in the particularly difficult mid-infra-red region (3).

Channels 1 and 2

In the visible waveband, the flux levels are reasonably generous, and the task is simply to demonstrate that there is no problem. Identical silicon photodiodes are used for both visible channels, driven by a −15-V bias supply. As we will be discussing later in the chapter, these devices generate a small current (the 'dark' current) in the absence of illumination, but they generate a much larger current which is proportional to the incident irradiance. The noise sources considered for these channels are signal shot noise (that due to the wanted signal) shot noise due to the dark current and thermal noise in the 4 MΩ feedback resistor. A 'degradation' factor is included to cover other noise sources such as $1/f$ noise, electronic pickup etc.

The following derivation is for Channel 1 (0.58–0.68 μm). Equivalent figures for Channel 2 (0.72–1.05 μm) differ by up to about 40%.

First it is necessary to estimate the irradiance falling on the detector. This must either be specified or derived from a source such as the LOWTRAN model (Chapter 2). LOWTRAN gives a spectral flux, incident upon the top of the atmosphere, of 158 W/m^2 over the waveband of Channel 1. LOWTRAN also gives a wide range of atmospheric attenuations over this waveband, depending on the atmospheric conditions assumed. ITT derive a combined figure of 42.4 W/m^2, incident upon the detector from 'Thekaekara's tables'. This figure assumes 100% albedo on the part of the ground patch, and so the actual minimum albedo specified (0.5%) represents an additional factor to be incorporated into the final equation. The transmittance of the optics is likely to be a significant factor. For Channel 1, ITT estimate 4.6%, including an allowance for in-flight degradation (readers who find this figure difficult to believe should consult Section 12.5). The spectral irradiance is then the product of:

— the incident flux, in watts per square metre (42.4),
— the optical transmission (0.046),
— the minimum spectral albedo specified (0.005),
— the square of the telescope aperture radius, in metres (0.1015^2),
— the square of the IFOV, in radians (1.31^2);

leading to a spectral irradiance, over the band, of 1.73×10^{-9} W.

The average 'responsivity' of the detector, over the waveband, is given as 0.37 A/W.

The signal shot noise under worst case (minimum signal) conditions is computed from the minimum signal irradiance incident on the detector (for this computation, ITT used 1.68×10^{-9} W) and the DC detector current thus generated (6.22×10^{-10} A); leading to a shot noise of 1.70×10^{-12} A, rms. The detector dark current is given as 17×10^{-9} A, maximum, leading to a dark-current shot noise of 8.88×10^{-12} A rms. The thermal noise of the feedback resistor, computed at 300 K and over an electronic bandwidth of 14.5 kHz, is given as 7.75×10^{-12} A rms. Thus:

$$\text{NEP} = \frac{\sqrt{(\text{signal shot noise})^2 + (\text{dark current shot noise})^2 + (\text{thermal noise})^2)}}{\text{responsivity}}$$

$$= \frac{\sqrt{((1.70 \times 10^{-12})^2 + (8.88 \times 10^{-12})^2 + (7.75 \times 10^{-12})^2)}}{0.37}$$

$$= 3.22 \times 10^{-11} \text{ W}.$$

The degradation factor is estimated to be 1.6. It includes stray noises as well as 'degradations in operation of the system'.

$$\text{The SNR is } \frac{\text{spectral irradiance}}{\text{NEP} \times \text{degradation factor}} = \frac{1.73 \times 10^{-9}}{3.22 \times 10^{-11} \times 1.66},$$

$$= 33 .$$

The specified SNR at minimum albedo is 3. Clearly the instrument is wildly over-specified for the visible channels. The reasons for this will however become abundantly clear very shortly.

Channels 4 and 5
The thermal channels use mercury–cadmium–telluride (HgCdTe) photoconductors. These are effectively resistors whose resistance is dictated by the level of incident illumination. (The mixes are individually tailored for the two wavebands.)

The irradiance in the thermal bands (10.3–11.3 and 11.5–12.5 μm) is interpreted directly in terms of the temperature of the radiating surface (ground or cloud top). The safety margin in these bands is less generous, however, and it is necessary to show that the users' requirement is met directly in terms of NEΔT. ITT's strategy here is to define the wanted signal in terms, not of its actual level, but of its rate of change with temperature at the specified temperature of 300 K (dR/dT). In ITT (1977), this is stated to be 0.264 W/m^2/s/K.

A 'noise equivalent radiance' is next arrived at. The only detector noise mechanism considered in this analysis is $1/f$ noise. Although the source of $1/f$ noise is not fully understood, its value may be computed from a complex formula involving the electronic bandwidth of the system, and the '$1/f$ knee frequency'. The latter is described as 'the frequency at which the $1/f$ detector noise is power equals the generation–recombination white noise', and is given as ≤ 1 kHz. For $1/f$ noise, it is important to consider carefully the minimum frequency to be taken. This is defined (in ITT (1977)) as 'lower than the frequency at which the electronics is re-zeroed' (Chapter 6 describes the re-zeroing procedure). Re-zeroing is done approximately once per second, and so a minimum frequency of 0.1 Hz is used. From this a degradation factor is generated, which can be used as a simple multiplier in the final equation. The figure derived for Channel 4 was 1.29.

An additional degradation factor is generated to account for noise generated in the electronics. This last was measured (and it is worth noting that remedial action was taken after it was discovered to be unduly high). The figure used was 1.4.

The degradation factor was converted to noise equivalent radiance, in watts per square metre per steradian, by taking into account the f-number of the optical system, the aperture of the telescope, the optical transmission efficiency (somewhat higher at these wavelengths), the detector IFOV, the detector D^* and the electronic bandwidth. The NER is then compared directly with dR/dT, to produce the expected $Ne\Delta T$. For Channel 4, $Ne\Delta T$ emerged as 0.05°C, at 300 K. This is comfortably below the maximum requirement of 0.12. The figure for Channel 5 was only slightly worse.

Channel 3
As we saw in Chapter 2, the signal levels in the region of Channel 3 (3.55–3.93 μm) are particularly low. A photovoltaic photodiode is used because, as will emerge, this is the mechanism best suited to low levels of illumination. The semiconductor is indium antimonide (InSb) because this material best suits the wavelength.

Because of the low signal levels, signal shot noise is a major factor, as also is background noise. Also considered are shot noise from the preamplifier input transistor, thermal noise from the feedback resistor and stray pickup.

The wanted irradiance falling on the detector is evaluated from LOWTRAN, 'Thekaekara's tables' or from another available source. It is given by ITT (1982) as 0.17 W/m^2. The 'system throughput', which covers the speed of the detector optics and the optical transmission losses, is given as 1.83×10^{-8} m^2 sr. From the specified quantum efficiency of 0.75, a responsivity for the detector of 2.26 A/W is arrived at. The detector output is the product of these three figures, namely 7.03×10^{-9} A, leading to an rms noise current, assuming an electronic bandwidth of 14.5 kHz, of 5.71×10^{-12} A.

The background flux noise is given as the product of:

— The background photon flux, which is assumed to be a uniform 300-K scene (i.e. the nominal temperature of the spacecraft). From Lowan & Blanch 'Tables of Plancke Radiation and Photon Functions', this is given as 3.22×10^{18} photons/m^2/sr/s.
— The 'optical filtering factor' which takes account of the precautions taken (see Chapter 12) to minimize the background irradiance reaching the detector. It is given as 1.80 sr.
— The detector area (2.98×10^{-8} m^2).
— The detector quantum efficiency (0.75).
— the charge on the electron, e, at 1.6×10^{-19} Coulomb.

This leads to a background signal of 2.07×10^{-8} A and to an rms noise signal of 9.8×10^{-12} A.

The preamplifier is considered to generate two types of noise, namely transistor shot noise and feedback resistor thermal noise. A figure of 4.42×10^{-12} A was obtained experimentally for the total preamplifier noise.

Finally we have stray pickup. As before this is covered by introducing a degradation factor. The figure evaluated for the visible channels (1.6) is chosen on the basis that the Channel 3 electronics appeared to be less noisy than those of Channel 4. A 4% allowance for $1/f$ noise is included.

The three individual noise figures are combined, as per Equation (5.3), and multiplied by the degradation factor (1.6) to yield an overall noise signal of 1.95×10^{-11} A rms.

Instead of invoking dR/dT, the above noise signal is compared directly with the change in signal current generated by a 1°C increase in wanted scene temperature. By the Lowan & Branch tables already referred to, this emerges as 1.37×10^{17} photons/m^2/sr/s (i.e. from 3.22×10^{18} to 3.36×10^{18} photons/m^2/sr/s).

The specification required an NEΔT of 0.12 which, by simple scaling, is equivalent to a noise-equivalent photon flux of 1.64×10^{16} photons/m^2/sr/s. This is converted to a *detector* photon flux (of 3.00×10^8 photons/s) by means of the system throughput already referred to. This flux is converted, as before, to a noise-equivalent signal of 3.60×10^{-11} A.

The specified NEΔT is thus achievable, with:

$$\text{SNR}=\frac{3.60\times10^{-11}}{1.95\times10^{-11}}=1.85\ .$$

Alternatively, the actual NEΔT is predicted to be 0.065°C.

This is obviously an uncomfortably narrow margin and, if it was a vital measurement, the wise designer would go through a further design cycle to see whether the situation could be improved.

5.2 RESOLUTION

Resolution is a term that has a precise meaning, but which is so much abused that it is no longer safe to use it in a precise way. The one thing that it does *not* mean, is the pixel spacing of digitized imagery. The term was originally coined by astronomers as a measure of the resolvability of bright point objects against a dark background. According to the 'Rayleigh' criterion, two points are defined as just resolved if

$$\alpha=\frac{2.44\times\lambda}{d}\ ,$$

where α is the angular separation, λ is the wavelength and d is the diameter of the Airy disc (see Chapter 3). If d is 1 mm and λ is 1 μm then α comes out at 0.22 seconds of arc.

The closest analogue of this, in the current context, is clearly connected with the discernment of small detail. However, a single object may be vanishingly small and still be seen if it contrasts sufficiently with its background. Two strongly contrasting small objects will normally have to be at least two pixels apart if they are to be resolved. Lower contrast objects may have to be further apart still. Fig. 5.3 illustrates the three main factors that control the detail content of digitized imagery.

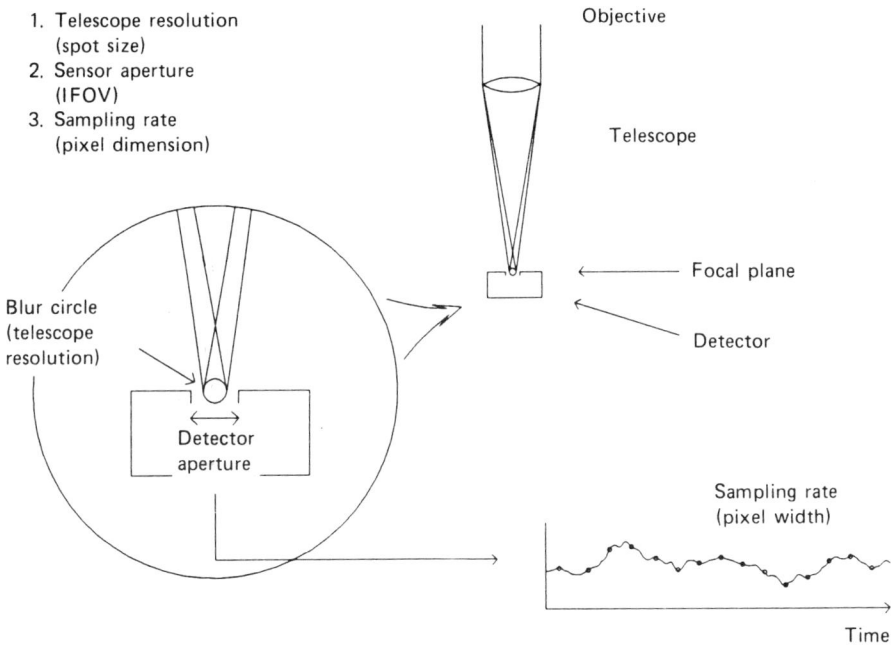

Fig. 5.3 — Factors contributing to 'resolution'.

The **instantaneous field of view (IFOV)** is often treated as synonymous with pixel size (or pixel spacing), and it should in fact be very similar in magnitude. But its basis is quite different. The IFOV is the area of ground that the detector sees at any instant. It is a function of the telescope magnification and detector aperture† only; whereas pixel size is mainly a function of sampling rate, and the forward motion of the spacecraft. IFOV and pixel size do not always match. Various studies were made of the actual IFOV for MSS (Simonett, 1983). The traditional value of 79 m was quoted as being reduced to 73.4 m (or to 76.2 m by a different worker) when an attempt was made to account for the effects of cladding round the fibre-optics. The pixel size quoted in the same results table was 79×56 m. The smaller dimension lies in the cross-track (along-scan) direction, and arises because the MSS detectors, unlike those of most later instruments, are scanned more often than once per IFOV. The

---

† It as apparently contrary to common usage to talk about *detector* aperture, the term normally being reserved for discussion of the aperture of the optical system. Approved terminologies are 'detector area' or 'detector response profile'. Neither seems to mean exactly what we want here.

implication of these results is that the radiance ascribed to each pixel emanates from an area which is not necessarily identical to that which the pixel represents. A similar situation arises on some of the bands of the Japanese MOS-1; where three scanning sweeps are made per IFOV. However, the spatial frequency response is 'slugged' to match the IFOV. Thus we have three rectangular pixels per IFOV which are not independent estimates. Only with every third sweep do we obtain the normal proportion of new information.

Pixel size is normally quoted in terms of metres, but this assumes a fixed altitude. Strictly, pixel size should be quoted in degrees or microradians, and converted to metres for a given altitude. A modified parameter, the **effective instantaneous field of view (EIFOV)** is a measure of the actual area on the ground whose intensity affects the reading at any instant. It takes into account telescope optical deficiencies, electronic filtration and any other limitations that there may be. EIFOV is not always particularly easy to measure.

5.3 SPATIAL FREQUENCY

Spatial frequency is a concept which may appear somewhat incongruous at a first meeting, but which is in fact a logical development of the better known 'temporal' frequency. The normal concept of frequency will be familiar to all who have listened to music. High frequencies emanate from high-pitched instruments, and generate rapidly fluctuating electronic signals in hi-fi equipment. If the signal created by a full orchestra is displayed on an oscilloscope, it will appear random and meaningless, with little sign of periodicity apparent. However the trained ear, being an analysis tool of matchless quality, will pick out all the main notes (i.e. frequencies) being played — and will alarm at once if any are out of tune.

The concept of spatial frequency relates to pictures and images, and is concerned with the way the intensity of any particular colour, or waveband, fluctuates along a given line. It is of course directly related to the signal produced by a detector scanned along the line. Fig. 5.4 attempts to illustrate the concept of spatial frequency, and the effect of sampling. (A two-dimensional spatial frequency is also used to describe the way an image varies in two orthogonal directions in one complex measurement.)

5.3.1 Spatial frequency response

Deficiencies in the optics, and in the electronics, combine to limit the detail that the detectors can pick out. If a detector is swept continuously across a rapidly changing terrain, such as a city, the finest detail will certainly be lost. This is partly because the finite resolution of the optics introduces a degree of blurring, and partly because the electronics will always incorporate a low-pass filter of some sort. The effect will be that the apparent contrast between the bright roofs and the dark streets, say, will get progressively less as we study areas where they are packed ever closer together. We can define a 'spatial frequency response' to quantify this fall off in response with the increase in spatial frequency. A 'cut-off' frequency might be defined at the frequency (street separation) which gives rise to a 50% reduction in measured contrast.

When (as is usual) the output signal is sampled instead of being measured continuously, then the additional danger of **aliasing** is introduced (Fig. 5.5). Aliasing is a general problem of all sampled-data systems, and occurs where the sampling is

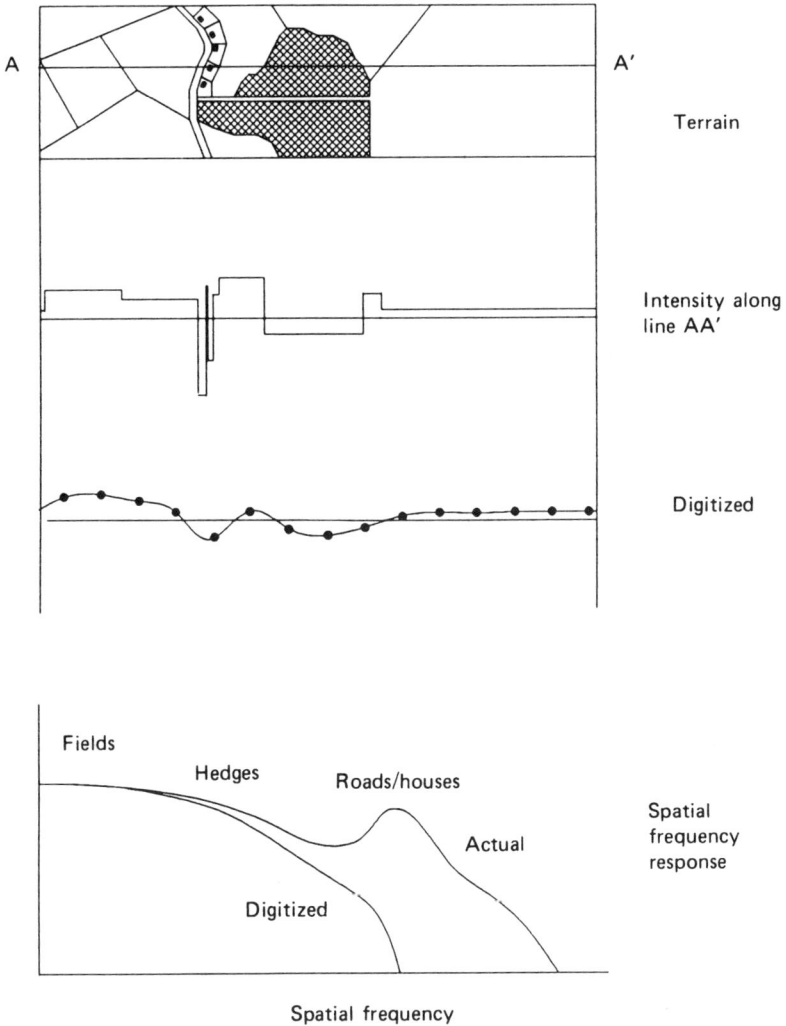

Terrain

Intensity along
line AA′

Digitized

Fields

Hedges   Roads/houses

Actual

Digitized

Spatial
frequency
response

Spatial frequency

Fig. 5.4 — Spatial frequency.

insufficiently rapid to capture the highest frequencies present. Frequencies above the
Nyquist frequency (half the sampling rate) can be misrepresented as being at a much
lower frequency. (Strictly, the output frequency is as far below the Nyquist
frequency as its input frequency was above it.) In the present context, a ghetto area
might appear as a leafy suburb. The only solution to this problem is to limit the
frequency response of the system to below the Nyquist frequency, so that data which
might be misinterpreted are not captured. In single-dimensional applications, the
problem can often be avoided by choosing what is perhaps a wastefully high sampling
rate. In multi-dimensional applications, such as imagery, however, this rapidly leads

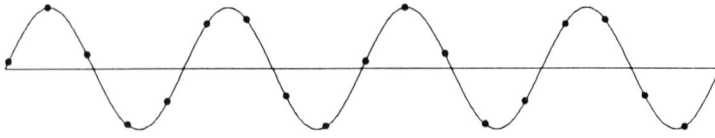

Overkill

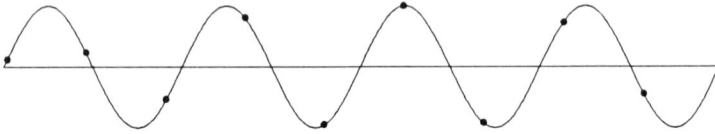

Approaching the theoretical limit
of two samples per cycle

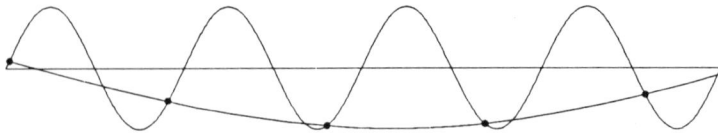

Aliasing

Fig. 5.5 — Data sampling.

to massive escalation of data rates and storage requirements, and it is more usual to
cut the margin of safety as fine as possible. To do this successfully requires a thorough
understanding of the underlying principles, and the use of sophisticated filtering and
analysis techniques. High-frequency information that is but moderately attenuated is
an additional source of noise, namely **aliasing noise**.

The purpose of evaluating the frequency response of any system is to measure
which frequencies are transmitted without loss, which frequencies are attenuated by
passage through the system — and by how much. In general, frequencies which are
attenuated also suffer a phase shift. In the case of an image, this implies a small
physical displacement, probably in the down-scan direction. The traditional procedure for evaluating the frequency response of one-dimensional systems used to be
to apply sinewaves at a range of frequencies, one at a time, and to measure the
amplitude and phase of the sinewave that emerged. This tedious procedure has been
largely replaced by what the author likes to call the PRBN method, in which a range

of sinewaves are superimposed and applied at the same time. (Pseudo-random binary noise is a particularly elegant tool for this purpose.) For the purist, it should be remarked that the cross-spectrum divided by the input power spectrum is the required frequency response. Genuine random noise may also be used. Indeed, the noise naturally occurring at the input to the system can be used if its amplitude and spectral characteristics are suitable. The principle requirement is of course that all frequencies of interest should be present, and at sufficient amplitude to give a clear-cut result.

There are many different ways of characterizing the frequency response of a system, and these have their parallels in remote sensing terminology. The **impulse response** is a particularly elegant concept, because an infinitely sharp impulse will excite all frequencies equally. Of greater use as a practical tool, however, is the **step response**, because it is easier to impart realistic energy levels using a step change of level. However a step input does not excite high frequencies well.

Photographic usage has developed a similar suite of concepts though via a different route. The term **modulation transfer function (MTF)** has been coined to describe what has heretofore been called spatial frequency response. **Modulation**, in this context, owes nothing to radio technology; the modulation of a photographic image is given by

$$M = \frac{E_{max} - E_{min}}{E_{max} + E_{min}} \; ,$$

where E_{max} and E_{min} are the maximum and minimum image intensities. If the object is a test card of black and white bars, then E_{min} might be defined as zero, forcing M to be unity. The modulation transfer function of an image of the test card is:

$$MTF = \frac{M_{out}(sf)}{M_{in}(sf)} \; ,$$

where M_{in} is the modulation of the test card and M_{out} is the modulation of the image.

M_{out} is clearly a function of spatial frequency (sf). M_{in} is also, because there must be a limit to the detail that can be reproduced on a test card. Thus is generality preserved.

Similar strategies are available to the designers of Earth observation instruments. Analysis can be of the two-dimensional spatial spectrum, or the along-scan and the across-scan spectra can be evaluated separately. Either way case, the responses in the two directions are likely to be significantly different. Optical effects are likely to be the same in both directions, but electronically induced attenuation is likely to be quite different. In the laboratory, test cards may be prepared to any design, and imaged using all the significant elements in the instrument chain. The result may then be compared with the original. Use of the PRBN method is probably not justified, because of the complex analysis procedure involved. Test cards might therefore be prepared whose reflectance varies sinusoidally between black and white at a range of different spatial frequencies. This is of course difficult, and it is understood that simple black and white bars or squares are more normally used. Such 'square-wave' test cards will introduce higher harmonics into the imager. If the harmonics are partially attenuated, and phase-shifted, then strange effects will occur. These must be correctly interpreted if unnecessary anguish is to be avoided. Markham (1984) gives a fairly comprehensive account of the derivation of square wave responses, with particular application to TM and MSS.

After the shock of launch, the entire instrument must be thoroughly re-checked, and this should include the MTF evaluation. The test-card method is inconvenient for in-flight checks, however, (though perfectly possible) and a variation of the naturally occurring noise method is normally resorted to instead. A logical extension of single-dimensional practice would be to choose an area with widely differing terrain, from densely packed city through to featureless desert, to evaluate the spatial spectra of its various regions, and to compare them with those of the imagery. Aircraft-borne imagers might be used, whose characteristics are as similar as possible to those of the spaceborne unit but with substantially higher resolution. If the spatial frequency spectra of the two images are evaluated, and the one divided by the other, then the result is a reasonable approximation to the spatial frequency response of the spaceborne imager. In practice, however, interest concentrates on the high-frequency end of the spectrum, where the response starts to fall off, and only cities are studied in any detail.

Other variations on the above theme will also commend themselves to various workers, depending on the precise nature of their problems. For Thematic Mapper, for example, comprehensive studies of MTF and 'square wave response (SWR)' were carried out, both at the component level and for the complete system (Markham (1984)).

SPOT used a photo-interpretive technique, based on aerial panchromatic photographs of selected cities (Leger *et al.*, undated, Begni *et al.*, 1985). Their basic concern was to check the focus of the telescope, and therefore only the most severe case, the 10-m panchromatic band, was chosen for study. The photographs were digitized at 3.33-m intervals, and then deliberately degraded by smoothing. A number of different images was prepared, with different degrees of smoothing; the best matching the MTF specification for the SPOT PA band. The images were then resampled to 10 m and printed. The flight imagery of the same areas was compared visually to the various prints. Unfortunately, no prints were prepared which were *above* specification, and HRV1 turned out to be better than the best of them. HRV2 was judged to be on specification.

Apart from its use in evaluating the performance of the complete instrument, the frequency response/MTF is a fairly powerful design tool, although to use it effectively, much more frequent sampling is required than the once-per-pixel norm. It is possible to evaluate the performance of individual constituents, plot them on the same axes, and quickly identify the 'pinch' points. For example, Fig. 5.6 shows MTF plots of two optical effects with, superimposed upon it, the frequency response of a particular design of low-pass electronic filter (Norwood & Lansing, 1983). Such a superposition requires the scanning rate to be known. In general, logarithmic scales are chosen for both axes of frequency response curves, because the curves then take on characteristic shapes which facilitate interpretation; although in this case that was

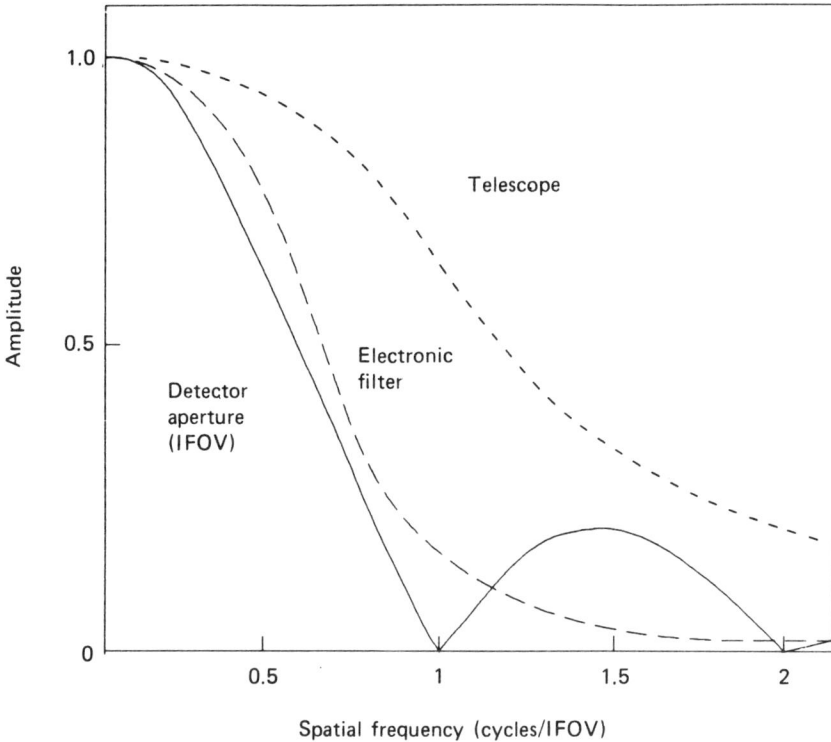

Fig. 5.6 — Spatial frequency responses (MTFs).

not done. The diagram shows that the instrument was field-stop-limited, rather than diffraction-limited, and that the filter appears satisfactory.

As we discussed in Chapter 3, the **80% blur circle** is another method of representing the performance of an optical system. It is the circle that encompasses 80% of the energy at the focal plane when the telescope is looking at a point source. The diameter of the blur circle is a simple, single-number descriptor of optical quality. It is slightly smaller than the **Airy disc**, which we also discussed in Chapter 3. Both are subsets of the **point spread function**, which is a complete characterization of the response of the optical system to a point source. For completeness, it might be mentioned that the MTF can be estimated by applying a Fourier transformation to the point spread function.

5.4   DETECTION MECHANISMS

To detect light, or its longer wavelength equivalent, radiant heat, it is necessary for photons to modify the physical, chemical or electrical state of some suitable material.

The traditional electrical detector is the vacuum-tube photo-diode. A **diode** is a device which can conduct electricity in one direction, but (in general use) not in the other (Fig. 5.7). A vacuum-tube diode requires electrons to be ejected from a **cathode**, by some means, into the surrounding vacuum; or there may be a trace of gas to generate **secondary emission**, and so increase sensitivity. There they are attracted to an **anode** by an applied **electric field**, generating a small current which can be measured by external circuitry. (An electric field results when a voltage is applied across a device such as a resistor, capacitor or a semiconductor. It is the force which attracts electrons towards the positive potential, so generating a current.) If there is no electric field, or if it is reversed, then there is no conduction. The standard way of providing a source of electrons is by heating the cathode. However, in a **photodiode** electrons are ejected, from a cold cathode, on impact from photons. Thus the current is controlled by the strength of the applied field, but also by the intensity of the incident illumination. A development of the photodiode is the **photo-multiplier tube**; in which secondary emission is used to provide a massive increase in sensitivity. The electrons ejected from the cathode are attracted towards a second electrode, called a 'dynode', which they strike with sufficient force to eject considerably more electrons. These are then attracted towards a second dynode, and so on until a large enough signal has been generated to drive traditional electronic circuitry. The use of secondary emission in this way could be equated with the function of a high quality pre-amplifier, placed so close to the detector as to be actually built into it. Photomultiplier tubes were invented in the days when good amplifiers were very difficult to make, were large, and generated a great deal of heat. MSS and GOES (at least up to GOES 7) both use photomultiplier tubes for the shorter wavebands.

The minimum energy that must be given to an electron to force it out of the cathode is called the **work function**, and is a function of the cathode material. Work function is measured in 'electron volts' (which is the work done on an electron when its electrical potential is increased by 1 V). The **threshold wavelength** is that of photons whose energy just matches the work function. The photodiodes and the photomultiplier are only sensitive to radiation whose wavelength is shorter (i.e. of higher energy) than the threshold wavelength. Note that, unlike the mechanism which gives rise to absorption bands in the atmosphere, this is not a resonance phenomenon. The probability of an electron escaping from the cathode increases continuously as the energy of the incident photon increases. The cathodes of traditional diodes are made of pure metals, whose work functions are high enough to exclude the red and infra-red regions of the spectrum. However, mixtures of alkali metals have been developed which extend the useful range of these devices well into the near infra-red. Only electrons within about 0.02 μm of the surface are likely to escape the cathode. Others will lose too much energy in collisions on the way. Since photons are likely to penetrate the material a good deal deeper than this, the efficiency of photo-emissive detectors is low. The term **quantum efficiency** describes the proportion of photons that give rise to ejected electrons. It is seldom better than 10–20% even for pure metals. For materials sensitive to wavelengths down to 1 μm, the quantum efficiency is as low as 0.1%.

Doping the cathode material in this way to improve the long wavelength response introduces another problem, that of **dark current**. This is the tendency of electrons to escape spontaneously under their own thermal energy. When the work function is high, the dark current is not a significant problem, and traditional vacuum tube photodiodes can work quite happily at temperatures as high as 200°C. But it is a

Symbol

Non-conducting   Conducting   Photodiode

Vacuum-tube diode

Symbol

Output

etc.

Negative ve EHT

Photomultiplier tube   Photoconductive device

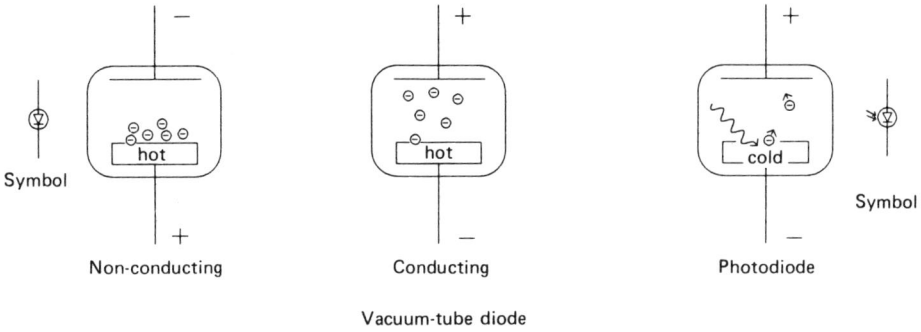

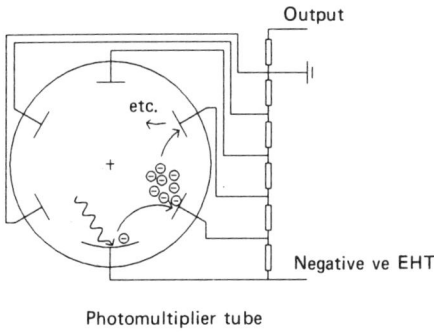

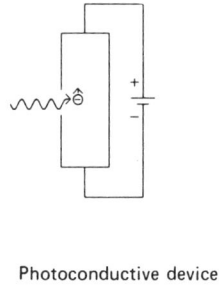

▨ = Depleted region

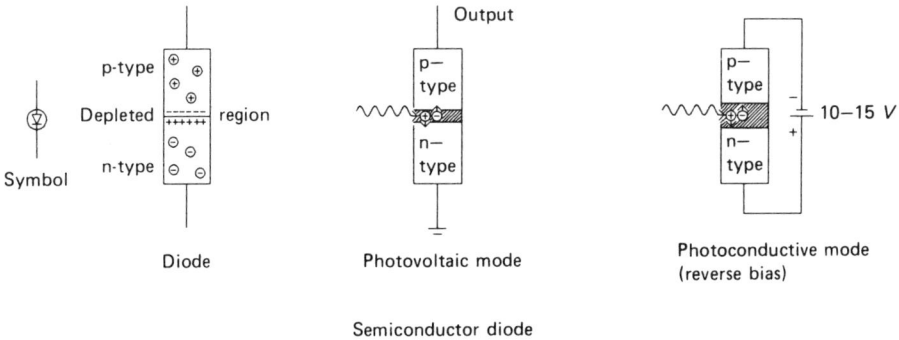

Symbol

p-type
Depleted ⸻ region
n-type

Output

p—
type
n—
type

p—
type
n—
type

10—15 V

Diode   Photovoltaic mode   Photoconductive mode
(reverse bias)

Semiconductor diode

Fig. 5.7 — Common detector types.

serious problem for the more advanced long-wavelength devices. This is partly because the photons are less energetic, so that more delicate energy balances have to be engineered within the material, but also because intensity levels tend to be substantially lower at the longer wavelengths, so there may well be fewer photons available to be detected. This increases the significance of a given level of dark current. The dark current of multi-alkali surfaces varies exponentially with temperature, in accordance with the Boltzman factor ($e^{-E_g/kT}$), and so the problem can be combatted by temperature control.

5.4.1 Solid-state devices

Most modern photodetectors are however solid-state devices. Here the incident photons alter the electrical properties of the material itself which, again, is sensed by its effect on external circuitry. The simplest solid-state detector is the **photo-conductive** device. A photoconductive solid-state detector is made of a poorly conducting material from whose atoms incident photons can eject electrons as before. However, this time the ejected electrons remain within the material, and render it slightly conducting. A voltage is applied across the device and the resulting current is a measure of the strength of the illumination. The energy required to eject an electron from an atom of pure bulk material is called the **band gap**. It is less than the work function would be, because the electron does not escape completely from the material. Photoconductivity induced in this way in pure materials is termed **intrinsic photoconductivity**.

The bandgap is invariably too large to render photoconductors sensitive to the longer wavelengths, and again, doping is resorted to to overcome the limitation. The addition of small amounts of a suitable impurity (such as mercury or copper to germanium) has been found to generate an intermediate energy level, at the site of the impurity, from which a photon can force an electron. This enables lower-energy photons to eject electrons from the bulk material, and extends the range of the device to longer wavelengths. This effect is called **extrinsic** photoconductivity. Unfortunately, impurities in the material are a source of dark current, for exactly the same reasons as before. Even nominally pure materials contain enough impurity to generate dark current, and in this case the effect appears to be considerably less temperature-sensitive than is the case with photoemissive materials.

The thermal-band channels, described in Part II, all use mercury–cadmium–telluride (HgCdTe) detectors and, as far as can be established, they are all photoconductors. This material may be thought of as an alloy of the semi-metal HgTe, and the semiconductor CdTe (Wilson & Hawkes, 1983). Depending on the composition of the alloy, the sensitivity can be varied between 5 and 14 μm. They are low-impedance devices and, because of their dark current, require to be cooled. Indium antimonide (InSb) photoconductors are also available, to cover the intermediate band, around 5 μm, but where the source material is sufficiently specific, the instruments that we cover use the material in photodiode form.

Semiconductors may be doped to extend their response down to around 100 μm. Such materials were used in the IRAS mission (Chapter 15). They need cooling to liquid helium temperatures, and are therefore well-suited to limited-life missions only.

The next device to discuss is the **solid-state photodiode**. All solid-state devices are photosensitive to a degree, and normally need to be well-protected by an opaque coating. However, purpose-designed devices can be made very much more efficient by aiming the incident radiation straight to the most sensitive region. In a photo-diode, the ability of incoming photons to enhance this conductivity is exploited, and increased by careful design. In a solid-state diode the metal/vacuum interface of the thermionic diode is replaced by two different types of semi-conducting material, p-type and n-type. These are an insulating material (traditionally germanium or silicon) which have been doped with different impurities to give complementary properties.

In an **n-type material**, free electrons exist, even in a totally unexcited state; thus the normally insulating material is rendered conducting. Note that the free electrons are part of the normal, non-ionized condition of the material. If the electrons are removed, then n-type material takes on a positive charge.

In a **p-type material**, there is a deficiency of electrons in the unexcited state. This means that there are **holes**, which are the locations within some atoms where these electrons should be. These holes can 'move', under the influence of an electric potential, as a result of atoms stealing an electron from a neighbouring atom. It is in fact still only the small, light electrons that do the moving, and so the holes in p-type material can be as mobile as the free electrons in n-type material. If the holes are 'removed', by filling them with electrons, then p-type material takes on a negative charge.

In between the two layers of conducting material a thin region, called the **depleted** region, quickly develops. This happens as the free electrons from the n-type region diffuse into the p-type region and fill the holes, leaving behind a positively ionized zone. The region where the holes have been filled becomes negatively ionized. An electric field is thus induced which discourages further migration. Equilibrium is established when the induced field just completely inhibits migration. As with a thermionic diode, a positive potential, applied to the p-type side, will cancel out the induced field and allow electrons to flow. A negative potential so applied will reinforce the induced field and further prevent conduction. The traditional appli-cation of a diode is, of course, to allow a current to pass in one direction only. However a potential applied deliberately as an aid to detection is called **bias**. **Forward bias** is the application of positive potential, to aid conduction in the absence of external excitation, while **reverse bias** is the opposite.

If photons of sufficient energy strike the material, electrons will be ejected, generating hole/free electron pairs. Those of the carriers that find their way to the depleted region will be attracted to its oppositely charged zone. If an external circuit is provided then a current will flow. Much the greatest sensitivity is of course achieved when the photons are aimed directly at the depleted region. This mode of operation is called **photovoltaic** mode, because the device itself generates the voltage that is observed. There is no dark current with this mode, because free electrons are only to be found in the n-type region; from there they are prevented from migrating by the induced field. On the other hand, the response of the device to changes in incident intensity is slow, because the only driving force is the relatively small induced field. If the circuit is closed with a low-impedance metering device then the electrons find their way back to the p-type side without significantly upsetting the induced field. However, if a high-impedance circuit is used, then build-up of electrons on the n-type side progressively reduces the induced field, and leads to a logarithmic response to incident illumination. Photo-voltaic mode, used open circuit in this way, generates a very sensitive detector; and Channel 3 of AVHRR uses an antimonide diode operated in just such a way.

An alternative approach is to use **photoconductive** mode. This is not the same as the use of a photoconductor. As we saw earlier, a photoconductor is a simple resistive element. A photodiode is operated in photoconductive mode by applying a reverse bias to enhance the induced field, as described above. Increasing the induced field increases the volume of the depleted zone, and thus increases the probability that hole/electron pairs will be generated within the zone itself, where they are much more effective. The applied voltage overcomes the tendency for the accumulation of electrons on the n-type side to interfere with the induced field. Altogether, photo-conductive mode is a much more effective method of operation, at least for reasonable levels of excitation. However, it re-introduces the problem of dark current, together with its thermal and shot noise.

Sensitivity is increased if an insulating layer is inserted between the n-type and p-type layers, to produce a p-i-n (or PIN) type photodiode. The visible channels of AVHRR use silicon PIN diodes, in photoconductive mode, with 15 V of reverse bias.

A solar cell is (at the time of writing) basically a silicon photodiode, such as we have discussed, operated in a regime that maximizes its power output rather than its signal fidelity.

A **light-emitting diode (LED)** is effectively a photodiode operated in reverse. Forward bias as used, which encourages the 'majority carriers' from both sides of the junction to migrate across it and increase the population of 'minority carriers' on the other side. This increases the probability of an electron meeting a hole, and hence the incidence of recombinations. Some, at least, of these recombinations generate an emitted photon. The flow of carriers across the junction, and hence the photon flux, is controlled by the bias voltage. For short periods, very high intensities can be obtained. This property is used in the calibration facility on *IRS*-1 (Chapter 18). The reversal is not exact, however. Silicon, which detects in the visible waveband, emits in the infra-red. At the time of writing, gallium arsenide (GaAs) is the usual material from which LEDs are made. It has a band gap of 1.44 eV, which corresponds to a nominal emission wavelength of 0.86 μm. However, for various reasons, the actual wavelength emitted is normally somewhat shorter than this (Wilson, 1983).

5.4.2 Thermal methods

Until recently, thermal wavelengths were too long to be detected by direct photon detection, as described above; and other methods had to be resorted to. These methods are still described in current literature, such as Pinson, 1985. The normal method of detecting thermal radiation was to focus the radiation onto a fully absorbing target, and to measure the temperature rise of the target. Two temperature-measuring devices are in regular use for this purpose, the tried and trusted thermocouple, and the more recently developed thermistor (see below).

The basis of the **thermocouple** is the **Seebeck** effect. This is that, if a pair of wires of two dissimilar metals are connected together at both ends — and if the ends are maintained at different temperatures — then an EMF (voltage) will be produced. The EMF can be measured by breaking into one of the wires and inserting a meter.

This EMF is a function of the temperature difference, and is unique to the particular pair of materials chosen. If it is measured, and the calibration curve consulted, then the temperature difference can be calculated. In normal operation, one junction (the 'reference' junction) is maintained at a known temperature, and the other (the 'active' junction) is attached to the test piece. It must be remembered, however, that a thermocouple measures the temperature difference between its own two junctions. There should normally be no difficulty in fixing the reference junction at a known value, but it can be quite difficult to ensure that the active junction itself is at exactly the same temperature as the test piece. The EMF produced by the thermocouple can be magnified many times by connecting a number of thermocouple pairs in series, thus producing a **thermopile**.

In radiation sensing applications one wishes to know, not the actual temperature of the test-piece/target, but the temperature rise generated by the incoming radiation. So in this case the reference junction would be mounted in the same assembly as the active junction, but shielded from any incoming radiation. Very careful design and calibration is required to ensure that any temperature difference between the two junctions is only that induced by the incoming radiation.

A **thermistor** is a semiconductor device, formed from suitably mixed metal oxides, whose resistance varies inversely with temperature. A thermistor **bolometer** is made from a matched pair of thermistor elements, which are used in the same manner as a thermocouple (although the electrical circuitry required is of course quite different).

There are also **pyroelectric** detectors. These are made from suitable crystals (of lithium tantalate, triglycine sulphate, strontium, barium niobate etc.) of which the capacitance varies with temperature. Thin wafers of crystal are placed between two electrodes to form a capacitor, whose capacitance therefore also varies with temperature. They are high-impedance devices, and require something like a field-effect transistor mounted on the same wafer to act as a source follower. Pyroelectric detectors are also slightly piezo-electric; and so care must be taken in designing the crystal mount to avoid any mechanical stress.

5.5 CCDs

The CCD is basically a technology for shifting arrays of numbers along an 'analogue shift register', manipulating them as necessary, and reading them out sequentially at one end. CCD technology has a wide range of applications, one of which is photodetection. In remote sensing terminology, the term CCD normally implies a linear array of around 2000 photocells (the devices were originally developed for photocopiers).† The array is backed by a CCD transfer register built into the same chip. The CCD register is used to read all the photocells, effectively instantaneously, and to output the readings sequentially, in the form of an analogue signal. A preamplifier would normally be built on to the output end of the chip to amplify the signals to a level suitable for routine processing (see Chapter 6).

The requirement that CCDs meet is for a long line of photodetectors packed

---

† Since the period in history that we cover, two-dimensional CCD packages have been developed. These are potentially able to capture an entire image in one snapshot.

closely together with a minimal gap between. The use of such an array makes it possible to cover a wide swath by scanning the detectors in one direction only. It would be perfectly possible to create a long array of conventional photodiodes. The problem that CCD technology solves is that of reading them all quickly and economically. A single CCD chip effectively takes the place of the conventional photocell array, a series of carefully matched preamplifiers, and a considerable amount of sampling electronics.

At the time of writing, CCDs are only used in reflective-band applications. No doubt, however, their use will be extended to longer wavelengths in due course. A CCD array comprises a strip of p-type semiconductor material, backed by an insulating strip. Metal electrodes are attached to the insulating surface. Thus we have a parallel array of capacitors, connected by one common (semiconductor) electrode. As shown in Fig. 5.8, a single photocell comprises a strip of this sandwich encompassing typically three of the electrodes (although other arrangements are possible). Note, however, that there is no actual barrier between the cells. During the acquisition phase, every third electrode is raised to a positive potential, which generates an electric field (or, in the jargon of the trade, a **potential well**) that attracts any free electrons generated in the locality. The remaining electrodes are held at a low potential; and so the sphere of influence of each activated electrode extends nominally half way to the next *activated* electrode. A mask may be added externally, to exclude light from a small area at the boundary. Each attracted electron reduces the field available for attracting more electrons, and so the 'depth' of the well is limited, being a function of the applied voltage. The dynamic range of a CCD detector is thus also limited, and is in fact very much less than that of the single detectors already described. Thus, at the time of writing at least, CCDs are at their best in 'production' environments where the requirements are already well understood. If a single detector is overloaded, perhaps by a bright point source, the well will overflow; and the extra electrons will be attracted to adjacent detectors. This will increase the apparent size of the source, but it may still be possible to estimate its brightness by summing the excess outputs from all the affected detectors.

At the end of each acquisition period, the charges are all transferred, in sequence, to the collection point at one end of the chip. This is done by manipulating the electrode potentials, for example, as shown in the figure. Any photons which strike during the transfer process will be misinterpreted, and add to the noise. However, the operation is done extremely rapidly, and the contamination so produced is negligible.

What emerges is a 'quasi-sampled' signal representing a snapshot of the state of the detector array at the time of sampling. It is in fact a continuous analogue signal, and is handled by ordinary analogue electronics. We describe it as quasi-sampled because only those parts of it which represent the reading of a detector are meaningful. In between will be a smooth transition between the voltages representing adjacent detector readings. The digitization process must be synchronized with the charge transfer rate, so that only the true readings are digitized.

Note that the sampling of a CCD system effectively takes place at the focal plane. This is where the high spatial frequency data are rejected, and where anti-aliasing filtration must be carried out. Filtration is carried out by blurring the image. The resolution of the telescope is degraded, to match the detector IFOV. Although at

Incident radiation

Mask

p—type
semiconductor

Insulating layer

Electrodes

13 μm (typical)

13 μm (typical)

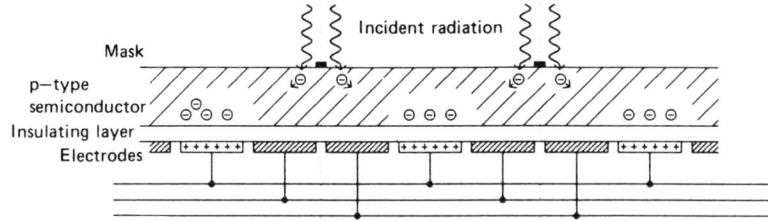

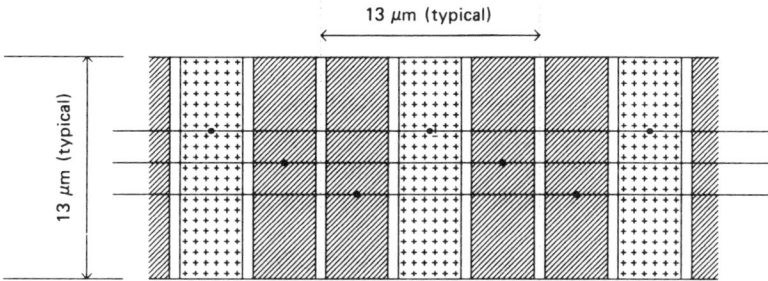

Acquisition phase

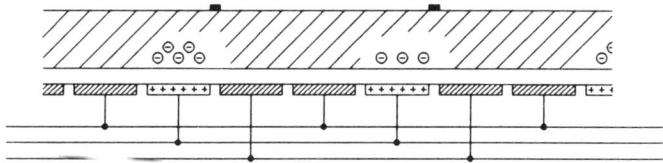

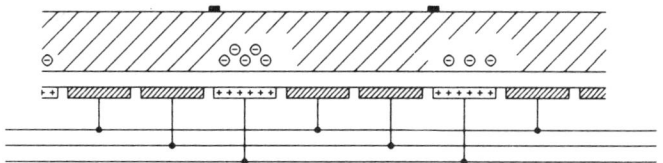

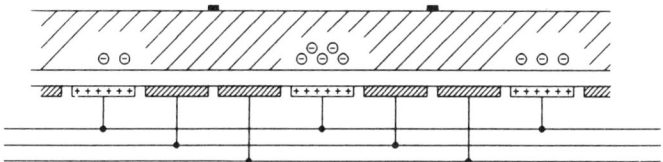

Transfer phase

Fig. 5.8.

this point we are still in the spatial domain, the effect is exactly the same as acquiring a high-definition analogue signal and passing it through the traditional anti-aliasing filter.

REFERENCES

Ahmed, H. & Spreadbury, P. J. (1973) *Electronics for Engineers*, Cambridge.

Begni, G., Boissin, B. & Perbos, J. (1985) 'Spot Image quality & Post-launch Assessment' *Adv. Space Res.* **5**, 5 1985.

Hughes (1984) *Thematic Mapper: Final Report*. Santa Barbara Research Center, Hughes aircraft company, October 1984.

ITT (1977) *INSAT VHRR proposal for the design and manufacture of the INSAT imaging instrument*. ITT Aerospace/Optical Div. Fort Wayne, Indiana, USA.

ITT (1982) *Advanced Very High Resolution Radiometer: Technical Description* ITT A/O division, Indiana, USA, September 1982.

Leger, O., Leroy, M. & Perbos, J. (undated) *Spot MTF performance evaluation*, CNES.

Markham, B. L. (1984) *Characterisation of the Landsat Sensors' Spatial Responses*, NASA Technical Memorandum 86130.

Norwood, V. T. & Lansing, J. C. (1983) *Manual of Remote Sensing*, Chapter 8. American Society of Photogrammetry,

Pinson, L. J. (1985) *Electro-optics*, John Wiley.

Robinson, B. F. & DeWitt, D. P. (1983) *Manual of Remote Sensing*, Chapter 7. American Society of Photogrammetry.

Simonett,D. S. (1983) *Manual of Remote Sensing*, Chapter 1. American Society of Photogrammetry.

Wilson, J. & Hawkes, J. F. B. (1983), *Optoelectronics*, Prentice Hall International.

6

The detector electronics

Consultant: **S. Braithwaite,** B.Sc, MIEE, C. Eng†

The task of the detector electronics is, first, to receive the minute electrical signal generated by the detector, and then to amplify it, so that it can be digitized, multiplexed and transmitted to Earth with as little as possible further degradation. The task must be done with considerable accuracy, so that the volts (or digital counts) that emerge from the electronics remain a direct measure of the microvolts generated by the detector. As we saw in the previous chapter, it must also be done without introducing a significant amount of additional noise.

6.1 THE TRANSISTOR

The transistor is the basic component of all modern electronic systems. It is normally described as an 'active' component to distinguish it from resistors and capacitors, which are purely 'passive'. However, after perusal of the paragraph on field effect transistors (below), the reader may well feel this distinction to be more apparent than real. He may well wish to regard the transistor simply as a resistor whose resistance is controlled by the voltage applied to a third electrode. This view does in fact give the non-specialist reader a qualitative understanding of the way transistors are used. Transistors are normally used as part of a 'potentiometer' chain, often connecting a 'supply' voltage to earth. A typical such chain is shown in Fig. 6.1(a). The voltage V_{out} is:

$$15 \times \frac{R_T}{R_1 + R_T} \, ,$$

and can be adjusted at will by moving the slider. A typical transistor circuit is shown

† University of Southampton, England.

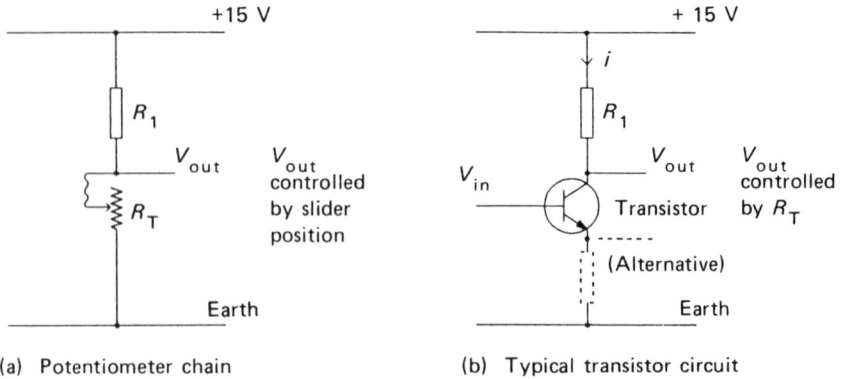

(a) Potentiometer chain   (b) Typical transistor circuit

Fig. 6.1 — Transistor usage.

by Fig. 6.1(b) (a practical circuit will contain additional resistors, which serve various secondary purposes. Alternative sites for both R_1 and V_{out} are shown dotted). R_T is now the resistance of the transistor. If R_T is varied, this time by varying V_{in}, then exactly the same effect is achieved. This is not, however, the full story because a linear relationship between V_{out} and V_{in} is generally required, and the relationship between R_T and V_{in} is not normally linear unless additional steps are taken. The main step is the application of 'feedback', which we will be meeting again in later sections.

It should perhaps be mentioned that the transistor is in fact better modelled as a 'controlled current source'. It is the current through the device (i), rather than the voltage across it, that is controlled by V_{in} (Braithwaite, 1990).

As we saw in the previous chapter, our application requires R_T to be large. However, we might also mention power transistors, because the reader will certainly meet them in other applications. Their task is to control large currents passing from the power supply to the 'load' (loudspeaker, servomotor etc.). Such a transistor must have a working resistance very close to zero, otherwise it will both waste power and overheat. Power transistors can also be used as high-speed switches. If a large change in V_{in} is applied, then R_T can change from effectively zero to a large value very rapidly indeed. Thus heavy currents can be switched by a relatively small transistor. These devices enable trailer owners who get tired of constantly replacing so-called heavy duty flasher units to 'transistorize' their trailer flasher systems very easily.

Within limits, the design of a transistor may be considered as a logical extension of the solid-state diode discussed in the previous chapter. Fig. 6.2 gives a schematic representation of a normal (bipolar) transistor. Typically, it is a very thin layer of p-type material sandwiched between two thick layers of n-type material — or vice versa — leading to NPN or PNP transistors. The 'meat' layer is termed the **base**. It is lightly doped, and therefore produces but a sparse supply of carriers. The bread layer that generates the carriers is the **emitter** and, in the symbol, is represented by the diagonal line incorporating the arrow. It is heavily doped, and produces a copious supply of carriers. The other diagonal line is the **collector**. It is less heavily doped, because its function is to receive carriers.

C = collector
B = base
E = emitter

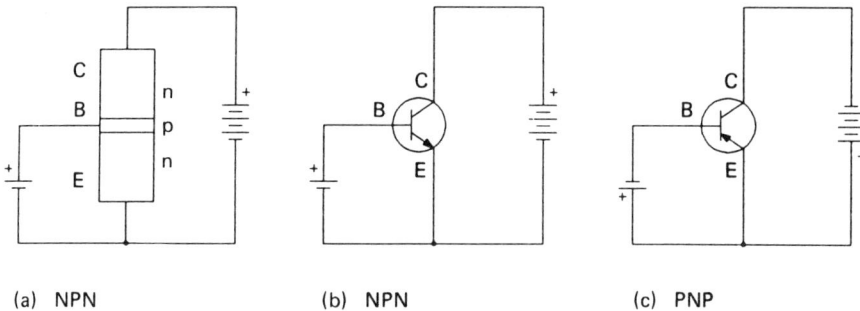

(a) NPN   (b) NPN   (c) PNP

Fig. 6.2 — Normal (bipolar) transistor (textbook circuit).

The operation of the transistor is extremely subtle, and authorities differ in their explanations of it. The device is sometimes described as 'two diodes placed back-to-back'. While this is in fact true, the concept omits the main features of the transistor, and does not enable its behaviour to be predicted. Two descriptions are offered here: one that the author finds helpful, and one that the consultant feels to be closer to the truth of the matter. Other explanations will be found in any good electronics textbook, of which Ahmed & Spreadbury (1973) is a typical example.

Crudely, therefore: if an external field is applied across the device then carriers (electrons or holes) will try to flow between the two 'bread' layers of the sandwich. However, the carriers generated in the 'meat' layer will be of the opposing type, and they will form a barrier tending to inhibit this flow. The trick is to bleed these opposing carriers away, in a controlled manner, by applying a suitable connection to the meat layer. Controlling the population of the opposing carriers controls the flow of the main carriers. It has proved possible to design the device so that a relatively small change in the 'bleed' current generates a large change in the main current, and so we have an amplifier.

More accurately: the carriers are attracted into the base region at a rate dependent on the voltage between the base and the emitter. The collector then collects these carriers as they pass across to it by diffusion. The doping of the base, emitter and collector are such that the current flowing in the base connection to establish the base–emitter voltage is very small compared with the current flowing into the base between the emitter and collector. Thus we have an amplifying action (Braithwaite, 1990).

An NPN transistor is the closest to the original thermionic triode in that the emitter region generates *electrons*, which are attracted to the collector by an applied *positive* voltage. Unlike a triode, however, a (smaller) *positive* voltage is also applied to the base, to bleed away the *holes* that are generated therein. The symbol for an NPN transistor has the arrow pointing *away* from the base, illustrating the direction of the electron flow. In a PNP transistor, the *quantities* are reversed. The term 'bipolar' is applied because it operates through the courtesy of both electrons and holes.

A device that is used extensively as the input device in high-performance amplifiers is the **field effect transistor**, or **FET** (Fig. 6.3). It is a 'unipolar' device,

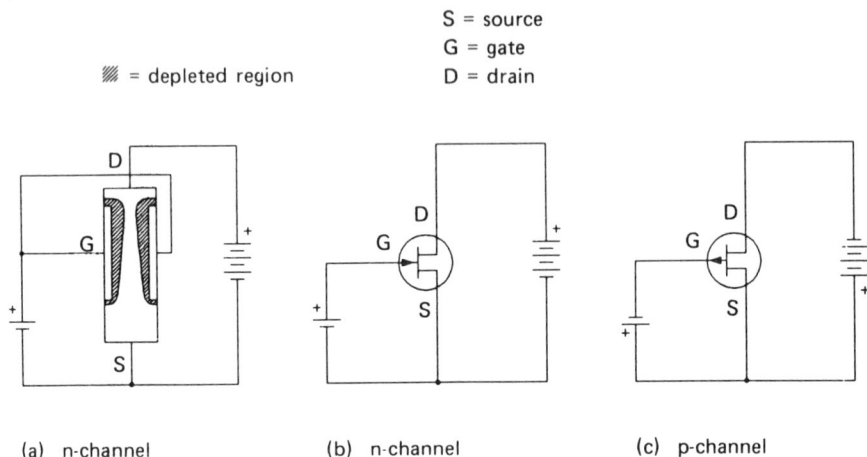

S = source
G = gate
▨ = depleted region   D = drain

(a) n-channel   (b) n-channel   (c) p-channel

Fig. 6.3 — JFET (textbook circuit).

because it relies upon either electrons or holes (depending on its doping) but not both. It is used because of its low noise characteristics and its high 'input impedance' (the significance of the latter feature will emerge in due course). The traditional FET (the **junction FET**, or **JFET**) is basically a bar of silicon, called a 'channel'. It is lightly doped to generate a relatively poorly conducting 'n-channel' or 'p-channel'. To control the conductivity of the channel, two heavily-doped regions ('gates') of the opposite type are diffused into its sides. They generate large depletion regions (see the previous chapter) which effectively constrict the channel and lower its conductivity still further. The magnitude of the constriction depends on the bias voltage applied to the gates. Hence we have a device which particularly closely resembles the simple model described at the beginning of this section.

The high input impedance, and low noise generation, of the FET makes it an ideal component for the first stage of a sensitive pre-amplifier. In particular, the availability of the FET is the reason why portable VHF radios are so sensitive, despite their puny whip aerials.

A more recent development is variously described as the **IGFET** (insulated gate FET), **MOSFET** (metal oxide semiconductor FET) or **MOST** (as MOSFET). Its mode of operation will not be described, but its characteristics are improved gain (strictly 'gain–bandwidth product') and reduced $1/f$ noise. The MOSFET was not available when many of the instruments that we discuss were designed. The limited performance of available FETs is the reason for, for example, the complex input circuit for the Meteosat infra-red electronics shown in Fig. 11.5.

6.2 THE AMPLIFIER

The **amplifier** is the backbone of most electronic systems, and the peak of perfection in amplifier theory and design is the **operational amplifier**. The ideal operational amplifier is a black box with infinite **gain** and infinite 'input impedance' (of which more later). The infinite gain feature requires that, in normal use, the voltage applied at the amplifier's input can never be other than zero. **Feedback** circuitry has to be built round the amplifier to ensure that this condition is met. (The principles of feedback are considered further in Section 6.4.) In a simple ('inverting') operational amplifier, the gain is actually arranged to be *minus* infinity to facilitate the design of the feedback circuit. As we will be discussing, the output impedance of an amplifier is normally designed to be reasonably low, to allow it to drive a wide range of following circuits.

It might perhaps be mentioned that many amplifiers are perfectly capable of being 'driven against their stops' by the application of a finite input. Sharp-edged pulses are then generated which find frequent use in logic circuits, which are another large application area.

Fig. 6.4(a) shows an operational amplifier, with the minimum circuitry required for effective operation. We will call the operational amplifier itself the 'chip', and the complete circuit the 'complete amplifier'. It should be noted that the symbol of a 'naked' amplifier is a triangle with a curved end. The use of a straight-ended triangle implies that any essential feedback components that are not specifically shown are assumed included. This makes it possible to indicate only those components which help to explain the function of the circuit.

In the figure, R_1 is the **input resistor**, and R_2 is the **feedback resistor**. Because the input impedance of the chip is infinite, no current can flow into or out of the amplifier. Therefore the current flowing through the feedback resistor must be identical with that flowing through the input resistor. And, because the system can only be stable when the voltage at its input is zero, the point v_0 is described as a **virtual earth**. Effectively, it is held at earth potential by the feedback signal. Therefore, notwithstanding that the gain of the chip is infinite, that of the complete amplifier is simply the ratio of the two resistors, namely

$$\text{gain} = \frac{R_2}{R_1} \,. \tag{6.1}$$

Any good operational amplifier will have a gain of many hundreds of thousands, and an input impedance of about a megohm. If an FET input is used, then the input impedance will be tens of megohms. With such an amplifier, Equation (6.1) will apply almost exactly. However, it also represents a reasonable working approximation when much simpler amplifiers are wired up in the same way.

The operational amplifier used to be a major constituent of analogue computers

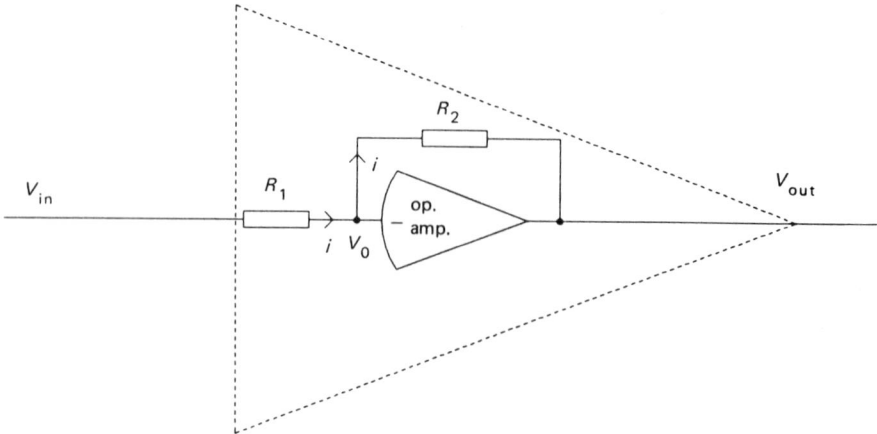

(a) simple inverting

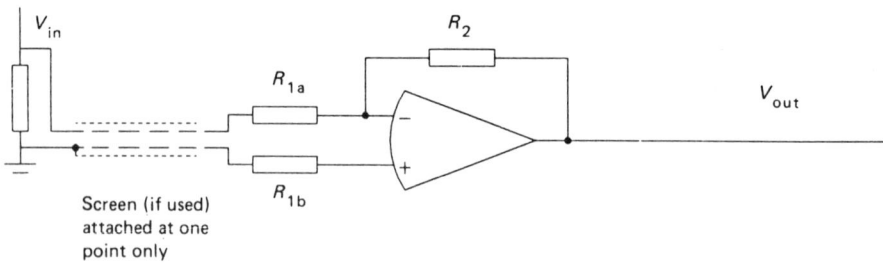

(b) differential

Fig. 6.4 — Operational amplifier.

because of the elegant manner and the accuracy with which the response of the system can be manipulated. It is still a vital component of any circuit where the gain or the frequency response needs to be accurately controlled. Because of their cheapness and reliability, lower grade 'op. amps' are now used almost everywhere where voltage (as distinct from power)amplification is required. As we will be seeing in (for example) Chapter 12, they can also be used as a component in a more exotic amplifier.

Amplifiers can be described as 'voltage amplifiers', 'current amplifiers', 'trans-impedance amplifiers' and no doubt others. The differences are quite superficial, and relate to the precise manner in which the input and feedback circuits are arranged.

The **input impedance** of a circuit element is its resistance (strictly its 'impedance') to earth at the input connection of the element. It dictates the current that the element will draw from the preceding element. Thus although the input impedance of the chip is normally very large, that of the complete amplifier is R_1. Similarly the **output impedance** of an element is the resistance to earth of the output terminal. It is a measure of the current that the element can supply without 'loading' (i.e. without having its output voltage dragged down). In many circuits, the output impedance is effectively that of the chip or other device. However, the output voltage may well be 'potted down' by a chain of resistors, in which case that between the output terminal and earth represents the output impedance of the circuit element. The general rule of thumb is that, to avoid loading, the input impedance of an element must be a nominal ten times larger than the output impedance of the previous element.

6.2.1 The differential amplifier

In practice, most modern operational amplifiers have two inputs, one inverting and the other non-inverting, as in Fig. 6.4(a). The signal that is amplified is the difference between the two. If the non-inverting input is earthed, then we effectively have the previous situation. However, it is possible to combat electrical 'pick-up' in long leads by doing this earthing close to the previous element. If both wires are run close together then the spurious signals that they pick up will be more or less identical, and the one will cancel the other out. This use is illustrated, for Thematic Mapper, in Fig. 14.5. Two other uses to which the differential inputs may be put are indicated in Fig. 12.5. The figure shows the three different arrangements used in AVHRR. The detectors for the visible channels comprise two identical photodiodes fabricated onto the same chip. Thus they are as near identical, both in characteristics and environment, as may reasonably be contrived. Only one, however, is subjected to the incoming illumination. Although the noise from the two diodes will be uncorrelated, and will therefore add, the two dark currents will be almost identical. Dark current will thus be almost completely cancelled out. The other two channels use discrete 'long-tailed pairs' to drive the main amplifier. This circuit does a similar job in relation to changes in the transistor characteristics due to temperature drift or power supply fluctuations. It would also minimize the effect of long leads should these be necessary. The transistors, if they are not actually fabricated onto the same chip, have to be very carefully matched.

6.2.2 The pre-amplifier

The pre-amplifier is normally the first item that the signal sees after leaving the detector; its performance is therefore particularly crucial. In a sensitive application such as interests us, low noise characteristics and/or high input impedance are often vital, and so most of those that we discuss feature FET input stages. Most of them however were developed before good MOSFETs became available.

The design of the pre-amplifier is particularly difficult when the output impedance of the detector is high. The input resistor must then be very much higher. If significant gain is required, then the feedback resistance must be very much higher still, perhaps leading to severe problems with thermal noise. To overcome this problem, the pre-amplifier may be designed as a unity-gain 'source follower'. This device simply serves to bring the impedance of the signal down to manageable levels. The rest of the chain may then be conventional.

Fig. 12.5 also shows three approaches used in the design of AVHRR's pre-amplifiers. As is discussed in Chapter 12, the signal strength for AVHRR's visible channels is comfortably high, and the detector outputs are taken direct to the two inputs of an off-the-shelf operational amplifier. The signal levels for Channels 4 and 5 however are much lower. The HgCdTe photoresistors employed are low-impedance devices, and so conventional bipolar transistors are used in the long-tailed pair that drives the main pre-amplifier. The signal levels for Channel 3 are very much lower still, and an (indium antimonide) photovoltaic detector is resorted to in order to handle the regime. These are high-impedance devices, and so FETs are used in an otherwise similar circuit. Note the difference in the values of the two feedback resistors. The 16 MΩ resistor used for Channel 3 is very large, and a significant potential source of noise problems; whereas the 3 kΩ resistor used for Channels 4 and 5 is comfortably small. (The feedback is taken to the 'inactive' side of the transistor pair to maintain the correct polarity.)

6.3 FILTERS

Operational amplifiers can also be wired up as integrators, differentiators, or as frequency-shaping networks to suit any practical requirement. These features are fundamental requirements of analogue control systems.

Fig. 6.5(a) shows an amplifier wired up as an integrating circuit. We can see that, within the range of operation of the amplifier, the voltage V_{out} will always be the integral of the voltage V_{in}. Similarly, a differentiating circuit can be produced by exchanging the capacitor and R_1.

Fig. 6.5(b) shows a simple 'low-pass RC' filter. The 'impedance' (complex resistance) of a capacitor is inversely proportional to frequency. Thus, at low frequencies, the capacitor C will have little effect, and the gain of the amplifier will be approximately as given by Equation (6.1). At high frequencies, the capacitor takes over control. It will dictate that the amplifier gain reduces progressively as frequency is increased. The 'response' of this filter is shown in Figure 6.5(c). The reader may prove for himself, or accept, the following statements:

(1) The DC gain of this circuit is R_2/R_1.
(2) At high frequencies, the response falls off at 6 dB (a factor of 2) per octave or 20 dB (a factor of 10) per decade.
(3) The 'break' point (or 'corner' frequency), at which the two asymptotes meet, occurs at a frequency of $1/(2\pi CR_2)$ Hz.
(4) At high frequencies, the output sinewave lags behind the input sinewave by 90°. At the corner frequency, the output lags by 45°.
(5) The **frequency response** of the circuit can be given in the form:

$$\frac{V_{out}}{V_{in}} = \frac{R_2}{R_1} \times \frac{1}{1 + j\omega CR_2} ,$$

where ω is equal to 2π times the applied frequency.

The frequency response, when quoted in this form, may also be referred to as the **transfer function**. However purists then prefer to replace the Fourier operator, $j\omega$,

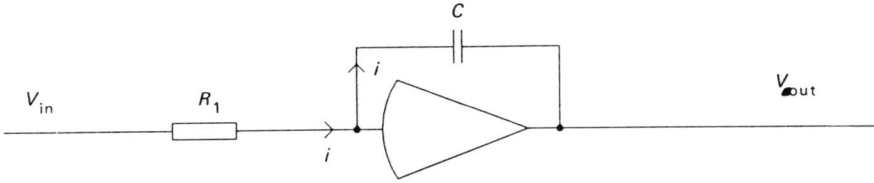

(a) integrator

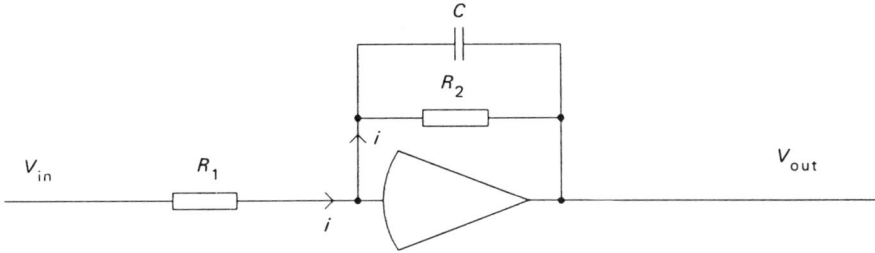

(b) simple low-pass filter ('simple lag')

Angular frequency (frequency $\times 2\pi$)

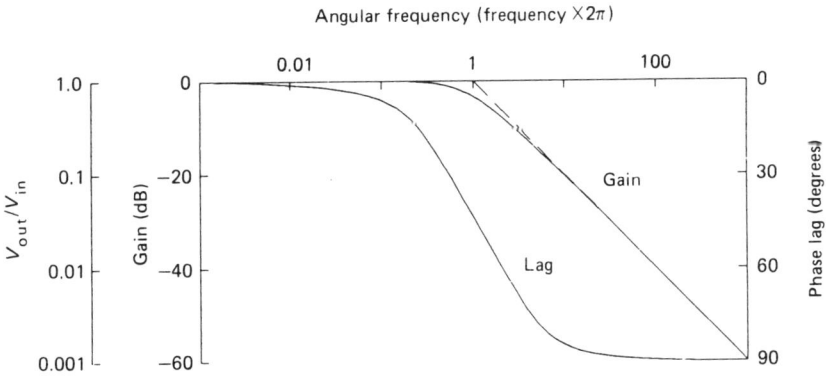

(c) frequency response of (b)

Fig. 6.5 — Low pass filter.

with the Laplace operator, p (where $p = \alpha + j\omega$). The practical difference is slight. In essence the Laplace transform is able to handle transient solutions, while the Fourier transform is not.

A wide range of different networks can be inserted in either the feedback circuit (to replace R_2) or the input circuit (to replace R_1) or both. With these one may produce simple high-pass, low-pass or band-pass filters. In each case the 'roll off' will be the same, namely 6 dB per octave, and a phase lag or lead of up to 90° will be produced. Six decibels per octave is not a great deal, and much faster roll-offs are often required. Any degree of improvement can be obtained by, effectively, chaining these simple filters together. This is the basis of the popular **Butterworth** design, which figures much in later chapters. However fast roll-off must be accompanied by large phase lags close to the corner frequency, and in the Butterworth design significant phase lags extend well down into the pass band. The higher the 'order' of the filter (i.e. the steeper the roll-off) the more serious this effect gets. Alternative designs are available which make use of 'tuned circuits' (electronic resonators). These cannot reduce the phase lag at the corner frequency, but they can prevent it from extending so far down into the pass band. However, being based on tuned circuits, they then introduce 'ripple' into the pass band, just as does the equivalent technique in optical filter design (see Chapter 4). As we will be seeing in later chapters, there is much scope for optimizing a filter for a given application.

6.4   DRIFT CONTROL

A number of the radiometers that we cover employ a strategy for short-term drift control that comes from the armoury of the controls engineer. Control practice is basically about detecting an 'error' between the desired state and that which exists at any instant, and introducing a correction to reduce the error.

Consider first Fig. 6.4(a). Suppose that the circuit is at rest, with V_{in}, V_{out} and V_0 all at zero. Now raise V_{in} to a small positive value. This will generate a current, i, as already discussed, which will tend to raise the voltage at V_0. The amplifier will react by starting to generate an infinite negative voltage at V_{out}. However, as the output voltage falls, it will generate a current through R_2 flowing in the direction of the arrow. The effect will be to tend to cancel out the voltage generated by the input circuit. In fact the system can only stabilize when the two influences cancel out exactly and V_0 is returned to zero. Thus we have the principle of feedback, and the basis of control theory and practice. In physics it is known as Lenz's Law, and in chemistry as Le Chatelier's Principle†. If a system in equilibrium is disturbed, then forces are unleashed which attempt to return the system to equilibrium. This simple discussion ignores the time lags that are an inevitable part of any practical system. If a sufficiently sluggish response is accepted, then the response of the control circuit can be slowed down until these time lags do not matter, and almost any system can be controlled with extreme ease. All the problems of control engineering occur when a rapid response is required from a system that is naturally less rapid.

The drift control problem is basically this. Variations in detector dark current, and drift in the electronics, will conspire to generate significant changes in the output voltage of the system independently of any genuine changes in illumination. Some of the instruments that we cover are **radiometers**, that is true measuring instruments,

---

† Hopefully, the reader will detect an element of tongue-in-cheek at this point. There is however a genuine analogy here.

from which absolute readings are required. These are basically the meteorological instruments, such as AVHRR, GOES and Meteosat. It will become clear that the Earth observation instruments (Thematic Mapper, SPOT etc.) are significantly less particular in this respect. The radiometers cannot tolerate this drift, and are provided with a single point calibration between every scan sweep to enable it to be corrected out. This is done automatically by an electronic feedback circuit, the operation of which is indicated in Fig. 6.6. During this calibration period, the detectors receive a

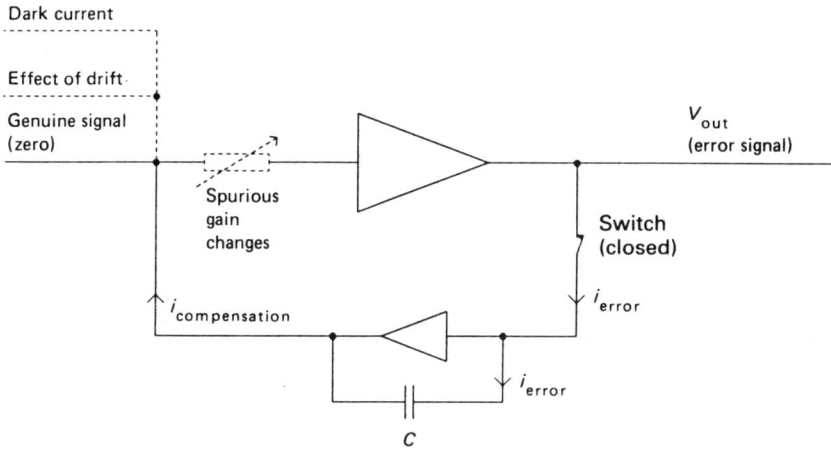

Dark current

Effect of drift

Genuine signal (zero)

Spurious gain changes

$i_{compensation}$

V_{out} (error signal)

Switch (closed)

i_{error}

i_{error}

C

(a) correction phase (detector viewing cold space)

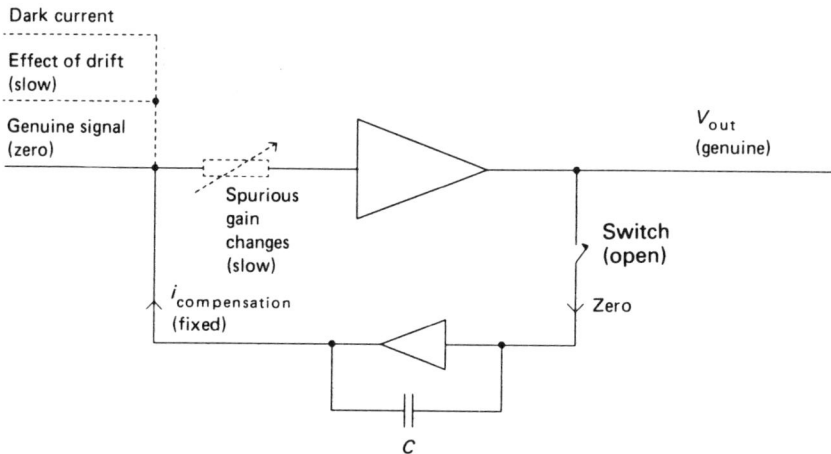

Dark current

Effect of drift (slow)

Genuine signal (zero)

Spurious gain changes (slow)

$i_{compensation}$ (fixed)

V_{out} (genuine)

Switch (open)

Zero

C

(b) data gathering phase (detector viewing target)

Fig. 6.6 — Drift correction.

brief view of cold space, at which time their output ought to be zero. Therefore the output of the electronic chain should also be zero. Any output voltage that it produces at this time represents an error signal. A current, proportional to the output voltage, is fed back, via an integrating circuit. As we discussed in Section 6.2, the current generated by the compensation circuit will continue to increase until it exactly matches the spurious input currents, and the error signal (the output of the main amplifier) is reduced to zero. This whole process takes place while the detector is still viewing cold space. Before the view of the detector changes, the switch shown is opened. The output of the integrator is thus frozen, for as long as the capacitor C holds its charge, which is of course for at least the duration of one scan. During this period all spurious inputs that existed at the start of the scan are corrected out, leaving only any changes that might occur during the scan. If these are small (which they are) then drift correction is accomplished.

The digital processes that the signal next meets are beyond the scope of this book.

REFERENCES

Ahmed. H. & Spreadbury, P. J. (1973) *Electronics for Engineers* Cambridge.
Braithwaite, S. (1990) Corrections in draft.
Millman, J. & Halkas, C. C. (1972) *Integrated Electronics* McGraw-Hill.

7

Data quality

Data quality may be something of a misnomer for this chapter, because the term 'quality' will mean different things to different users. Many will regard freedom from 'striping' and other visual artifacts as a major requirement for good image data. However, for current purposes, we regard a remote sensing instrument as just that — a device for measuring (or at the very least, estimating) the properties of patches of ground, water or cloud. We therefore concentrate on the following properties.

7.1 RADIOMETRIC PERFORMANCE

While the general currency of instrument performance specifications tends to be terms such as 'linearity' and 'dynamic range', by far the most important property of most remote sensing instruments is its **signal/noise ratio (SNR)**. This is because it is a vital part of the design optimization strategy to ensure that the SNR is just high enough but no more.

As we discussed in Chapter 5, noise tends to be introduced into an instrument when small incoming signals have to be greatly amplified. Noise (or uncertainty) is also inherent in all physical processes from atomic physics through acoustics to politics and economics. In the present case, noise is introduced by the quantization of radiation. Individual photons arrive with a degree of randomness, which is only averaged out to a smooth and progressive signal after a large number have been detected. If SNR is to be eliminated as a problem then each reading must indeed be based on the accumulation of a large number of photons.

However, collecting photons is expensive, in terms of instrument size and weight, or in terms of complexity and cost (and reliability) or in terms of performance penalties in other areas. For example, Thematic Mapper (Chapter 14) buys its high performance by the use of as many moving parts (relatively speaking) as a sewing machine. AVHRR (Chapter 12) buys its rugged simplicity, and its potential for easy development, at the cost of extremely poor spatial resolution. Pushbroom systems trade the avoidance of moving parts for a large number of very small detectors (usually CCDs, which introduce problems of their own, including a rather limited dynamic range), and for the problems associated with a wide viewing angle. IRS-1 eliminates the losses associated with beam splitting by quadrupling the entire instrument. And so on. In each case the same strategy is used to minimize the additional problems introduced. The strategy is to base each flux measurement on the minimum number of photons that the specification will tolerate.

The limiting noise source may not actually be fluctuations in photon arrival. If detectors receive little incident energy then they generate small output voltages (or currents), and massive amplification is required to generate usable signals. How much additional noise this produces will depend on how much money is spent, and on the state of the electronic art at the time that the design decisions were made. It can also depend on how much cooling is made available for the vital first stage of amplification.

The task of producing an acceptable performance is much easier in some wavebands than others, and this is reflected in the approach to the specification of channel performance. In general, life is reasonably easy in the reflective band, more difficult in the thermal band, and verging on the impossible in the $3–5\,\mu$m band.

7.1.1 Reflective band

In the reflective band, incident flux levels are generally relatively high. Even for instruments which cover only this band, it is normally possible simply to demonstrate that noise effects are negligible. The reflective band serves two main user communities; Earth observation groups and meteorologists. To the former it is their principle tool. They are interested in differences in the reflectance of a wide range of ground features. They therefore demand a given maximum **noise equivalent ground reflectance variation** (NE$\Delta\rho$) for each reflectance band. This effectively defines the maximum permissible uncertainty in the veracity of any particular measurement. Unfortunately, however, the supplier must deal in absolute radiance values. By some means, therefore, both an assumed spectral radiance incident upon the top of the atmosphere and a spectral atmospheric transparency must also be supplied.

The strategy for arriving at a usable specification varies with the project. In the case of Thematic Mapper, much of the responsibility was laid·upon the supplier. The customer supplied NE$\Delta\rho$ figures for each band (from 0.5% to 2.4% of full scale depending on band), but the user input concerning the incident radiances against which these were to be set appears to have been simply that they should take account of 'anticipated high reflectance ground features' (Hughes, 1984). The supplier then undertook a considerable investigation in order to derive realistic incident at-instrument radiances (Norwood & Lansing, 1983).† By contrast, for the later IRS-1, the 'design-goal' NE$\Delta\rho$ figures, together with the equivalent saturation radiances, are given clearly in the Users' Handbook (NRSA, 1986).

Meteorologists use the reflectance band mainly as a qualitative tool, for example for feature identification. (Although it has to be said that they would probably make greater use of it if ways could be found of providing dependable absolute calib-

---

† The dynamic range of TM is often quoted as extending to negative radiance values. This is obviously nonsense. The situation arises because dynamic range is quoted strictly in terms of a standard analogue to digital conversion, a conversion whose calibration gives rise to this unfortunate effect.

rations.) The meteorologists tend to approach the supplier half way, and to specify the performance of their reflective bands in terms of the SNR at some minimum **albedo**. In the case of AVHRR, the requirement is for an SNR of at least 3 at 0.5% albedo (ITT, 1982), with $400 \, W/m^2/\mu m/sr$† (approximately) being assumed to be incident upon the top of the atmosphere. Note that this specifies the albedo of the total atmospheric light path, which will not generally be the same as that of the cloud or sea patch of interest to the meteorologist. From these two figures the supplier computes a minimum incident flux in watts. From a knowledge of the major sources of instrument-generated noise, he can then compute a **noise equivalent power (NEP)**, which allows an SNR, at this flux level, to be derived. For AVHRR, this emerges as around 35:1, which is an excellent figure. It shows that the minimum reflectance measurable is in fact dictated wholly by the resolution of the 10-bit A/D converter. (We have already remarked that this superb radiometric performance is bought at the expense of a low spatial performance.)

SPOT was specified in terms of an NEΔρ of 0.5% of full scale at the ground (Begni *et al.*, undated). This is then converted into a 'noise equivalent radiance' at the instrument, by taking into account assumed atmospheric losses, by the customer (CNES).

The builders of IRS-1 were given a design goal of an NEΔρ of 0.5% of full scale for all spectral bands (NRSA, 1986); based on saturation radiances of about $300 \, W/m^2/\mu m/sr$ (except for the 0.8-μm band which is somewhat lower). Corresponding SNR (measured) levels are also given, at around 150.

The Insat requirement is based on an incident radiance of $324 \, W/m^2/\mu m/sr$. It demands an SNR of 8 at 2.5% albedo, although the actual ratio expected pre-launch as a result of design calculations was 17 (ITT, 1977).

7.1.2 Thermal band

By contrast with the reflective band, thermal-band radiance levels are relatively low. As we will be discussing, it is invariably necessary to cool the detectors and their preamplifiers down to around 100 K in order to minimize electrical noise. Moderately exotic preamplifier designs also have to be used, but that is discussed elsewhere. Even so, the noise levels remain distinctly borderline, and it is necessary to do a reasonably careful analysis to show that the user's minimum requirements will in fact be met.

The thermal band is the main band used by meteorologists, and they require it for precise measurements. They require both accuracy, which demands absolute calibration to high standards, and precision, which means good SNRs. As we have already discussed, flux levels in the thermal region are much lower, and the SNRs may be expected to be less good, than those of the reflective bands, in spite of the best that the electronic engineers can do. In consequence it is not possible simply to prove that the noise level is small enough to be ignored, because it will not be. It is necessary to show that acceptable measurements can be made *in the region of main interest*, and to take 'pot luck' as it were in other parts of the radiometric range. The performance is specified in terms of the **noise equivalent temperature difference (NEΔT)**, more or less as before. What is of course different is that the specific temperature quoted represents the zone of principle interest rather than at some minimum. In the case of

† The steradian (sr) is a unit of solid angle. There are 4π steradians in a complete sphere.

AVHRR, the thermal-band NEΔT is required to be no greater than 0.12°C at 300 K (27°C), which is near the top of the range of interest. No limits appear to be placed upon their performance at other parts of the range. For Thematic Mapper, no fixed temperature is specified, which is perhaps something of a puzzle. However the NEΔt requirement is somewhat more relaxed, at 0.5°C.

Only Meteosat appears to specify the performance of its thermal channel at more than one point on the range. It demands an NEΔT of no greater than 0.4°C at 300 K and no greater than 1.2°C at 200 K (Péraldi, 1978).

The INSAT requirement is for an NEΔT of 0.2°C at 300 K, although the actual value expected as a result of design calculations was 0.11°C.

7.1.3 Mid-infra-red band

This region is well down the 'dip' slope of the solar radiance curve (see Fig. 2.1), so that solar irradiance is a factor of approaching 10^4 smaller than for the reflective bands. However, it is also well down the 'scarp' slope of the Earth's radiation band. In general, this band is used for night-time thermal measurements, and this makes photon noise a particularly severe problem. Not only are the flux levels low, but the photons themselves are more energetic than they are in the thermal band proper, and so the available radiant energy is shared among fewer of them. As a result, photon noise is almost certain to be the limiting factor, and particular precautions have to be taken to ensure that noise levels from other sources are lower than this. If, in spite of all precautions, analysis does not show that an acceptable performance can be achieved, then the sights of the user must be lowered or the entire instrument redesigned.

Of the two instruments covered that include this band, in-depth information is available to the author only for AVHRR. The calculation is much as for the thermal band, but the results inevitably make less happy reading. Photon noise is indeed the limiting factor, and the 'quantum efficiency' of the detector itself therefore assumes an importance (because failure to detect a photon is akin to receiving fewer). It is computed that the specification of NEΔT will be met, but by a much smaller margin than for the thermal band proper. It is, for example, computed that a single digitizer step is about equal to the estimated noise level. This implies that, unlike other channels, the user can have but a moderate confidence that a given reading is not out by more than one count. However, a user who is taking too much account of a single reading is in danger from other deficiencies such as spatial frequency response (to be discussed later in the chapter). Over larger areas, this particular weakness will appear as visible noise, or speckle. If a nominally uniform area comprising more than a few pixels is under study then averaging should be an effective remedy.

Readers wishing to delve more deeply into the subject are invited to study the numerous publications of the Optical Science Center, University of Arizona, possibly starting with Slater *et al.*, (1987).

7.2 SPECTRAL PERFORMANCE

The function of the spectral filters is to pass all the incident energy that falls within the nominal band limits, and to reject all that which falls outside them. In practice it is never possible to provide 100% transmission, and a considerable loss, even in the pass band, must be accepted. More seriously, it is not possible to produce a filter with both a flat top and steep sides either in electronics or in optics, and for very similar mathematical reasons. As we saw in Chapter 4, it is not always easy to provide a high degree of rejection even well outside the pass band. Each of these deficiencies, if too severe, could destroy one of the main bases upon which an imaging remote sensing instrument operates. An additional complication is that the overall spectral response of the instrument is the sum of the responses of all the elements in the chain. Although the response of the primary optics is likely to be flat, that of the beam splitters may well not be. Detectors are quite often used in spectral regions where their response is far from flat. This is particularly likely when silicon detectors are used in the visible band. In principle it is the task of the filter designer to take all these responses into account, and to produce a filter which delivers an overall profile which meets the specification. In practice however, at least during the period in history that we cover, it appeared to be more usual to demand flat-topped filters, and accept that the detector profile would impose a slope on it.

This procedure was clearly intended to ease design and contractual problems. However, it has to be said that it is moderately useless as a specifier of the performance of the overall instrument. There is no way that a user in the field can correct out the effects of a sloping top unless he knows the precise spectral profile of his target patch and of the incoming radiation (Schueler, 1983a). Lest the above comment should be regarded as overly critical, however, it must also be pointed out that the real killer is not a moderately rough top to the pass band, but poor rejection of out-of-band energy.

Not surprisingly perhaps, the specification of an optical filter is less than perfectly straightforward. An electronic engineer would expect the pass band profile to be defined in terms of pass band ripple, as a percentage, and band edge roll off in decibels per octave. (He might also expect to see a maximum phase shift specified, but the author has only come across one report that mentions this in the current context.) He will not find much that is familiar in the specification of, for example, the TM filters (Markham, 1983). The specification parameters are illustrated in Fig. 7.1. The figure shows the overall response of a typical Band 1 channel. It includes the sloping response of the silicon detectors. The 'filter flatness' is intended as a measure of the flatness of the *filter* pass band, as estimated from the *overall channel* characteristic. It is defined as the proportion of the nominal passband which may be represented within specified limits by a sloping straight line. The limits are 10% (or 20% for the thermal band). The quantity of interest to the user is the 'channel' flatness. It is defined as that proportion of the nominal passband which can be represented (within the specified limits) by a *horizontal* straight line.

The band edges of the Thematic Mapper filters are defined as the points at which the response drops to 50% **relative spectral response (RSR)**, which is the response related to the peak response of that particular filter. Peak response, as we have said, is likely to be considerably less than 100%. The unit is watts (per square metre per micrometre), which is a unit of power. Therefore the 50% RSR points are the same as the '3 dB down' points familiar to readers versed in electronic terminology.† The

---

† The **decibel** is defined so that 3 dB represents (almost exactly) a factor of 2 in power terms, or a factor of $\sqrt{2}$ in volts, speed, quanta per second, etc. Six dB then represents a factor of 4 in power, of 2 in voltage etc. and so on. A factor of 10 in power is 10 dB. A factor of 10 in voltage is 20 dB.

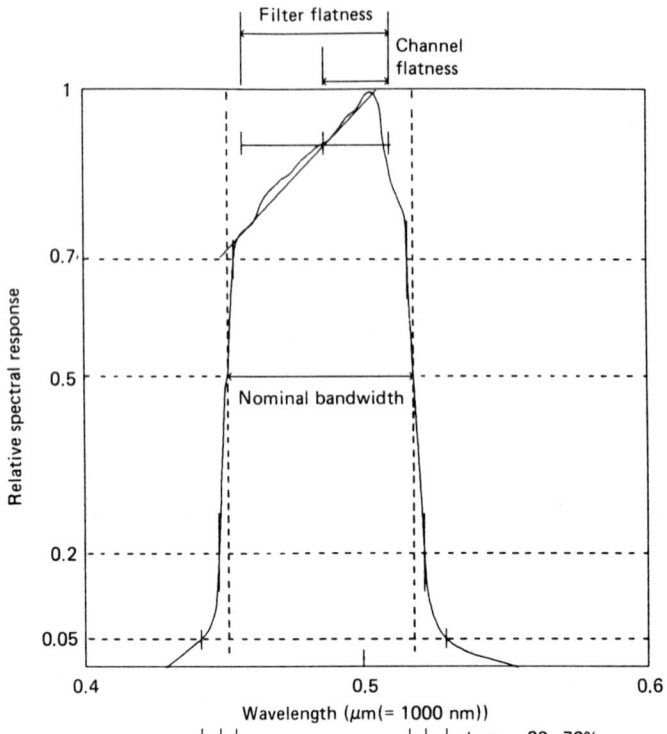

TM

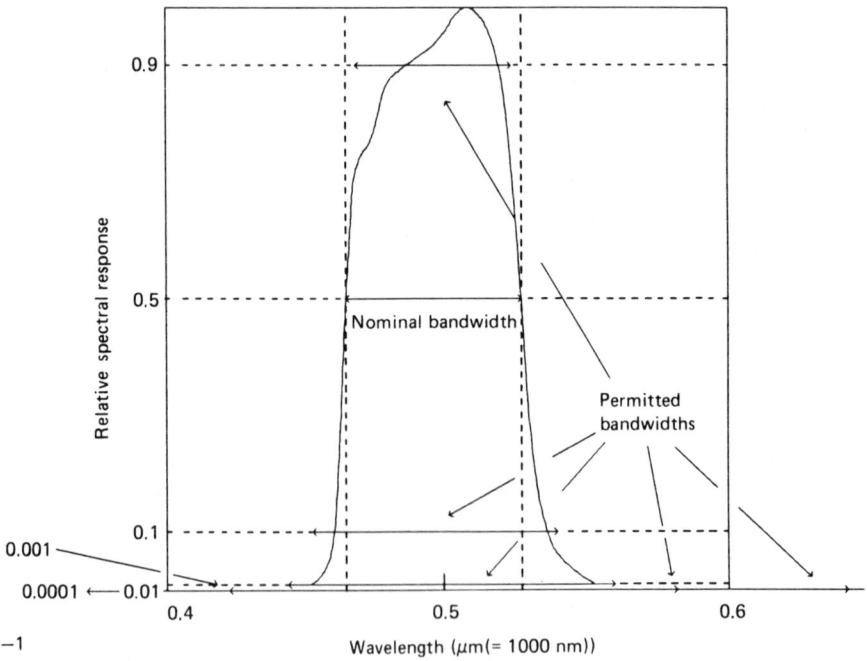

IRS–1

Fig. 7.1 — Filter performance criteria.

slope of the skirts is defined as the separation in nanometres ($0.001 \mu m$) of the 20% and 70% RSR points and, separately, the 5% and the 70% RSR points. This method of specification reflects the fact that the skirts tend to broaden out at the base more rapidly than would an electronic filter, and that the effect of this could be serious if it should be too pronounced. Thus although the steepest part of the skirt is very steep, it gets very much less steep further 'out' (away from the band edge). To cover the tendency of optical filters, particularly interference filters, to pass radiation in zones well beyond the desired passband, it was required that the total out-of-band energy should be less than a certain fraction of the in-band energy.

IRS-1 uses high performance dichroic (interference) filters. The specification procedure is slightly different, and somewhat more generalized (Jouan *et al.*, 1989). Again the filter specification does not take account of the response of the detectors. This is perhaps not surprising since the telescopes and filters were made in France, and were integrated with the focal plane in India. Flat filter tops were specified, with ringing or ripple of no more than 5% RSR and in-band transmission to be no less than 50%. The slope of the skirt was specified by defining the width of the peak at a number of RSR values, ranging from 90% to 0.01%. The width was specified as a fraction of the nominal bandwidth (i.e. at 50% RSR). An attempt has been made to illustrate these criteria with the aid of a curve from Jouan *et al.* (1989). As is indicated in Fig. 7.1, the bandwidth at 0.01 RSR was permitted to be about 4 nominal bandwidths. Jouan's profiles are clearly those of the overall instrument, because the tops presented would not meet the flatness specification. This has affected the illustration of the 90% bandwidth point.

AVHRR uses bulk absorption for the short-wave skirt and interference techniques for the long-wave skirt. Filter, and overall channel, spectral performance is discussed in some detail in ITT (1982). A spectral profile of the silicon detectors used for the reflective bands is also supplied (see Fig. 12.6). Also supplied are profiles of the raw filters and of the overall system. The profiles of the longer-waveband detectors are reasonably flat over the region of interest, and only overall spectral response curves are shown. Again, the bandwidth is defined in terms of the 50% RSR points; however, top flatness does not appear to be covered. Skirt steepness is specified in terms of the spread, as a percentage of the nominal passband wavelength, over a certain region of the skirt. In addition it is specified that the integrated spectral transmission, over specified out-of-band zones, shall be less than 0.1%. In-band transmission must average out at at least 75%.

7.3 SPATIAL FREQUENCY RESPONSE

If a detector is swept continuously across a rapidly changing terrain, such as a city, the finest detail will certainly be lost. This is partly because the finite resolution of the optics, and of the detector, independently introduce a degree of blurring, and partly because the electronics will always incorporate a low-pass filter of some sort. The effect will be that the apparent contrast between the bright roofs and the dark streets, say, will get progressively less as we study areas where they are packed ever closer together. We can define a 'spatial frequency response' to quantify this fall-off in response with the increase in spatial frequency. A 'cut-off' frequency might be defined at the frequency (street separation) which gives rise to a 50% reduction in measured contrast.

The (temporal) frequency response of a non-imaging instrument is a measure of its ability to respond to high frequencies in the electronic or other input signal. It could equally well be thought of as representing its ability to pick out fine detail in the incoming signal, because fine detail is generated by high frequencies.

Similarly, the **spatial frequency response** of an imaging instrument represents its ability to pick out the fine detail in the scene that it is imaging. In a well-designed instrument, the spatial frequency response is mainly dictated by the IFOV of the detectors. However, as we discussed in Chapter 5, this will be because the other major factors (the resolving power of the telescope and the response of the electronics) will have been matched to it. The response of each of these components can be measured, or computed, as shown in Fig. 7.2. The response of the complete system may then be estimated by multiplying together the individual attenuation figures. (Readers versed in electronics will be used to seeing such curves plotted on log–log scales, which would have enabled the overall response curve to be obtained particularly easily.)

Total system response can also be measured, at least on the ground. A standard technique is to prepare a series of cards, each with a series of black and white bands painted on it. The bands on each card represent a different spatial frequency, as depicted in Fig. 7.2. If the cards are presented in turn, then the complete spatial frequency response can be built up. Although the use of black and white bands is usual, there should ideally be a sinusoidal variation in grey level between black and white. Sinusoidal cards are in fact sometimes used, and may be expected to give a slightly differently shaped curve.

Once the instrument is in orbit then measurement, in its true sense, is virtually impossible. However, checks of a sort can be made by evaluating the spatial frequency content of imagery and comparing the result with high resolution imagery of the same scene obtained from a different source (Chapter 16).

Readers of remote sensing literature will often find the spatial frequency response presented in terms of the **modulation transfer function (MTF)**, which we have already discussed in Chapter 5. The term 'transfer function' implies a Laplace or Fourier transformation. It therefore also represents a (spatial) frequency response. In this context, modulation effectively means contrast, and so we are talking about the 'contrast transfer function' or the loss of contrast as the fineness of the detail increases. It is intuitively obvious, for example, that where the separation of black and white bands, or city streets, is much greater than the sampling interval (which, for illustrative purposes, we might equate with IFOV), then they will be picked out in reasonably full detail. At these low spatial frequencies, the MTF will approach its maximum value (often defined as unity). Where the separation is exactly equal to the sampling interval (IFOV) then each detector will see the average radiance of one black and one white band, or of one street combined with that of its lining of houses, and the MTF will approach zero. In between, the MTF will take on an intermediate value, and we will end up with a curve which looks somewhat like the solid curve shown in Fig. 7.2. Note however that as the spatial frequency is increased still further the 'cancellation' effect that we have just described becomes less than complete — and we are into the 'aliasing' region.

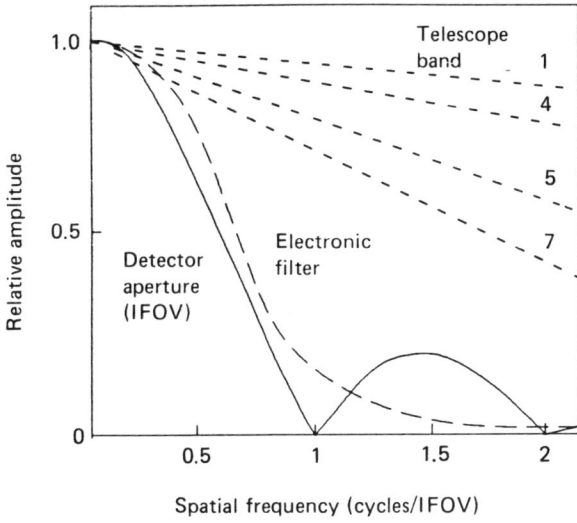

(taken from TM)

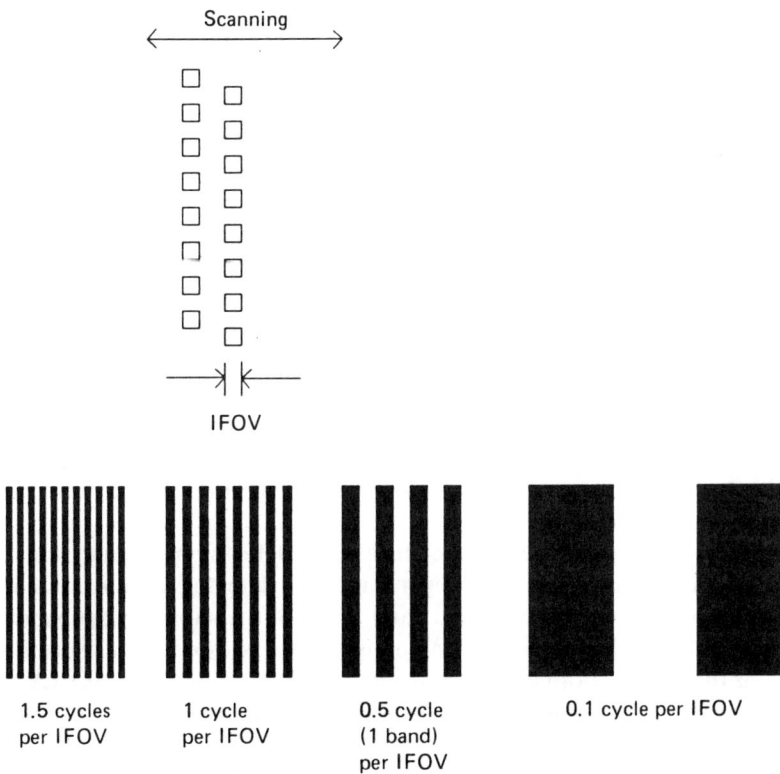

Fig. 7.2 — Spatial responses.

As we have seen, the MTF is a frequency response, and is therefore a continuous curve† covering the entire range of interest. However, it is often quoted at a single spatial frequency only, namely the **Nyquist frequency** (Markham, 1984). This is the frequency which is sampled exactly twice by the A/D converter. In the current context it normally represents a spatial period of 2 IFOVs, although it must be emphasized that this only means the same thing if the sampling rate is matched to the IFOV. The Nyquist spatial frequency is also often described as 'one bar per IFOV'

As might be expected, the spatial response of TM was evaluated with great thoroughness (e.g. Markham, 1984). Comprehensive studies of MTF and 'square wave response' were carried out, both at the component level and for the complete system. It is of course important that the spatial performance of the system should be re-evaluated after the shock of launch. This is unfortunate, because to do so is not without difficulty. However Schueler (1983b) proposed a strategy based on the derivation of the 'line spread function' that could facilitate simple checks. The lines in question would be a series of North/South (or East/West) linear features of varying widths. The expected MTFs (which are a 'frequency-domain' representation) could be converted to expected line spread functions ('spatial-domain' representations) by use of the inverse Fourier transform. These could then be compared relatively easily with what was actually observed.

The photo-interpretive technique used for SPOT is discussed in Chapter 5.

7.4 INTERBAND REGISTRATION

A major application area for remotely sensed imagery is the derivation of the 'spectral signature' of the ground surface. From this, a great deal may be inferred about the nature of the terrain and/or of its ground cover. However, to make such inferences with confidence requires that the pixels for all the bands line up with each other with considerable precision. For single-detector (per band) low-definition instruments, such as AVHRR, an adequate degree of precision is relatively easy to achieve. For a pushbroom high-definition system such as SPOT, the task is much more severe. As we shall be seeing in later chapters, the design of the focal plane is affected quite substantially by the importance, or otherwise, of this criterion.

The interband registration for AVHRR is generated/measured by aligning each of the five detectors to within the prescribed circle (see Chapter 12). Apart from making the structure sufficiently strong to survive launch without distortion, this is all that requires to be done.

The Thematic Mapper focal plane comprises 16 detectors per band, whisked to and fro across the ground track. The longer waveband detectors require cooling, and are mounted on a separate focal plane from those of the shorter wavebands. The registration quality of the instrument on Landsat 4 was assessed by measurements at '29 locations throughout both the forward and reverse scan profiles' (NOAA, 1984). Registration was slightly better in the along-scan direction than across-scan. It was within about 0.1 pixel within one focal plane, but the average misregistration was approximately doubled across the focal planes.

---

† If evaluated by means of the so-called fast Fourier transform, the curve will of course be discrete. However the real Fourier transform, like the Laplace transform, is a continuous funcion.

For a pushbroom system, registration accuracy could vary widely over the length of a scan line, depending on the quality of the optics and where the beam splitting takes place. In general, when it comes to evaluating or describing the registration accuracy, only statistical methods are realistic. On the ground, this accuracy may be measured with precision; however Boissin, (undated) outlines a strategy whereby interband registration may be estimated from flight imagery. The strategy involves assuming that most ground features will show up in all three bands. There will therefore be a high degree of correlation between equivalent lines and columns on all three multispectral bands. If registration is perfect, then a cross-correlation between two bands should peak at exactly zero spatial displacement. Random mis-registrations would be averaged out by this technique. However, a systematic mis-registration, even of much less than one pixel, should show up as an asymmetry of the principle peak. As we will be discussing in Chapter 16, SPOT's 6000-pixel pushbroom arrays are made up from four shorter CCD chips, butted together 'optically' by means of a complex beam-splitter arrangement. Discontinuities may therefore be expected at the chip boundaries. Boissin & Gardelle outline how the relative alignments of the individual chips were investigated, by carrying out cross-correlations over parts of the scan lines only. By this means it was possible to estimate the degree of mis-registration in the region of each end of each chip. Band 1 was used as a reference (i.e. assumed to be perfect) and the alignment of each chip of the other two bands was compared with it. From this, the degree of mis-registration at the ends of each array was estimated. In general the results were within 0.15 of a 20-m pixel.

The strategy was also used to reveal any timing problems over the scanning of adjacent chips. Here, it was assumed that most features would be large compared with the pixel size (i.e. that the spatial frequencies are mainly low), and that therefore adjacent lines of imagery should also be highly correlated. Again, the correlation should peak at zero displacement. Even a small mis-timing of the scanning of adjacent chips should show up as an asymmetry of the correlation peak, although to eliminate completely the effects of oblique linear features, it might be necessary to repeat the analysis over several images.

With SPOT, it will emerge that the beam splitting is done close to the focal plane, and so the effects of optical distortion are, to a first approximation, eliminated. With IRS-1, however, (Chapter 18) the single CCD chip for each waveband was mounted at the focal plane of a separate telescope. We therefore have two potential sources of mis-registration to evaluate. The first is relative mis-alignments of the individual telescopes, which would give rise to a mis-registration of a complete band, which might vary linearly along the imagery line. The second source is due to optical distortion. The designs of the telescopes are identical and so, again to a first approximation, distortion effects should cancel out. However, second order effects due to manufacturing differences, and to the effects of wavelength, could remain and would generate mis-registration which might be worse towards the ends of an imagery line. As far as it known, all these effects are within the specified 0.25 pixel.

REFERENCES

Begni, G., Leroy, M. & Dinquirard, M. (undated) 'Spot radiometric resolution performance evaluation: preliminary results'. CNES & CERT.

Boissin, B. & Gardelle, J. P. (undated) 'Spot Localization Accuracy and Geometric Image Quality', CNES & Matra.

Hughes (1984) 'Thematic Mapper, Final Report'. Santa Barbara Research Center, Greenbelt, Maryland, USA, Oct. 1984.

ITT (1977) 'INSAT VHRR' proposal for the design and manufacture of the INSAT imaging instrument. ITT Aerospace/Optical Div. Fort Wayne, Indiana, USA, Nov. 1977.

ITT (1982) 'AVHRR Technical Description' ITT Aerospace/Optical Div. Fort Wayne, Indiana, USA, Sept. 1982.

Jouan, J., Martinuzzi, M. M., Mallinge, N., & Coussot, M. (1989) 'Wide field high performance lenses' Matra Space Branch.

Markham, B. L. (1983) 'Spectral Characterisation of the Landsat Thematic Mapper sensors' Landsat-4 Science Characterisation Early Results Symposium, Goddard Space Flight Center, Greenbelt, Maryland, USA February 1983.

Markham, B. L. (1984) 'Characterisation of the Landsat Sensors' Spatial Responses'. NASA Tech. Mem 86130.

NOAA (1984) *Landsat 4 (& 5) Data Users Handbook* U.S. Geological Survey.

Norwood, V. T. & Lansing, J. C. (1983) *Manual of Remote Sensing* (American Society of Photogrammetry); Chapter 8.

NRSA (1986) *IRS Data users Handbook.*

Péraldi, A. (1978) *The Meteosat Radiometer* 15th European Space symposium, Bremen, June 1978.

Schueler, C. (1983a) 'Sensor Limitations' Landsat Short Course, presented at University of California, August 1983.

Schueler, C. (1983b) 'Thematic Mapper Protoflight Model Line Spread Function' 17th International Symposium on Remote Sensing of Environment, Ann Arbor May 1983.

Slater, P. N., Biggar, S. F., Holm, R. G., Jackson, R. O., Mao, Y., Moran, M. S., Palmer, J. M. & Yuan, B. (1987) 'Reflectance and radiance based methods for the in-flight absolute calibration of multispectral sensors' *Remote Sensing of Environment* **22**.

Part II

8

Multispectral scanner

8.1 INTRODUCTION

The multispectral scanner, or MSS, was the first of NASA's mass-market instruments. On reading this chapter, in the 1990s, it is particularly important for the reader to remember that the design concept of MSS dates from about 1965. It is a remarkable tribute to the pioneers who created it that MSS data were still popular at the time of writing (1990). This is largely because of their economy both in first cost and in their demands on computing and data storage capacity. MSS imagery created something of a revolution when it first became available in 1972.

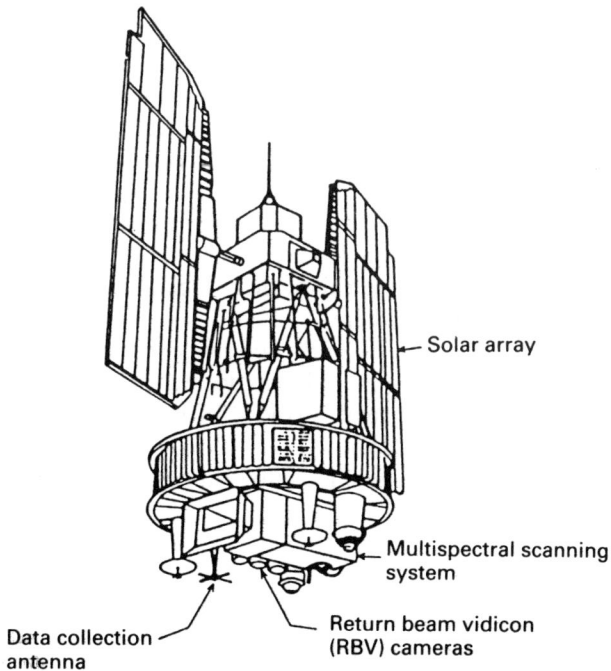

Fig. 8.1(a) — Landsats 1–3. Courtesy: NASA.

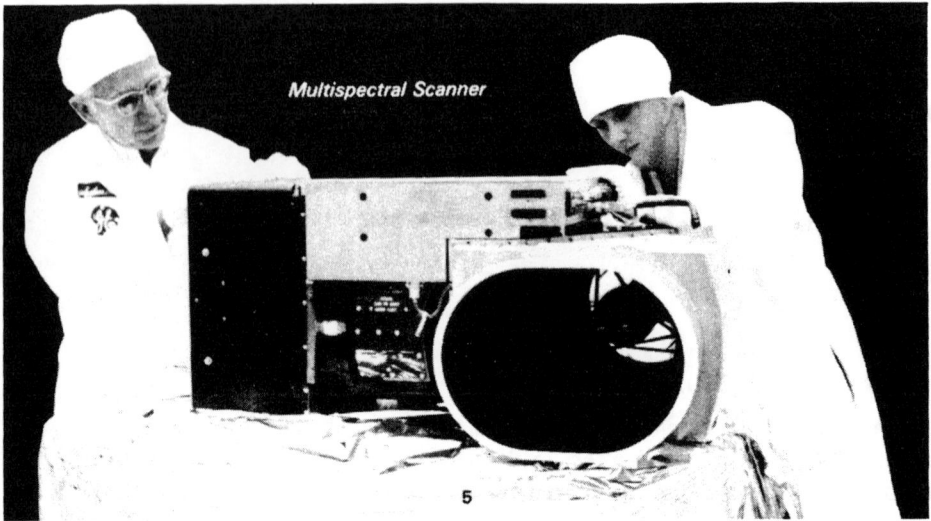

Fig. 8.1(b) — MSS. © General Electric.

MSS was flown on its own on Landsats 1 to 3, and in parallel with the much more advanced Thematic Mapper (TM) on Landsats 4 and 5. Subsequent Landsats will provide MSS-quality data by data reduction from TM output. For Landsats 4 and 5, the orbit was lowered by some 200 km, partly to facilitate the design of Thematic Mapper (see Chapter 14) and partly with the idea of making maintenance by Shuttle easier. This resulted in a change to the ground-track pattern and to the world reference system. The old pattern is described in this chapter; and the new in Chapter 14.

MSS operates in four bands in the 'reflective' region of the visible and near infra-red. On Landsats 1 to 3, the channels were numbered 4–7; from Landsat 4 onwards, however, they have been numbered in the more logical sequence of 1 to 4. This latter numbering sequence is adopted here.

On Landsats 4/5, the geometry of the telescope was altered slightly to keep the ground IFOV the same when the altitude was decreased. Apart from the orbit, all the quantitative data in this chapter relate to the MSS instruments flown on Landsats 4/5. An exception, however, is the orbit.

On account of its age, little detailed information has been made available about the design of MSS.

8.2 THE ORBIT

We describe here the orbit of the first three Landsat missions, upon which MSS was the flagship. The orbits of the later missions, where MSS took second place to TM, are described in Chapter 14.

Table 8.1

Spacecraft — Landsat 3			
Mass — satellite (kg)	—	Launched	1972
— MSS (kg)	65		1975
			1978
		Target	Land

Orbit — Landsat 3			
Altitude @ Equ. (km)	920	Orbits/day	13.94 ($13\frac{17}{18}$)
Semi-major axis (km)	7285.8	Orbits/cycle	251
Eccentricity (%)	—	Days/cycle	18
Period (min)	103.15	Shift/orb. (°)	28.82
Inclination (°)	99.12	(km) @ Equ.	2875
Descending node	09:31	Gnd track spacing (°)	1.43
Ground speed (km/s)	6.45	(km) @ Equ.	159.38

Telescope — Landsat 4/5			
Type	Ritchey Chretien		
Focal length (m)	0.826	80% blur c.d. (μrad)	—
Aperture (m)	0.229	F.P. dia. (mm)	—
f-number	3.6	Field-of-view (°)	—
1y mirror dia. (m)	0.229	IFOV (μrad)	117.2
2y mirror dia. (m)	0.094	(μm @ F.P.)	96.75
Total obscuration (%)	50	(m on ground)	82.6
Clear area (m^2)	—	EIFOV (m on ground)	77.7†
		Inter-band reg. (pix.)	—

Scanning — Landsat 4/5			
Method	Raster scan		
Rate (scans/s)	13.62	Mirror swing (°)	±7.45
(rad/s)	8.05		
Period (ms)	32.75×2 (approx.)		
Active scan time (ms)	32.75		
Efficiency (%)	50 (nominal)		
Sampling interval (ms)	0.01		
Integrating time (ms)	—		

Channel details (nominal) — Landsat 4/5

Band	Wavebands (μm)	Detectors	Dynamic range (W/m^2/sr/μm)	Noise equiv. reflectance (% F.S.)	MTF @ pix. rate trk/scan	Cooled?
1	0.495–0.605	PM tubes	. . .−225/75		0.57	No
2	0.603–0.698	PM tubes	. . .−211/71		0.57	No
3	0.701–0.813	PM tubes	. . .−156		0.65	No
4	0.808–1.023	PM tubes	. . .−186		0.70	No

Electronics	
Consumption (kW)	0.335
Tape recorder?	Yes (Landsats 1–3 only)
Compression	No

Downlink	
Data rate (Mbits/s)	15
(GHz)	S-band

Data	
Word length (bits)	6
Grey levels	64
Image width (pix)	2700
Image width (km)	185

† Scanning rate is not synchronized to IFOV. Pixel width is 68.5 m.

The orbits of the first three Landsats were nominally 920 km, Sun-synchronous, with a cycle of 251 orbits in 18 days. This gives rise to a k (orbits per day) of $13\frac{17}{18}$ or 13.94 (see Chapter 1); and the ground track pattern as shown in Fig. 8.2. Tomorrow's orbits are one ground track to the west of today's.

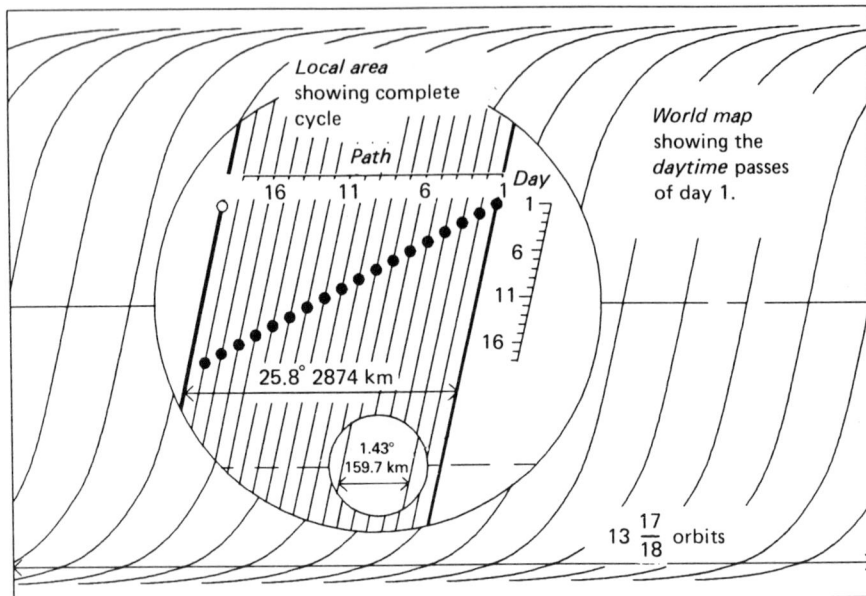

Note that Day 2's orbits are
1 track to the *west* of those
of Day 1.

The figures quoted have been computed from the
orbital cycle of Landsats 1–3. They will not necessarily
agree perfectly with the published figures.

Fig. 8.2 — Ground-track pattern for MSS.

NOAA (1984) gives the orbits of the first three Landsats as all being slightly different, but all geared to the same orbital cycle of 251 orbits in 18 days. As we saw in Chapter 1, this is impossible, and the varying orbits must be seen as gradually improving approximations to the ideal. As the achieved orbit approaches closer to the ideal, then the accuracy with which the specified ground tracks are followed improves, and the frequency and magnitude of the corrections required reduces.

Table 8.2 gives the orbital parameters as derived from the equations in Chapter 1, as well as those actually quoted. The 'nominal' figures are those computed directly from the published orbital cycle which, being integer data, must be strictly correct. The achievement of these nominal figures would result in perfect Sun-synchronism

Table 8.2

	Landsat			
	Nominal	1	2	3
Altitude (km)	934	920	920	920
S/M axis (km)	7291.24	7285.44	7285.99	7285.78
Period (°)	103.27	103.14	103.16	103.15
Incl. (°)	99.10	99.91	99.21	99.12
Desc. node	—	08:50	09:08	09:31
Precess (°/dy)	0.9856	1.0719	1.0034	0.9937
Error (%)	—	8.8	1.8	0.82

without recourse to systematic adjustment. The 'precession' column gives the daily precession rate, in degrees, obtained by applying these equations to the published data. The 'error' column gives the percentage discrepancy between the computed precession rate and the ideal figure of 0.9856 °/day. This discrepancy is a relative measure of the amount of systematic orbital adjustment that would be required to maintain the published ground track pattern.

8.3 SCANNING

The scanning system for MSS is a modification of the traditional television technique of raster scan, but with one important difference. On a television camera, scanning is electronic and flyback can be virtually instantaneous. This enables the next live scan to proceed almost immediately the preceding scan is complete. With remote sensing instruments such as MSS, however, scanning is achieved through the use of a moving mirror. A mirror, being other than weightless, is not amenable to this treatment, and therefore flyback must be reasonably slow. On MSS, flyback is at forward-scan speed. This effectively halves the light-gathering efficiency of the system. However, the radiometric specification of MSS is relatively relaxed and so no insuperable problems arise. (Compare the MSS system with that of TM, where considerable additional design complexity is employed to recover this light-gathering time.) Nevertheless, in order to provide adequate illumination, it is necessary to scan six rows of imagery at once so that the scanning rate may be reduced by this factor.

The pixel size of an instrument is dictated by the satellite ground speed (which is fixed by the orbital height) and the scanning and sampling rates. In the more recent instruments discussed in later chapters, the pixel size is normally both square and matched to the IFOV of the detectors. With MSS, however, neither is the case. Landsats 4/5 move forward at a ground speed of 6.79 km/s, which is equivalent to about 82 lines of imagery per second. The mirror oscillates at a rate of 13.62 scans per second, scanning 6 lines at a time. This makes the pixel height 83.1 m. The cross-track image speed is nominally 6.82 m/μs, while the sampling interval is 9.958 μs (NOAA, 1984). This makes the pixel width 68.5 m (or 67.9 m, as given in NOAA,

1984), which is significantly different from the IFOV figure of 82.6 m square, derived from the telescope and detector geometrices (Section 8.5). Note, however, that elsewhere NOAA quotes a 'ground IFOV' of 82.7 by 57 m, without explanation.

8.4 THE TELESCOPE

The MSS telescope is shown in Fig. 8.3. It is a Ritchey Cretien design — the first of many to be described in this book (see Chapter 3). The focal length is 0.826 m, which is quite short by the standards of later instruments, but which is appropriate to the relatively large pixel size of MSS.

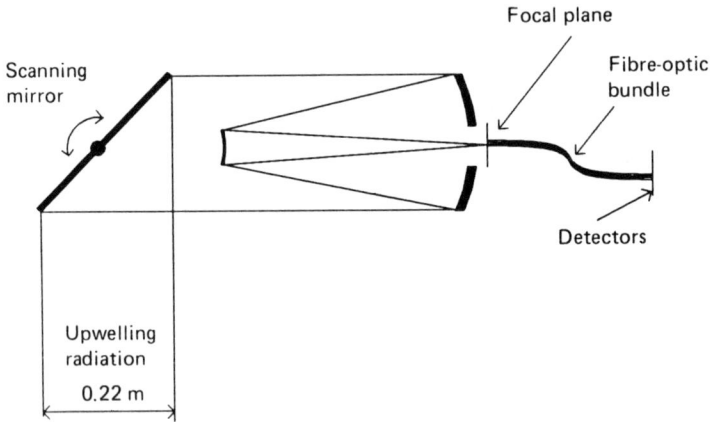

Fig. 8.3 — The MSS telescope.

8.5 THE FOCAL PLANE

The MSS Focal plane is shown in Fig. 8.4. It represents four arrays of six detectors each, as already discussed, ensuring that six lines of imagery, in each of the four bands, are scanned for each cycle of the mirror. The 24 apertures shown are not, in fact, the detectors themselves, but the ends of light pipes, which lead the radiation off to the detectors situated elsewhere. This arrangement enables the apertures to be placed much closer together then the relatively bulky detectors of the time would have permitted.

Unlike more recent instruments, the scanning rate of MSS is not synchronized with the IFOV of the detectors (NOAA 1984). Whereas the IFOV is given as 83 m square, the movement of the line-of-sight within a scanning period of 9.958 μs is only 68.8 m. Thus there is significant overlap, in the East/West direction, between the areas imaged for adjacent pixels. This will have the effect of a degree of low-pass filtering in the across-track direction.

Similarly, the inter-band registration is not as good as on later instruments such as

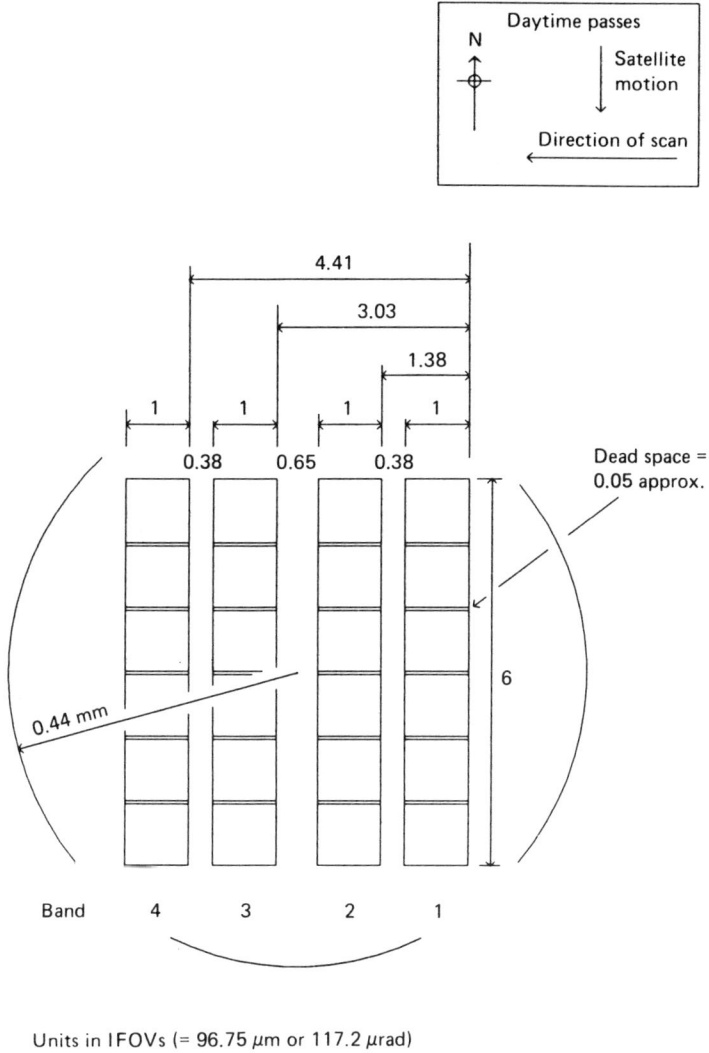

Units in IFOVs (= 96.75 μm or 117.2 μrad)

Fig. 8.4 — The MSS focal plane.

TM and SPOT. As is shown in Fig. 8.4, the effective apertures for the six lines that are scanned at one time are separated by exactly one IFOV. This is in line with later instruments. However, unlike more recent designs, sample-and-hold amplifiers are not used to enable all six detectors to be scanned at the same time. There is a time delay between the sampling of the first and least detectors of 3.98 μs, or 22.3 m 'for a non-rotating Earth'. To obtain correct inter-line registration, therefore, without resamplng, it is necessary to regard the lines acquired during a single sweep as being offset by some 3.7 m each.

It can be seen from the figure that the apertures for the different bands are positioned at intervals along the in-scan direction. Therefore the different bands will image a given ground patch at different times. This situation is inevitable if a single telescope is used to cover multiple bands without beam splitters. (See, for example Chapter 16 for a beam-splitting system and Chapter 18 for a multiple-telescope system.) In principle, this offset could be corrected for by reading the detectors at slightly different times; however, on an early instrument such as MSS such sophistication was not perhaps to be expected. It must be remembered that reliability is all-important for any satellite-borne instrument. Instead the detectors covering the four bands of any one line are all read at the same time, and the mis-registration is corrected for by 'inserting the appropriate number of dummy bytes prior to the data during CCT production' (NOAA, 1984). Bands 1–3 are delayed by this means to match Band 4.

8.6   THE DETECTORS

The exposure given to an individual detector is a function of the telescope f-number, the effective 'shutter' speed (integrating of dwell time), the effective detector aperture or capture area, and the optical efficiency of the light path. Reflecting telescopes can have quite a large central blockage, and the light pipes of the 1960s had quite significant losses. So the efficiency of the MSS light path is not particularly high. Whereas reflecting telescopes remain common, at least at the time of writing, the use of light pipes has not persisted, except for GOES, which has particular dynamic balancing problems (Chapter 9).

The first three bands are sensed by vacuum-tube photomultipliers, while Band 4 uses silicon photodiodes. No information has been made available on the operation of the photomultiplier tubes.

8.7   THE DETECTOR ELECTRONICS

The electronics system comprises the usual sequence of amplification, track-and-hold and DC restoration, before A/D conversion and multiplexing in to a single 15-Mbit/s data stream. However, a linear and a non-linear amplification system is available, selected by ground command. The non-linear system employs four-segment quasi-logarithmic amplifiers to compress the high radiance levels, and thus enable the lower radiance levels to be expanded. This characteristic matches the quantization noise more nearly to photomultiplier noise, and thus suits bands 1–3. Linear amplification better suits the noise characteristics of solid-state devices, and is therefore used for Band 4. Two alternative gain levels are available for Bands 1 and 2, differing by a factor of 3 (10 dB). This is because some scenes are expected to provide relatively low irradiance levels in these wavebands.

Quantization is into 6 bits, leading to 64 grey levels. Mirror synchonization pulses and a time code are multiplexed into the telemetry signal. The latter is used to identify the position of the imagery, and to control the splitting of the swath into scenes in accordance with the world reference system. The time code is based on Greenwich Mean Time.

8.8   THE DOWNLINK AND DATA COMPRESSION

No data compression is used although, as discussed above, a non-linear amplification characteristic is used for Bands 1–3.

The downlink arrangements, for Landsats 4/5 only, are considered in Chapter 14.

8.9   CALIBRATION AND IN-FLIGHT CHECKS

Every other flyback period is used for calibration purposes. An on-board calibration lamp illuminates the detectors through a rotating density-wedge optical filter.

8.10   DATA QUALITY

8.10.1   Spatial frequency response

The anti-aliasing filter on MSS is a three-pole Butterworth filter cutting off frequencies above 42.3 kHz. Butterworth is one of the most common filter designs, combining a reasonably sharp cut-off characteristic with minimum induced phase lags. A three-pole filter cuts off at 10 dB per octave, or a factor of 10 per doubling of frequency (see Chapter 6). This filter has no effect on the along-track spatial frequency response but, as with other scanning instruments, is responsible for a significant reduction in the across-track direction.

8.11   THE HOST SPACECRAFT

8.11.1   Attitude control

Unlike later vehicles, Landsats 1–3 only controlled attitude about one axis, resulting in a pointing accuracy of 0.7°, as against 0.1° for Landsats 4 and 5.

REFERENCES

Markham, B. L. (1984) 'Characterisation of the Landsat Sensors' Spatial Responses'. NASA Tech. Mem 86130.

NOAA (1984) 'Landsat 4 (&5) Data Users Handbook' U.S. Geological Survey.

NOAA (1984) 'A Prospectus for Thematic Mapper Research in the Earth Sciences' NASA Tech. Memo 86419.

9

GOES (-E and -W)

Consultant: **Frank Malinowski†**
Sources: **NASA** (1980), **Hughes** (1990)

9.1 INTRODUCTION

The Amercan National Oceanic and Atmospheric Administration (NOAA) makes two contributions to the global network of geosynchronous satellites. They are the GOES spacecraft (Geostationary Operation Environmental Satellites). GOES-E views the Eastern and Midwest United States and the Western Atlantic, while GOES-W views the Eastern Pacific and the Western United States. This is in addition to the NOAA series of low-altitude satellites which fly in Sun-synchronous orbits as described in Chapter 12. Other contributors are the European Meteosat (Chapter 11), which basically covers Europe, Africa and the Middle East from a point over the Greenwich Meridian; the Indian INSAT (Chapter 13), which is stationed over the Indian Ocean; and the Japanese GMS (Chapter 10), stationed over a point north of Australia. At the time of writing, a gap in launchings has resulted in only one fully operational GOES satellite being available. It is currently positioned midway between these two stations.

Meteorological satellites are truly operational tools, upon which a massive international network of weather forecasters depend for their day-to-day operations. Furthermore, very large sums of money hang on the forecasts that are produced, and which are increasingly dependent of the timely arrival of their imagery and other measurements. The design and operational philosophy is thus geared to dependability and continuity of service, to an extent that is entirely unmatched by the Earth-resources instruments such as Landsat and SPOT. Tried and tested technology is therefore the order of the day, and this shows very much in the detailed designs discussed in this book.

The function of the geosynchronous satellite is, effectively, to hang over a fixed

† Chief Scientist, Systems Division, Santa Barbara Research Center, Hughes Aircraft Company.

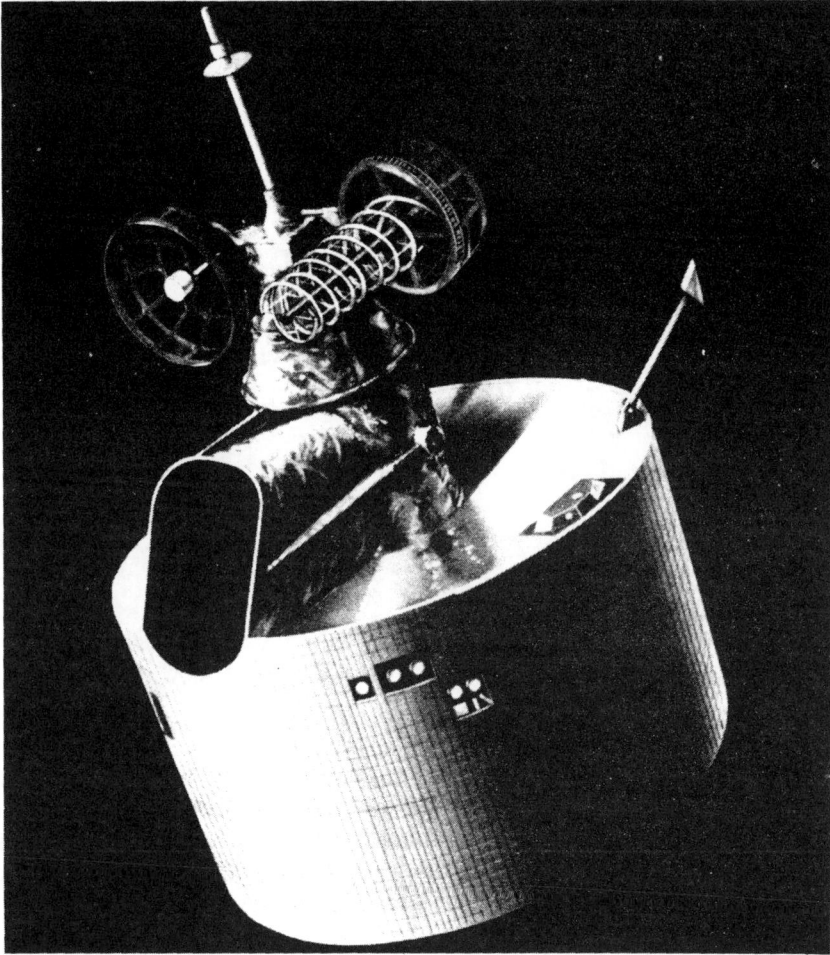

Fig. 9.1(a) — GOES. Courtesy: NASA.

point on the Equator, and to view nearly half the globe at any one time. This enables it to produce frequent, low-definition, imagery to complement the less-frequent, but higher-definition, imagery provided by the NOAA series. The Earth resources missions, of course, produce much higher definitions still, but at a coverage rate measured in weeks rather than the hours required for weather forecasting.

The United States pioneered the use of the geosynchronous meteorological satellite with the ATS program, first launched in 1966 (Hughes, 1990). This was followed by the SMS program, and the GOES series, the first of which was launched in 1975. The first three GOES satellites (launched 1975 to 1978) were more or less identical, carrying, as their main instrument, the 'visible and infra-red spin scan radiometer' (VISSR). An upgrade to the instrument was introduced with GOES 4, including the incorporation of atmospheric sounding facilities (atmospheric sound-

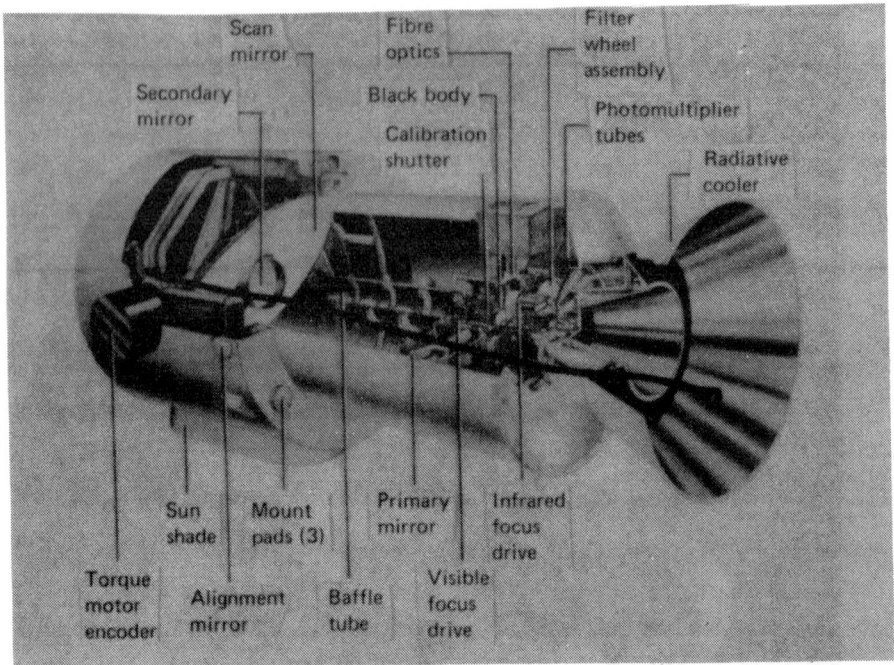

Fig. 9.1(b) — The VAS radiometer. Courtesy: NASA.

ing, notwithstanding its importance, has regretfully had to be excluded from this book.) The new instrument is the 'visible and infra-red spin scan radiometer and atmospheric sounder' ('VISSR Atmospheric Sounder') or VAS. GOES 4 was launched in September 1980, with GOES 5 and 6 following in May 81 and April 83 respectively. GOES 7 was launched in March 1987 and, at the time of writing, is the only satellite fully operational.

A **radiometer** is an instrument which is capable of delivering an absolute estimate of the radiance reaching it, to a reasonable level of accuracy. Use of the descriptor implies good, absolute calibration facilities (not always easy to provide), regularly operated and applied to the data on a routine basis. In general it is the meteorological fraternity that require accurate absolute radiance measurements, and meteorological instruments that get accorded the accolade of radiometer. Thematic Mapper, for example, is not normally so described.

Of the geosynchronous satellites that we cover, all but one scan by spinning the entire satellite at some 100 r.p.m. The gyroscopic effect of spinning this relatively large mass improves the stability of the platform considerably, and it simplifies the design of the scanning mechanism (see Chapter 13). However, part of the vehicle has to be **despun** to allow the antennae to beam to a fixed point on the ground. This can cause difficulties, particularly if, like INSAT, the spacecraft must cover a number of other duties such as communications and broadcasting. The current GOES series uses actual despinning, as is described in Section 9.9, whereas Meteosat (and some of GOES' predecessors) achieves the same effect electronically (Chapter 11).

The introduction of a more advanced main instrument on GOES 4 posed a similar problem to that posed by the introduction of Thematic Mapper on Landsat 4 (see Chapter 14), namely that of providing continuity of service to users of data to the old standard. With Landsat 4/5, the problem was solved by flying the old instrument in parallel with the new. However the VAS instrument on GOES was designed to be operated, for routine operations, as a VISSR lookalike. On GOES 4–6, the sounding features were intended to be experimental (Hughes, 1990). With GOES 7, the facility is effectively operational, although the term 'demonstrator' is still used in relation to it.

GOES also carries a number of other instruments for determining the near-Earth Space environment. These include an Earth's field magnetometer, a solar X-ray sensor, and an energetic particle sensor covering electrons, protons and α particles. In addition, it acts as a relay station for transmitting raw observation data from unmanned weather stations, and for transmitting processed imagery and other data to real-time users.

GOES is a relatively old design, and the company has clearly moved on to more advanced projects. The detailed final report that has proved such a mine of information on some other instruments has not been made available in the case of GOES. In consequence this chapter also is less detailed than otherwise it might have been.

9.2 THE ORBIT

Considerations of orbital cycles and ground track patterns are all more or less irrelevant to a satellite in geosynchronous orbit. The orbital cycle is one orbit in one sidereal day, and the ground track is, ideally, a spot exactly on the Equator. (A **sidereal** day is the time that the Earth takes to make one complete revolution against the fixed star background. The sidereal day is slightly shorter than the more familiar solar day.) In practice, of course, an orbit is never perfect. Residual eccentricity will cause the ground point to execute a sinusoidal excursion along the Equator, and residual inclination will cause it to execute a small figure of eight, with a period of one sidereal day, as illustrated, schematically, in Fig. 11.2. In the case of GOES, inclination is kept with $\pm 1°$ (Hughes, 1990). There is, however, an additional untoward movement caused by the varying thickness of the Earth's crust. This results in a small additional pull in the direction of two fixed potential minima. The first is directly above 79°, south of India. The second, not surprisingly perhaps, is almost diametrically opposite over 252.4°, or 20° west of the Galapagos Islands. The satellite is normally allowed to drift by up to about 0.5°, when it is given a kick sufficient to cause it to drift a double amount in the opposite direction. A second sinusoidal excursion along the Equator is therefore executed, this time with a period measured in months and an amplitude which is controlled to within the limit described above. Additional, though irregular, perturbations are generated by the attractions of the Sun and Moon. Fig. 9.2 shows the specified stations of both GOES satellites.

9.3 SCANNING

As we have discussed, the GOES series are spin-stabilized. The entire vehicle spins at 100 r.p.m., or 1.67 revolutions per second, about a north/south axis, which provides a natural scanning mechanism for the telescope. Scanning is from west to

Table 9.1

Spacecraft

Mass — satellite (kg)	396	Launched	1975, 1977, 1978
— VAS (kg)	75		1980, 1981, 1983
			1987
Spin speed (r.p.m.)	100	Target	Cloud, sea surface temp.

Orbit

Altitude @ ;Equ.(km)	35 786	Orbits/day	1 (/sidereal day)
Semi-major axis (km)	42 164	Orbits/cycle	1
Eccentricity (%)	—	Days/cycle	1
Period (min)	1436	Shift/orb (°)	0
Inclination (°)	0	(km) @ Equ.	0
Descending node	n/a	Gnd track spacing (°)	0
Ground speed (km/s)	0	(km) @ Equ.	0

Telescope

Type	Ritchey Chretien		
Focal length (m)	2.92	80% blur c. d. (μrad)	12/226
Aperture (m)	.406	F.P. dia. (mm)	—
f-number	7.17/1.0	Field-of-view (mrad)	0.005/5.635
1y mirror dia. (m)	0.406	IFOV (μrad)	25/192†
2y mirror dia. (m)	0.10	(μm @ F.P.)	73/78
Total obscuration (%)	16	(m on gnd @ nadir)	900/6900
Clear area (m²)	0.109	EIFOV (m on ground)	—
Exit pupil (cm)	1.3/13.5	Inter-band reg. (pix)	—

Scanning

Method	Spinning vehicle/rotating mirror		
Rate (revs/s)	1.67	Mirror swing (°)	360
(rad/s)	10.47		
Scan repeat time (ms)	600		
Time for 1 line (ms)	33		
Efficiency (%)	5.5		
Sampling interval (ms)	0.0020/0.0040		
Dwell time (μs)	0.0024/—		

Channel details (nominal)

Band	Wavebands	Detectors	Dynamic range	Noise equiv. ref-lectance	MTF @ pix rate
	(μm)		(W/m²/sr/μm)	(% F.S.)	trk/scan
1	0.55–0.75	PM tubes	— — —	—	0.35
8	10.3–12.1	HgCdTe	— — —	—	—
2–7 and 9–12	Various	InSb or HgCdTe	Used for atmospheric sounding and beyond the scope of this book		

Electronics

Consumption (kW)	—
Tape Recorder?	—
Compression	—

Downlink

Data rate (Mbits/s)	28
(GHz)	—

Data

Word length (bits)	6/8
Grey levels	64/256
Image width (pix)	n/a
Image width (km)	n/a

†Visible pixel size is 21×25 μrad

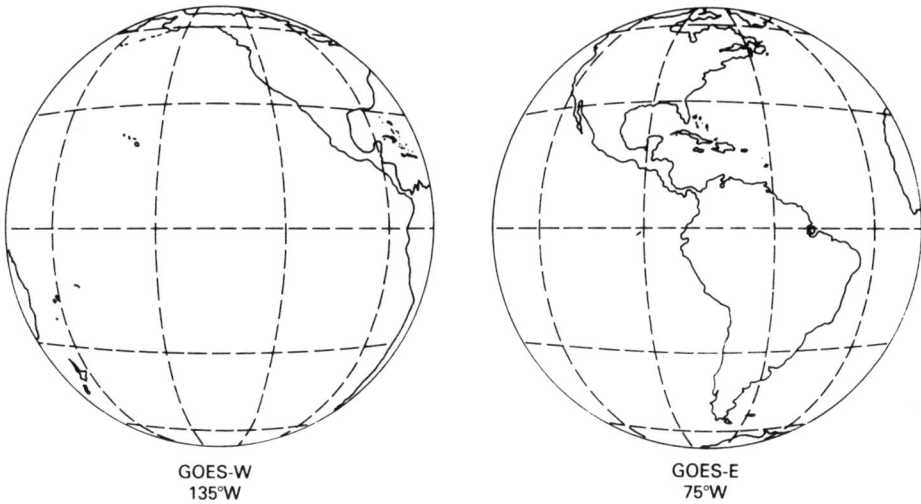

Fig. 9.2 — Ground coverage of the GOES satellites (when in their standard positions). These maps show views of the Earth as seen by GOES. Each extends out to about 80° from the nadir point.

east. (It will emerge in due course that Meteosat scans from east to west, and also from south to north.)

At the geosynchronous altitude (35786 km), the Earth's diameter (12735 km, mean) subtends an angle of 17.2°, as can very easily be shown. However, at that altitude, the Earth's true edge is some 10° below the horizon, and it is slightly more difficult to show that the angle subtended by the visible edge is in fact 18°. Thus the maximum efficiency of a circular scanning motion, at geosynchronous altitude, is about 5%.

North/south scanning is provided by a plane scanning mirror, which is stepped from north to south. The optical beam is stepped in units of 192 μrad, or 6.9 km at nadir, which is equal to the quoted IFOV for the infra-red bands. A complete pole to pole scan takes 15.7 min (Hughes, 1990), which is significantly faster than Meteosat. Each scan commences on the hour or on the half-hour.

9.4   THE TELESCOPE

The GOES telescope is the traditional Ritchey Chretien instrument (Fig. 9.3), constructed entirely from beryllium. Its focal length is 2.92 m. The telescope is mounted down the centre of the satellite, i.e. along the spin axis, to minimize centrifugal effects, and to facilitate the balancing of the rotating mass.

Compare this with the Meteosat telescope (Fig. 11.3) where the telescope is mounted across the spin axis. This is made necessary by the absence of a scanning mirror, which means that the telescope itself must face outwards. Note also that this

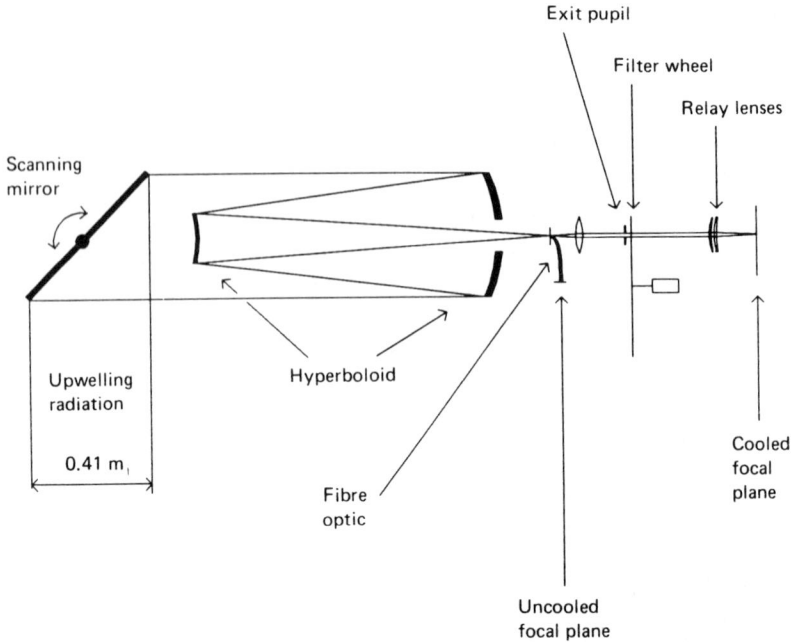

Note

The uncooled focal plane, in fact, comprises
eight photomultiplier tubes mounted in
different radial positions around the telescope
axis.

Fig. 9.3 — The VAS telescope on GOES.

feature means that the Meteosat telescope must be physically quite short, in spite of its great focal length, whereas that on GOES can be as long as conventional design techniques dictate.

The focal plane is split, to enable the infra-red detectors to be housed in a cooled area. (This strategy, and the reasons for it, are discussed in more detail in later chapters.) The relay optics are fully refracting, and comprise a field lens and two relay lenses, all of germanium. The multiplicity of bands required for the sounding function are provided by a rotatable filter wheel, described in more detail in Section 9.6. Uniquely, however, except for MSS (Chapter 8), the visible detectors are also housed remotely, and the radiation is channelled to them through fibre optics.

9.5   THE FOCAL PLANE

The GOES focal plane is shown in Fig. 9.4. The fibre-optic apertures are nominally 25×21 μrad, or 73×62 μm. This leads to a nadir ground IFOV, in the visible band, of 0.89×79 km, or roughly one-eighth of that of the cross-track swath width. The apertures are mounted in a row across the scan track, so that eight lines of visible imagery are acquired per scan. There is a dead space equivalent to 4 μrad between adjacent apertures.

The infra-red detectors are solid-state, and are mounted together on the cooled focal plane, which is installed at the heart of the radiative cooler (see Section 9.11). The relay optics provide further magnification of 7.17:1, enabling the physical size of the infra-red detectors to be reduced by that amount.

The pattern in which the detectors are arranged on the logical focal plane is designed to facilitate the two alternative modes of operation. Both modes involve the collection of visible data during all scans. The main operational mode is the 'VISSR' mode, in which only the two small infra-red detectors are used, together with a filter that passes the standard thermal band of 10.3 to 12.1 μm. The resulting imagery is identical to that produced by the earlier VISSR instruments.

9.6 THE DETECTORS

The visible detectors are eight photomultiplier tubes, as used in MSS. Being large and heavy, compared with solid-state devices, the installation of the photomultiplier tubes has had to take account of centrifugal and balancing considerations. They are mounted, equispaced, round a radial plate like the spokes of a wheel and, as we have seen, receive their excitation via optical fibres. MSS and GOES are the only known users of optical fibres in this connection. The sensitivity band of the detectors is between approximately 0.4 μm and 0.72 μm. Therefore the longwave cut-off is provided naturally by the photomultipliers. However, the blue response of the tubes is countered with a filter cutting on at 0.55 μm.

Filtration for the thermal channels is provided by a filter wheel built into the relay optics. This enables the effective bandwidth to be changed, if necessary after each east/west scan. At each filter wheel position, all the detectors view through the same filter.

The infra-red detectors are solid state devices. As shown in Fig. 9.4, two indium selenide (InSb) devices, are used for the shorter wavelengths required for the sounding function (3.9–4.5 μm). The rest use mercury cadmium telluride (HgCdTe) photoconductors, and cover the longer wavelengths.

As the figure shows, the HgCdTe detectors come in two sizes, to facilitate the dual mode of operation described in the introduction. When mimicking the operation of the VISSR instrument, only the small detectors are used. During this mode of operation, the filter wheel is normally fixed to give these detectors a spectral bandwidth of 10.3 to 12.1 μm. During 'multispectral scan' mode, however, selected pairs of detectors may be used; namely the two small HgCdTe, the two large HgCdTe or the two InSb. By altering the filter wheel position, any one of 12 infra-red filters may be chosen. By switching filters between scans, and by interleaving the scans, it is possible to obtain a full frame of scans in each of up to four infra-red wavebands.

In both modes, visible imagery is collected on every scan.

9.7 THE DETECTOR ELECTRONICS

No onboard recording facilities are provided on GOES. All data are transmitted in real time, at high data rate, direct to the ground receiving station at Wallops Island,

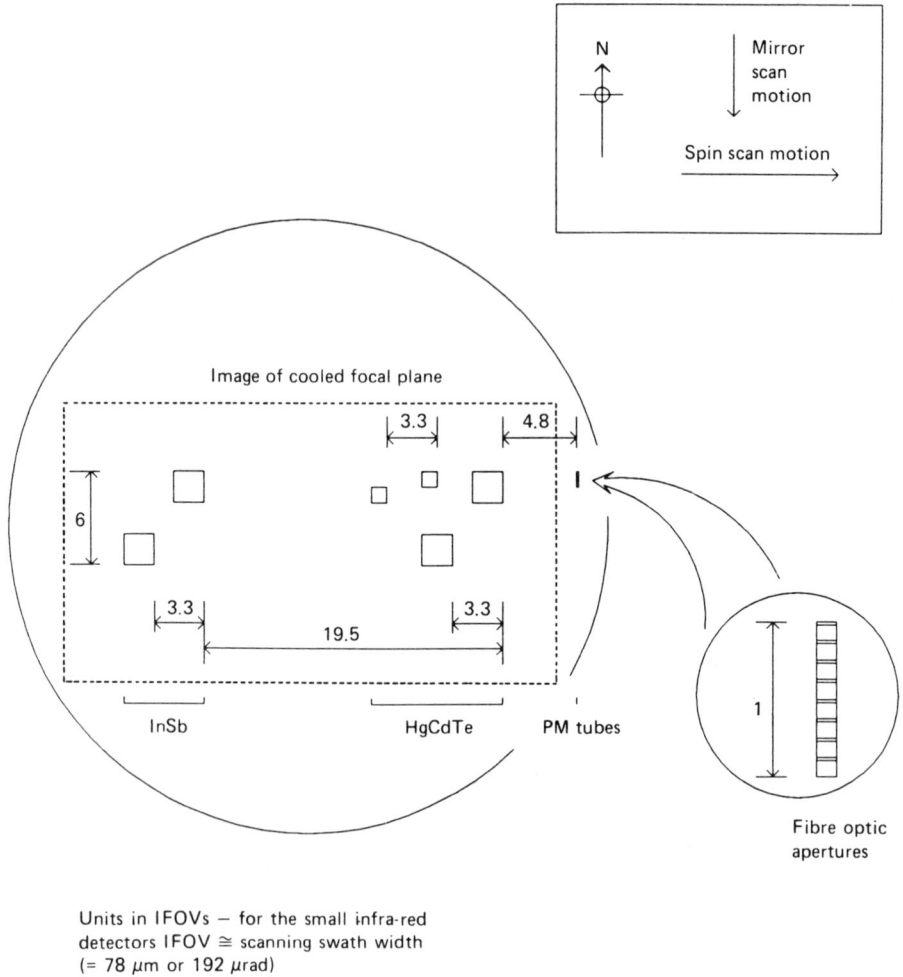

Fig. 9.4 — The VAS logical focal plane.

Virginia. (This is in direct contrast to Meteosat (Chapter 11), where the data for each scan are delivered to an onboard memory bank, and released to the downlink system at a rate which utilizes effectively the whole of the spin cycle.)

Geosynchronous vehicles occasionally go into eclipse for up to about two hours either side of local midnight. The eclipse phase occurs for about six weeks around the equinox periods. The electrical loads are divided into three categories, 'sunlight', 'eclipse' and 'essential'. The sunlight load comprises the entire VAS instrument, and is shed by the battery protection unit as soon as it detects a reduced charging current. The eclipse loads comprise other non-essential services, and are shed should the battery voltage drop below a certain level. The essential loads are simply those required to maintain a command link to the satellite. They go down with the battery.

9.8   THE DOWNLINK

Raw VAS data are transmitted, from both GOES, directly to the Command and Data Acquisition (CDA) Station at Wallops Island, Virginia. Transmission is in real time. This demands a high data rate downlink, but obviates the need for online storage. Data are processed and re-transmitted at a 'stretched' rate, utilizing the whole of each spin cycle, back to the satellite that originated the data. The data are then broadcast to all suitably equipped receiving stations.

9.9   THE HOST SPACECRAFT

As we have discussed, the current GOES vehicles are spin-stabilized. This means that the bulk, at least, of the mass of the vehicle spins about a north/south axis at approximately 100 rpm. This spinning action stabilizes the vehicle through its gyroscopic effect, and provides a natural scanning motion for east/west coverage. It also evens out the thermal loads generated by solar radiation. In the case of GOES, unlike Meteosat, a structure on the end of the rotating cylinder is despun, to provide a stable platform for mounting the antennae. The despun component has to be as light and as compact as possible to minimize dynamic interactions between the two.

The requirement to spin smoothly about a predetermined axis imposes severe constraints upon the mechanical design of the vehicle. As discussed in the next section, a mass will tend to spin about the axis having the largest moment of inertia (MOI). To ensure a smooth and stable spin, it is necessary to arrange that the MOI about the required axis is much larger than that about any other. This means that the vehicle must be designed to be short and fat, with plenty of weight near the periphery. This could well run counter to other design requirements. The mass of the vehicle and payload has to be very carefully distributed, and the structure accurately balanced, statically and dynamically. Balance has to be maintained, both with a full load of hydrazine and as the fuel is gradually used up.

Further constraints are imposed if it is necessary to 'fix' part of the structure, in relation to the Earth, as a platform for the antennae. It is in fact possible to design antennae that can rotate with the spacecraft, and remain strongly directional in a fixed direction. The first five prototype GOES missions sported electronically despun antennae (Greaves & Shenk, 1985). However, for the production series, this concept was abandoned in favour of conventional antenna design (see Fig. 9.1(a)). Hence the need to despin a part of the vehicle.

The size and weight of the GOES vehicles are designed to be within the payload constraints of both Shuttle and the Delta 3914 launch vehicles.

They are designed for a minimum active life of seven years. Full duplication is provided for all critical functions except for the mechanical parts.

9.9.1   Attitude control

As we have discussed *ad nauseam,* GOES obtains stability by spinning about the telescope central axis. Many satellites do this, and they can be identified on account of being cylindrical in shape. Spinning is extremely effective as a means of achieving the high degree of stability required for a high-altitude instrument. However, extremely careful dynamic balancing is required, much as is needed for the road wheels on a high-performance car.

The basic strategy for controlling the relative motion of the two components of GOES is to generate the required spin rate for the principal mass, and then, as it were, to spin the antenna assembly backwards at the same speed. Overall control of the antenna attitude (the outer control loop) is maintained by manual observation of the downlink signal (Hughes, 1990).

The satellite is initially spun up by means of a spin table on the third stage of the launch rocket, and it remains spinning throughout the orbital manoeuvres which transfer it from low orbit, via a highly eccentric transfer orbit, to the final circular high orbit. Adjustments to the spin speed are carried out, when necessary, by manual operation of the appropriate vernier thrusters (see below). The instantaneous attitude about the spin axis is monitored by means of a Sun sensor or one of two Earth sensors. The Earth sensors are considerably less accurate than the Sun sensor, because the atmosphere blurs the edges of the Earth's disc, so they are only used during eclipse or under failure conditions. The selected sensor provides a pulse which establishes a precise orientation, once per revolution. This pulse controls the frequency and phasing of an oscillator, which runs free until it receives the next pulse, to produce exactly 65536 (2^{16}) pulses per revolution, or a pulse every $0.0055°$. (The basic strategy is similar to that of the free pendulum clock, beloved of museum curators, in which a clock is synchronized, at relatively infrequent intervals, with a large pendulum which otherwise swings free.) The oscillator output is then used to control all the position-sensitive activities, including scanning and the despinning of the antenna assembly.

In normal operation, the oscillator is under the influence of the Sun sensor, and follows solar time, whereas to maintain long-term synchronism with the Earth it must follow sidereal time (see Section 9.2). The oscillator output is therefore passed through a variable delay line, controlled by a time-of-year clock, to accomplish the necessary conversion. When either of the Earth sensors is in use the delay line is switched out.

Every 16th pulse is picked off and used to indicate the required instantaneous antenna orientation (to the nearest $0.088°$). This is compared with a similar signal from the antenna position sensor, and the error is used to drive the despin servo.

To generate attitude adjustment, GOES effectively makes use of the station-keeping thruster system (see Section 12.11.1). Six small, 1-lbf (4.4-N), thrusters are mounted around the outside edge of the vehicle to enable the attitude and position of the satellite to be adjusted. Three of them can be seen in Fig. 9.1(a); they are the assembly mounted at the edge of the flat end-plate. The other three are diametrically opposite. The two thrusters exhausting through the tops of their respective assemblies provide pulsed thrusts parallel to the spacecraft axis. When fired together, they provide a torque in what might be termed 'yaw'; i.e. about an axis parallel to the telescope equatorial viewing sight line. If one is fired continuously on its own then, as the vehicle spins, it provides a steady impulse north or south. The four thrusters that exhaust through the bevelled sides of the two assemblies are mounted radially, i.e. with their axes passing through the centre of gravity. Continuous firing of pairs of radial thrusters produce torques about the other two axes, while carefully timed pulsed firings can produce a steady impulse east/west or north/south. As we discussed earlier in the chapter, an eastwards adjustment is required every few months. A north/south manoeuvre is required every year or two (Hughes, 1990).

9.9.2 Nutation control
As we have already discussed, a body spinning free will tend to spin about the axis of greatest MOI. Therefore spin-controlled satellites are designed so that the required spin axis has a greater MOI than any other. However, if any perturbation is applied from the thrusters, movement of fuel etc., then a wobbling motion, **nutation**, is generated about the spin axis. If the spin is basically stable, because it is about the axis of greatest MOI, then the amount of energy in the nutation is relatively small. It can be damped out by a simple passive device — perhaps incorporating a viscous fluid which is at rest during normal spinning, but which is disturbed by the nutation. The nutation damper on GOES comprises a hollow hoop which is filled with isopropanol and 'displaced unsymmetrically relative to both the spin axis and the centre of gravity'.

During orbital injection, however, the apogee boost motor is attached to the radiative-cooler end of the satellite, effectively as an extension to the spinning mass. This extra mass, at one end of the vehicle, increases the MOI about the transverse axes and makes the spin unstable. During this phase, therefore, nutation is a serious problem, and active nutation control is required using accelerometers and one of the thrusters.

9.9.3 Temperature control
The rotation of a spin-stabilized satellite simplifies the thermal control problem somewhat, because there are no areas that are exposed to continuous solar radiation, and only the ends are exposed continuously to the cold of space. On GOES, one of these ends is occupied with antennae, which tend to be relatively insensitivie to temperature, and the other carries the radiative cooler.

The solar panels are used as the main heat sink, experience with previous vehicles having shown that they tend to settle to the very comfortable temperature of 13°C, ±8°. Components that require closer control than that are operated at slightly raised temperatures, which are maintained by small heaters. During eclipse the heat balance is seriously upset, and protective measures are taken to prevent damage to sensitive items.

The VAS instrument is designed to maintain as constant a temperature as possible over the entire structure apart from the cooler. This is important both for improved infra-red calibration and to maximize dimensional stability. It is therefore designed to maintain its own internal environment, using its optical aperture to absorb and radiate heat. This poses interesting design problems, because the main heat sink is at the opposite end of the structure to the main thermal loads, represented by the eight photomultiplier tubes. Furthermore, this structure has to be maintained dimensionally particularly stable because it carries the optical elements of a very high powered telescope.

A standard two-stage radiative cooler is fitted. The operation of these is discussed in more detail in Chapter 12, because particularly detailed information on the working of the AVHRR cooler was made available. Briefly, however, the GOES cooler is designed to achieve a flat-out temperature, on the cooled focal plane, of 75–80 K (Hughes, 1990). The usual thermostatically controlled heater is used to generate a working temperature of 94 K.

The despun bearing assembly is a precision mechanism which is attached to the thermally uncontrolled antenna assembly, but which requires close thermal control itself. This is achieved by providing insulation on the antenna end, and a highly emitting/absorbing finish on the side facing the VAS instrument. This couples the bearing assembly thermally to the much more massive VAS assembly, fortunately at its controlling end.

9.10 APPLICATIONS

As with all similar instruments, GOES data are usable for quantitative measurements, out to about 60° from nadir. Beyond that, only qualitative interpretation is possible.

The visible bands are mainly used for following the movement of clouds, and thus estimating wind speed and direction.

REFERENCES

Greaves, J. R. & Shenk, W. E. (1985) 'The Development of the Geosynchronous Weather Satellite System', *Progress in Astronautics & Aeronautics* **97**, pp. 150–181.

Hughes (1990) Corrections, and additions, in draft. Staff of Hughes Aircraft Company.

NASA (1980) 'GOES D, E, F, Data book' NASA 06424B.

10

GMS-4

Source: **NASDA** (1987)

10.1 INTRODUCTION

The GMS series of geosynchronous satellites are Japan's contribution to the World Weather Watch. The design of the satellites and payloads is very similar to NOAA's GOES series, and they carry, as their main instrumentation a 'visible and infra-red spin scan radiometer' (VISSR) similar to that on the earlier GOES missions. GMS occupies a station between India's INSAT vehicles and America's GOES-W, at 140°E.

The first three GMS vehicles were created with, at the very least, the close co-operation of Hughes (the builders of GOES). GMS-4, however, appears to have a considerably greater Japanese content.

GMS-1 was launched in July 1977, aboard a US delta rocket, and was operated for just under four years, being shut down at the end of 1981. GMS-2 was launched in August 1981, aboard a Japanese N-II rocket, and operated for $2\frac{1}{2}$ years. GMS-3 followed in August 1984, again Japanese launched. GMS-4 was launched in September 1989.

GMS also carries a number of other instruments for determining the solar X-ray and near-Earth Space environment. In addition, it acts as a relay station for transmitting raw observation data from unmanned weather stations, and for transmitting processed imagery and other data to real-time users.

Very little information has been made available on the GMS series, and this chapter is included for completeness only.

10.2 THE ORBIT

As already mentioned, GMS occupies a station over the Eastern Pacific, at 140°E. Its field of view is shown in Fig. 10.2.

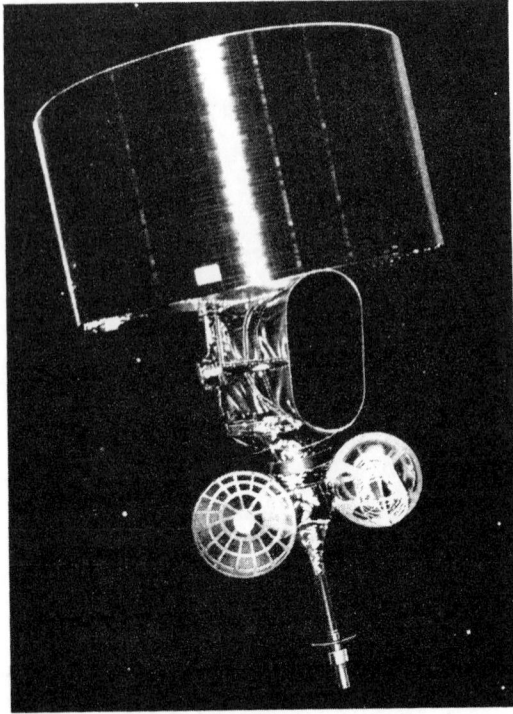

Fig. 10.1(a) — The GMS spacecraft. ©NASDA

10.3 SCANNING

Like GOES, the GMS series are spin-stabilized spinning at 100 r.p.m., about a north/south axis, which provides a natural scanning mechanism for the telescope. Scanning is from west to east. North/south scanning is provided by a plane scanning mirror, which is stepped from north to south. The optical beam is stepped in units of 140 μrad, or 5.0 km at nadir. This is slightly less than the 192 μrad quoted for GOES, but greater than the 125 μrad of Meteosat. Scans are repeated every half hour.

10.4 THE TELESCOPE

The telescope is clearly very similar to that of GOES, but the details were not made available.

10.5 THE FOCAL PLANE

Fig. 10.3 encapsulates the information that is available on the layout of the GMS logical focal plane.

The IFOV of the visible band, at nominally 35 μrad, is larger than that of GOES, whereas the thermal band IFOV, at 140 μrad, is smaller. Whereas it takes eight GOES visible band IFOVs to cover the swath width, set by the infra-red IFOV; on

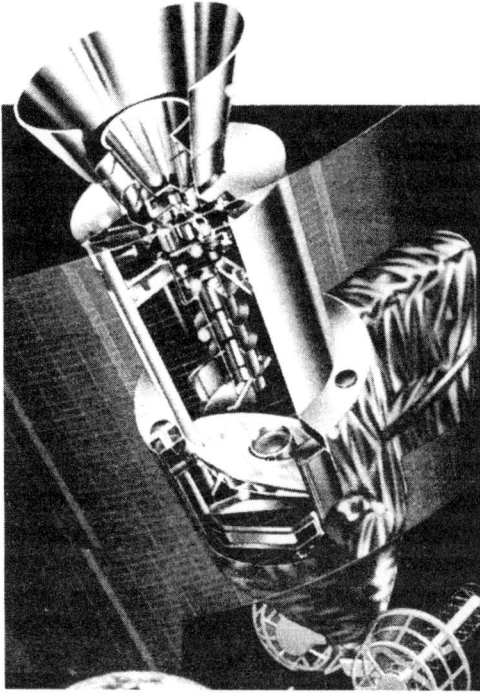

Fig. 10.1(b) — The VISSR instrument. ©NASDA

GMS it only takes four. Eight photomultiplier tubes are still installed, however; the second four being duplicates. As already mentioned, the fibre-optic apertures are nominally 35 μrad. This leads to a nadir ground IFOV, in the visible band, of 1.25 km, which is larger than the 0.9 km offered by GOES, but smaller than the 2.5 km of Meteosat. Each group of four apertures is mounted in a row across the scan track, so that four lines of visible imagery are acquired per scan.

The single infra-red detector is also duplicated. The IFOV of each is 140 μrad, or 5 km on the ground. The detector pair are solid-state HgCdTe devices, mounted together on the traditional 'cooled' focal plane, which is installed at the heart of the radiative cooler.

10.6   THE DETECTORS

As we have already discussed, the visible sensing is carried out by, nominally, four photomultiplier tubes. The measured visible band is from 0.50 to 0.75 μm; which is slightly wider than that of GOES, but narrower than the Meteosat equivalent, whose 3-dB down points are from about 0.55 to about 0.95 μm.

10.7   THE DETECTOR ELECTRONICS

No information has been made available on the GMS electronics.

Table 10.1

Spacecraft			
Mass — satellite (kg)	325	Launched	1977, 1981, 1984
— VISSR (kg)	—		1989
Spin speed (r.p.m.)	100	Target	Cloud, Sea surface temp.

Orbit			
Altitude @ Equ. (km)	35 786	Orbits/day	1 (sidereal day)
Semi-major axis (km)	42 164	Orbits/cycle	1
Eccentricity (%)	—	Days/cycle	1
Period (min)	1436	Shift/orb (°)	0
Inclination (°)	0	(km) @ Equ.	0
Descending node	n/a	Gnd Track spacing (°)	0
Ground speed (km/s)	0	(km) @ Equ.	0

Telescope (unspecified details probably similar to GOES)

Type	—		
Focal length (m)	—	80% blur c. d. (μrad)	—
Aperture (m)	—	F.P. dia (mm)	—
f-number	—	Field-of-view (mrad)	—
1y mirror dia. (m)	—	IFOV (μrad)	—
2y mirror dia. (m)	—	(μm @ F.P.)	—
Total obscuration (%)	—	(m on gnd @ nadir)	1250/5000
Clear area (m^2)	—	EIFOV (m on ground)	—
Exit pupil (cm)	—	Inter-band reg. (pix)	—

Scanning			
Method	Spinning vehicle/rotating mirror		
Rate (revs/sec)	1.67	Mirror swing (°)	360
(rad/s)	10.47		
Scan repeat time (ms)	—		
Time for 1 line (ms)	—		
Efficiency (%)	—		
Sampling interval (ms)	—		
Dwell time (μs)	—		

Channel details (nominal)

Band	Wavebands (μm)	Detectors	Dynamic range (W/m^2/sr/μm)	Noise equiv. reflectance (% F.S.)	MTF ∂=at. pix rate trk/scan
1	0.50– 0.75	PM tubes	— — —	—	—
2	10.5–12.15	HgCdTe	— — —	—	—

Electronics		
Consumption (kW)	—	
Tape Recorder?	—	
Compression	—	

Downlink		
Data rate (Mbit/s)	—	
(GHz)	S-band/UHF	

Data		
Word length (bits)	—	
Grey levels	—	
Image width (pix)	n/a	
Image width (km)	n/a	

140°E

Fig. 10.2 — Ground coverage of GMS. This map shows the view of the Earth as seen by GMS. It extends out to about 80° from the nadir point.

10.8   THE DOWNLINK

The communication arrangements appear to be similar to other meteorological systems, in that the satellite is also used as a relay station, both for other input data, and for the dissemination of processed products.

10.9   APPLICATIONS

GMS data are used for routine weather forecasting, as well as for the monitoring of typhoons, rain and snow storms. GMS imagery is used over a wide area of Southeast Asia and the Western Pacific.

REFERENCES

NASDA (1987) GMS-4 promotional material.

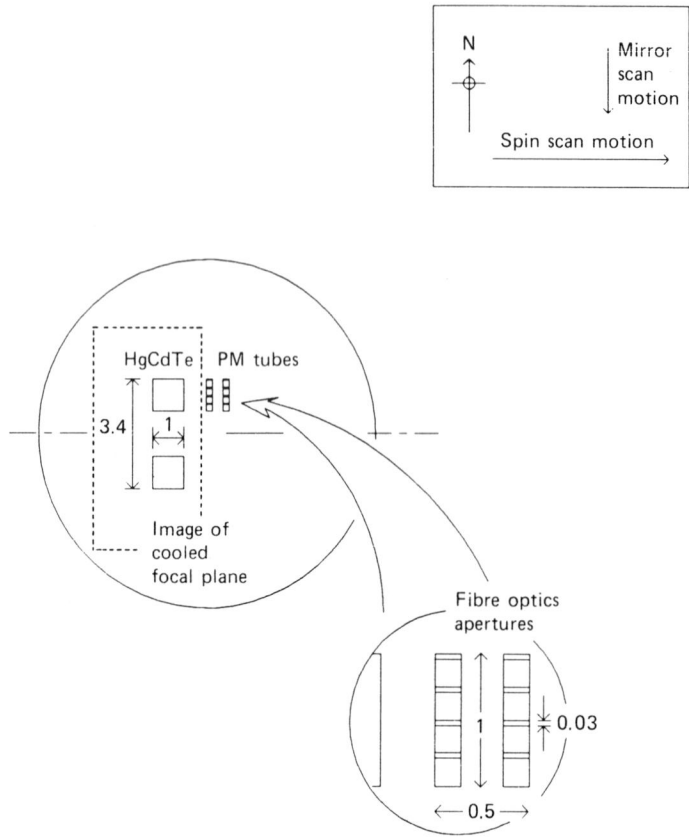

Units in IFOVs — for the infra-red detectors
IFOV ≅ scanning swath width (=140 μrad)

There is a degree of conjecture in this diagram

Fig. 10.3 — The GMS VISSR focal plane.

11

Meteosat

Consultant: **Mr Armand Péraldi†**
Sources: ESA (1981), Péraldi (1978, 1989)

11.1 INTRODUCTION

Meteosat is part of a global network of four or five geosynchronous satellites which

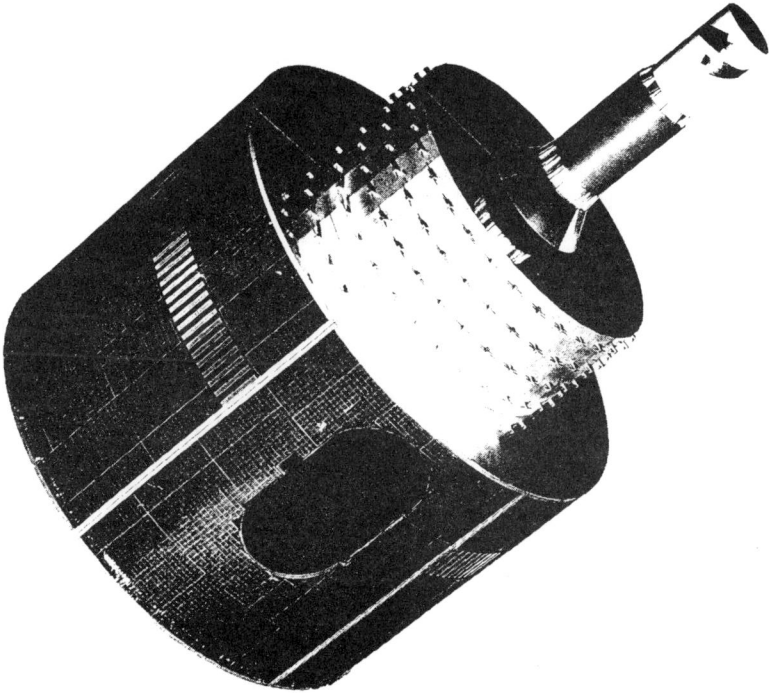

Fig. 11.1 — (a) Meteosat. Courtesy: MATRA.

together view almost the entire globe from an altitude of 35 786 km above the Equator. Their principle function is to provide an online imaging service for weather forecasters. However, the readily available up-to-the half-hour imagery also finds

† Mr Armand Péraldi was Scientific and Technical Director of Matra until his tragic death in late 1989.

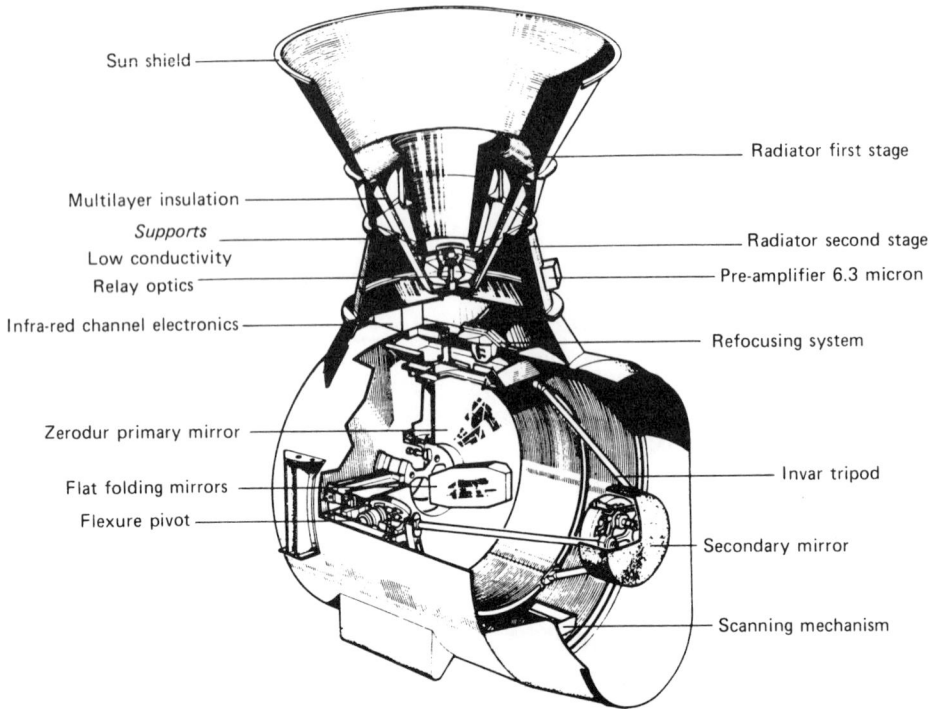

Fig. 11.1 — (b) The Meteosat radiometer. Courtesy: MATRA.

ready use in a wide range of other applications. Meteosat is the European contribution to the World Meteorological Organization's World Weather Watch. It is stationed above the Greenwich meridian, at 0° of longitude and, as Fig. 11.2 shows, views the Atlantic, Africa, Europe and the Middle East. America makes two contributions, GOES-E and GOES-W, which view the United States and the Pacific respectively. India flies INSAT, which views much of the Far East from a station over the Indian Ocean, and Japan has contributed GMS, which is stationed over a point North of Australia and views the Far East and the Western Pacific. There is thus considerable overlap between the coverage areas of the various satellites in the network. Coverage of the higher latitudes, both north and south, is of course increasingly poor, as the angle of view gets more oblique. That factor, and the extreme altitude of the geostationary orbit, have caused the worldwide meteorological community to demand a complementary network of lower-altitude instruments, flying in the near-polar Sun-synchronous orbits which we will be covering *ad nauseam* in later chapters (see Chapter 12). Very briefly, the geosynchronous instruments provide frequent, low-resolution, imagery, and are optimum for wind velocity measurements and 'local' forecasting applications. The near-polar orbiting instruments provide high-latitude coverage, and higher resolution, but their re-visit interval makes them more suited to longer-range forecasting (Pick, 1989).

All the meteorological satellite projects are in fact extended missions series, to an

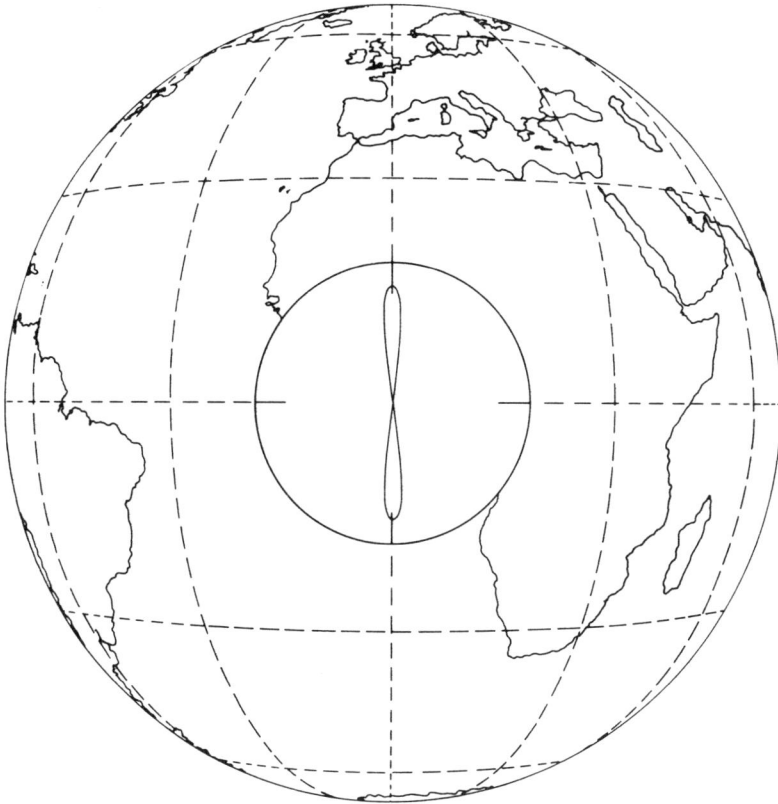

Fig. 11.2 — Ground coverage of Meteosat. This map shows a view of the Earth, as seen by
Meteosat. It extends out to about 80°, in all directions, from the nadir point. (The fact that the
full 90° cannot be seen is indicated by the slightly curved latitude lines.) The 'sub-satellite point'
is marked by the centre of the figure-of-eight pattern which indicates — much exaggerated —
the motion that would be induced by any residual inclination.

extent that Earth-resource projects such as Landsat are not. Most have several
prototype, or pre-operational, missions which precede the introduction of a fully
operational service. Meteosat is no exception to this. The first design studies for
Meteosat were carried out by France in the late 1960s. In 1972, the programme was
'Europeanized', by being taken over by the European Space Agency, ESA. (All
ESA member states, except Sweden, participate in the Meteosat programme.) The
ground segment is concentrated in or around Darmstadt, in Germany.

The first three Meteosats were pre-operational models, and were given design lives of three years. Meteosat 1 was launched in November 1977, and lasted almost exactly two years before a power supply failure destroyed its imaging function. In 1978, ESA took over full responsibility for the programme, including the operation of the satellite, the ground segment, and provision of replacements. Meteosat 2 was launched, under the auspices of ESA, in June 1981, commencing routine operations just under two months later, on 12 August 1981. Thus a gap in service was left of 21 months. Its mission was terminated at the end of 1988, shortly after the launch of Meteosat 3, when fuel depletion prevented accurate orbital control. Meteosat 2 thus remained on active service for some $7\frac{1}{2}$ years, which is claimed as a record for this type of instrument (Péraldi, 1989).

Although the design life was three years, the satellites were stocked with consumables for five years. After that a further year of free flight may be available (assuming that a small amount of hydrazine is kept for the smaller corrections) before the inclination of the orbit has built up to an unacceptable amount. Nevertheless, the fact that the satellite could be kept in service for so long is due in part to the excellent injection accuracies now routinely achieved by Ariane. This has saved most or all of the fuel allowed for initial orbital correction (see also Chapter 16).

The first fully operational Meteosat was originally intended to be launched in late 1988. However, delays led to the danger of a gap in service, and it was decided to refurbish the second prototype satellite (which had been in mothballs for about 12 years — possibly another record (Péraldi, 1989)) and fly it as Meteosat 3. As an additional economy measure Meteosat 3 was launched, free, in June 1988 aboard the Ariane 4 proving flight (thus possibly leading to a third record, in cut-price deployment). Ariane 4 was delayed for two years, and so it was particularly fortunate that Meteosat 2 proved so reliable. The operational Meteosats will have a design life of at least five years. The first of these, Meteosat 4, was launched in March 1989.

The objective is to have a spare Meteosat, stationed at about 10°W. However, Meteosat 3, which is still in full working order, was moved in October 1989 to a position over the Atlantic at about 40°W, to plug a gap left by a failure of one of the American GOES satellites. (This information was gleaned from European sources. Specialists involved in Chapter 9 have disputed it.)

A second-generation Meteosat is under discussion which will be considerably more ambitious in its scope. At the time of writing, however, the implementation has been put effectively on ice for the foreseeable future. This is partly for reasons of cost, partly out of caution in view of the problems that Matra believe to be being encountered by NOAA with their equivalent, which is to be three-axis-stabilized, and partly because the present system is in fact entirely satisfactory. It is likely that the existing operational Meteosat design will see the present century out (Péraldi, 1989).

The only instrument on Meteosat is the imaging radiometer. The visible region is covered by a single broad band covering the range 0.4 to 1.0 μm. The thermal band is covered by one 10.5–12.5 μm channel, while a third band, 5.7–7.1 μm, covers a region where water vapour absorbs strongly. The water vapour channel was a relatively late addition to the specification of the pre-operational instrument, inserted at the request of the meteorological user community. It was intended (initially at least) as an experimental feature only.

To minimize the changes needed, the water vapour channel stole downlink capacity from the visible channel, whose effective resolution was reduced accordingly. The water vapour channel was, therefore, switched out much of the time. On Meteosat 4, however, the water vapour channel is fully incorporated and is available full time as part of the online service.

No final report was made available on Meteosat. The information contained herein was derived from a number of sources, and the coverage is inevitably a little uneven.

11.2 THE ORBIT

The orbital considerations that apply to Meteosat are basically identical to those discussed in Chapter 9. Briefly, Meteosat's orbit is designed to position it over the point where the Greenwich meridian crosses the Equator. In practice, residual eccentricity causes the sub-satellite or 'nadir' point to execute a small figure of eight up and down the Equator (although its magnitude must be assumed to be small, since it is not discussed in any of the source material). More significantly, residual inclination will cause it to execute a figure of eight, with a period of one sidereal day, as indicated in Fig. 11.2. The prelaunch tolerance for inclination was specified as ±0.3°. However, in practice, this proved to be excessively stringent. North/south variations in the thickness of the Earth's crust generate an asymmetric pull on the satellite, and will tend to pull the orbit to an inclination at which the asymmetry is evened out. In the case of Meteosat, this introduces an increase in inclination of about 1° per year, and a regime of annual corrections has been introduced (Péraldi, 1989). As we have already discussed, it also means that a year's life extension is possible after the main supply of hydrazine has run out. The effect on the imagery can easily be corrected for on the ground. The excursion limit is, in fact, set by the beam width of the fixed ground station antennae.

The irregularities in the Earth's crust referred to in Chapter 9, pull Meteosat in the direction of the Indian Ocean. The satellite thus drifts slowly eastwards at a rate of just over a degree a month. When the error reaches 1°, small thrusters are fired. The firing gives the vehicle sufficient westwards impetus to generate a movement of 2° against this 'geopotential slope'. A second sinusoidal excursion along the Equator is therefore executed, this time with a period of some five months and an amplitude of ±1°.

11.3 SCANNING

In common with virtually all other geosynchronous satellites of its generation, Meteosat is spin-stabilized (see Section 11.11). The entire vehicle spins at a rate of 100 r.p.m. about a north/south axis, which provides a natural scanning mechanism for the telescope. Scanning is from east to west. As we have already seen, GOES scans from west to east, and from north to south.

South/north scanning is obtained by rotating the entire front end of the telescope, via reduction gears, by means of a lead screw and stepper motor. (The screw is wet-lubricated, and hermetically sealed using stainless steel bellows. The movement is transmitted to the telescope structure by means of flexible strips, thus eliminating

Table 11.1

Spacecraft

Mass — satellite (kg)	293	Launched	1977, 1981,
— single HRV (kg)	61		1988, 1989
Spin speed (r.p.m.)	100	Target	Cloud, sea
			surface temp.

Orbit

Altitude @ Equ. (km)	35 786	Orbits/day	1 (per sidereal day)
Semi-major axis (km)	42 164	Orbits/cycle	1
Eccentricity (%)	—	Days/cycle	1
Period (min)	1436	Shift/orb (°)	0
Inclination (°)	0–1	(km) @ Equ.	0
Descending node	n/a	Gnd track spacing (°)	0
Ground speed (km/s)	0	(km) @ Equ.	0

Telescope

Type	Ritchey Chretien		
Focal length (m)	3.65	80% blur c.d. (μrad)	—
Aperture (m)	0.4	F.P. dia. (mm)	—
f-number	—	Field-of-view (°)	±2.15
1y mirror dia. (m)	0.4	IFOV (μrad)	65/140
2y mirror dia. (m)	0.14	(μm @ F.P.)	270/70[†]
Total obscuration (%)	25.7	(m on gnd @ nadir)	2500/5000
Clear area (m^2)	0.1	EIFOV (m on ground)	—
Exit pupil		Inter-band reg. (pix.)	¼ IR pixel

Scanning

Method	Spinning vehicle/rotating instrument front end		
Rate (revs/s)	1.67	Mirror swing (°)	360
(rad/s)	10.5		
Scan repeat time (ms)	600		
Time for 1 line (ms)	30		
Efficiency (%)	0.05		
Sampling interval (ms)	0.012 (infra-red bands)		
Dwell time (μs)	—		

Channel details (nominal)

Band	Wavebands (μm)	Detectors	Dynamic range (W/m^2/sr/μm)	Noise equiv. reflectance (% F.S.)	MTF @ pix. rate trk/scan
1	0.50– 1.0	Si	— — —	>200[‡]	0.5
2	5.7 – 7.1	HgCdTe	— — —	<0.4@290K[‡‡]	0.5
3	10.5 –12.5	HgCdTe	— — —	<1.0@260K[‡‡]	0.5

Electronics

Consumption (kW)	0.069 (average)
Tape recorder?	No
Compression	No

Downlink

Data rate (Mbits/s)	—
Data rate (GHz)	—

Data

Word length (bits)	—
Grey levels	—
Image width (pix)	5000/2500
Image width (km)	n/a

† Actual, after further magnification.
‡ In fact SNR at 25% albedo.
‡‡ NEDT

lost motion or backlash.) After each scan line, the stepper motor rotates the front end assembly northwards by $125\,\mu$rad, or 5 km on the ground at nadir. The whole area of the disc visible to the instrument is imaged in 2500 scan lines, for the infra-red bands, representing a subtended angle of 18°. The sampling rate is chosen to generate 5 km square pixels (at nadir). Two half-size visible-band detectors produce 5000 scan lines during the same period. The sampling rate for these two channels is doubled, to generate $2\frac{1}{2}$ km square pixels (at nadir). A complete acquisition takes 25 min. The main mirror assembly is ramped back over $2\frac{1}{2}$ min. A final $2\frac{1}{2}$-min is left for the calibration of the detectors, and to allow the passive nutation damper (see Section 11.11) to have its effect. The next acquisition starts exactly half an hour after the current one.

It is also possible to reduce the latitude range scanned, and so increase the repetition rate. This facility has been used occasionally for reasearch purposes. However, operational users in Europe normally prefer full-disc images.

On Meteosats 1–3 the downlink capacity was limited to handling the outputs from three detectors, and a selection was made between the second visible channel and the water vapour band. When the water vapour was on therefore, the north/south pixel spacing for the visible band was increased to 5 km at nadir. There was no change to the dynamics of the remaining channel, and, therefore, strictly, every other line of data was lost. However the spatial frequency of meteorological phenomena is relatively low and so no serious deficiency resulted. The 5-km (at nadir) water vapour channel was sampled at the same rate as the $2\frac{1}{2}$-km visible channel that it replaced. Hence, in the east/west direction, it was oversampled by a factor of two. This, however, resulted in little improvement in resolution, because the spatial frequency response of the channel remained matched to a 5-km pixel. In normal practice, the water vapour channel was switched in and out at intervals during the day according to operational needs. The reason for this limitation arose because the decision to include the water vapour band was made at a late stage in the development of the instrument.

Scan assembly position is sensed by a potentiometer, measuring the position of the screwjack nut. A microswitch, mounted on the frame, detects end-of-scan. Another, positioned outside the norminal scanning range, switches off power to the motor in the event of malfunction. All three components are duplicated for reliability.

Because of the importance of the scanning unit, more than 20 years of simulated operation have been accumulated on the various development models. One model has been run through over nine years of operation without any degradation in its performance. Repeatability on ground test was within $\pm 2\%$ of step value, or $\pm 2.4\,\mu$rad (0.5 second of arc), a performance which is well within the specified limits.

Fourier analysis techniques were used (possibly for the first time in this application (Péraldi, 1989)) to identify faults in the leadscrew mechanism. Each component (gearwheel, drive nut or ball race) will have its own rotation rate and/or other characteristic frequencies, which can be measured or deduced. Each component will introduce periodic errors into the movement of the scanning assembly at one or more of these frequencies. During bench testing, the signature of any faulty components will show up as peaks on the 'error' spectrum, so that it can be readily identified and attended to.

11.4   THE TELESCOPE

The telescope is, in essence, a conventional aplanatic Ritchey Chretien instrument such as we will be meeting many times. (An **aplanatic** instrument is corrected for spherical aberration and coma (see Chapter 3.) However, its focal length at over $3\frac{1}{2}$ m, is exceptionally long on account of the very high angular resolution required. (Note that GOES, which has a similar basic requirement, is just under 3 m.) This of course places extreme demands on the accuracy and stability of positioning of the various components. The other remarkable feature of the design is that it is very highly folded, with sectors of the light path travelling in many different directions. The simple two-dimensional style of drawing employed elsewhere (e.g. Fig. 14.3) has proved incapable of depicting this system adequately. Fig. 11.3 is therefore taken from Péraldi (1989) with permission.

The basic design principle is that the primary elements (the primary and secondary mirrors, 'A' and 'B') should pivot, to provide the scanning action described in the previous section. The sensing stages of the instrument are fixed to the satellite structure. In between is the optical equivalent of a slip-ring. (This may be something of a misnomer because the mechanism only works over a relatively narrow angle of movement.) Folding mirror 'C' is also attached to the pivoting structure, and is mounted at the junction of the centreline and the pivot axis of the telescope. (It is also sited on the satellite spin axis, but this is to line it up with the cooler.) The rest of the optics are 'fixed', being attached to the radiometer structure. The infra-red detectors are installed in a zone cooled by a two-stage radiative cooler (see Section 11.11). As with other spin-stabilized satellites, the Meteosat cooler is symmetrical about the spin axis. This places the 'cold patch', and hence the infra-red detectors, naturally on the axis. The visible detectors are mounted close by.

The pinciple reasons for this unusual scanning mechanism are two-fold. The first is that it saves on a large and heavy mirror. The second is that scanning accuracy is improved, because the whole design is more compact. There is thus less scope for relative movement between components from either mechanical or thermal stresses. In spite of its optical complexity, the design has worked extremely well, and would certainly be used again in an instrument involving a small number of detectors (Péraldi, 1989).

As might be expected, the design of the structure supporting the optical elements posed considerable problems. The pivoting elements are supported by a structure made entirely of Invar, an alloy of iron and nickel, whose coefficient of thermal expansion is less than $2\times10^{-6}/°C$. (These days carbon fibre would probably be used, but it was not available when Meteosat was designed.) It comprises a cylindrical shell and a tripod, which is designed to be exceptionally stiff. All the mirrors are made of 'Zerodur', a vitreous ceramic with a coefficient of thermal expansion of less than $10^{-7}/°C$. It is not to be confused with ULE glass, much used in America, which is a form of doped fused silica with a coefficient that is significantly greater. The optical surfaces are coated with AG/ThF_4, which is a reflective coating of silver protected with an overcoat of thorium fluoride. The primary mirror was made from a blank

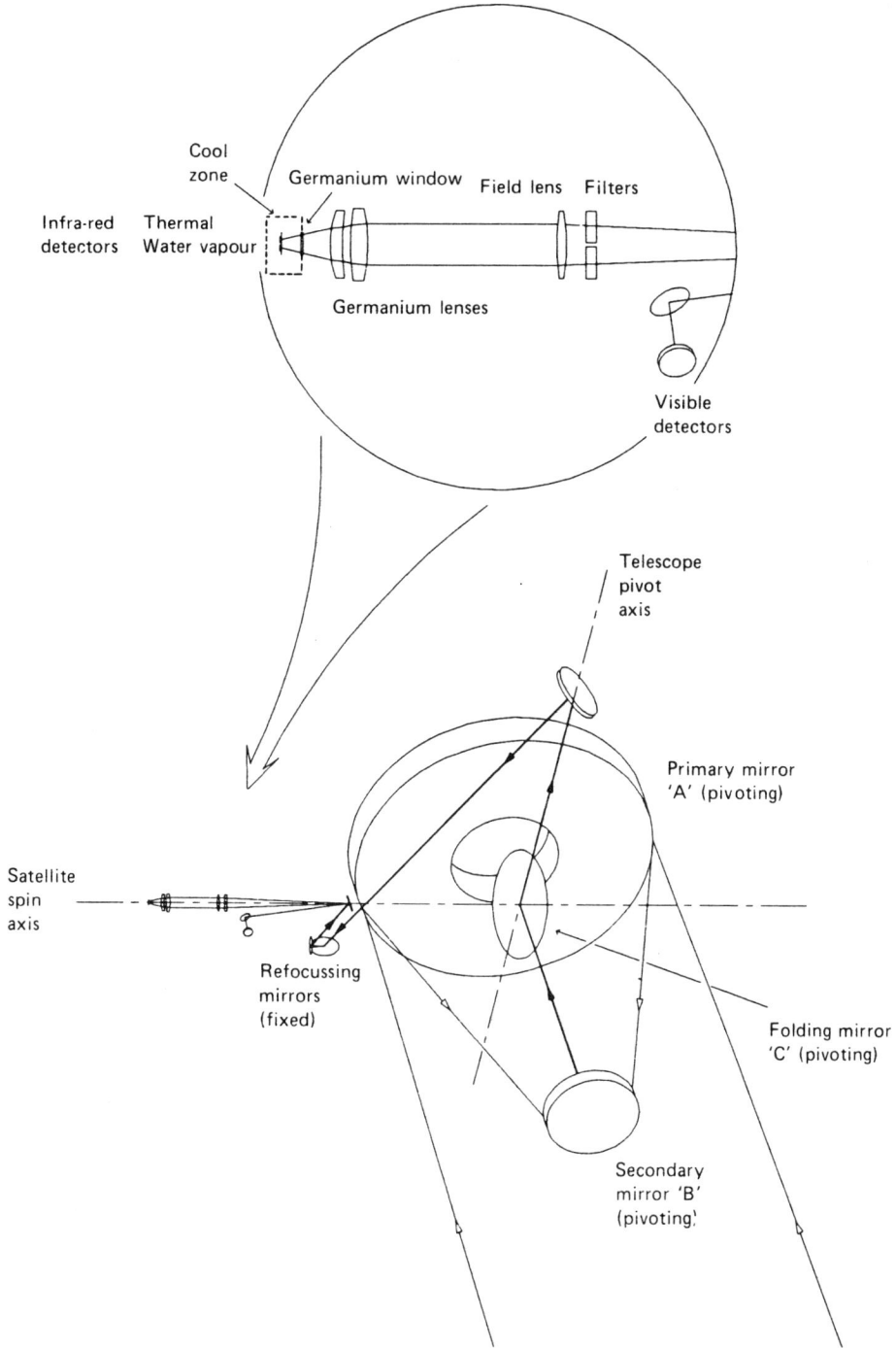

Fig. 11.3 — The Meteosat telescope.

already lightened by 65% by milling out the back but leaving a stiffening rim around the edge and also by boring out a central hole not needed for the optical path. The plane mirrors were made from blanks lightened by the use of an ultrasonic machining process. A similar process was later used on the SPOT mirrors.

The pivoting assembly is mounted on Bendix flex pivots, which allow a movement of ±10°. A system of cables (incorporating fully duplicated pyrotechnic cable cutters) was used to take the strain during launch.

Focussing facilities are provided by moving one of the folding mirrors. Because of the low thermal sensitivity of the system, refocussing was expected to be a rare event, and so a dry-lubricated system was chosen. (This is by contrast to the scanning mechanism, of which the sealing in of the lubricant was a significant problem.) In fact, there being no easy way of measuring the degree of defocus — if any — the facility has never been used (Morgan, 1989). (See Chapter 16 for an example of a focus-checking strategy.)

11.5   THE FOCAL PLANE

The focal plane (Fig. 11.4) is similar to that of AVHRR (and dissimilar to that of most of the other instruments covered), in being extremely sparsely populated. Indeed, again as with AVHRR, no single physical focal plane exists. Instead, the detectors are all placed at convenient points, and the radiation is directed to them.

Unlike AVHRR, however, beam-splitting is not employed. The detectors, including several spares (see next section), occupy different places on the logical focal plane, and observe different ground patches at any instant. This enables the folding mirror that deflects the rays destined for the visible detectors (see Fig. 11.4), to be fully silvered. The displacements are engineered to be exact numbers of infra-red IFOVs, to within about a quarter of an infra-red pixel, and so, by simple manipulations, the bands may be registered to acceptable accuracies.

Details of the individual detectors are given in the next section.

11.6   THE DETECTORS

The visible band is served by two silicon photodiode detectors, generating alternate half lines (see Fig. 11.4), which are duplicated from Meteosat 4. The thermal band is served by a single HgCdTe photoconductor, with its mix optimized to produce a sensitivity peak in the appropriate waveband (see Chapter 5). It is duplicated from Meteosat 1. The water vapour band is provided with a single HgCdTe detector, similarly optimized. Being less vital, however, it is not duplicated until Meteosat 4. The visible and infra-red detectors are mounted separately, as befits their differing environmental requirements. An additional, fully silvered, mirror directs light to the visible detectors.

The two visible detectors are fabricated onto a single chip to maximize performance homogeneity. The size of each is $250 \, \mu m$ square, which is appropriate to a published IFOV, at nadir, of 2.5 km. The nadir IFOV for the infra-red bands is 5 km. However, an additional field lens is added to each of the infra-red detectors to enable the active dimension to be reduced from 500 to $70 \, \mu m$ square. This strategy will become familiar as later chapters are perused. Its purpose is to compensate for the

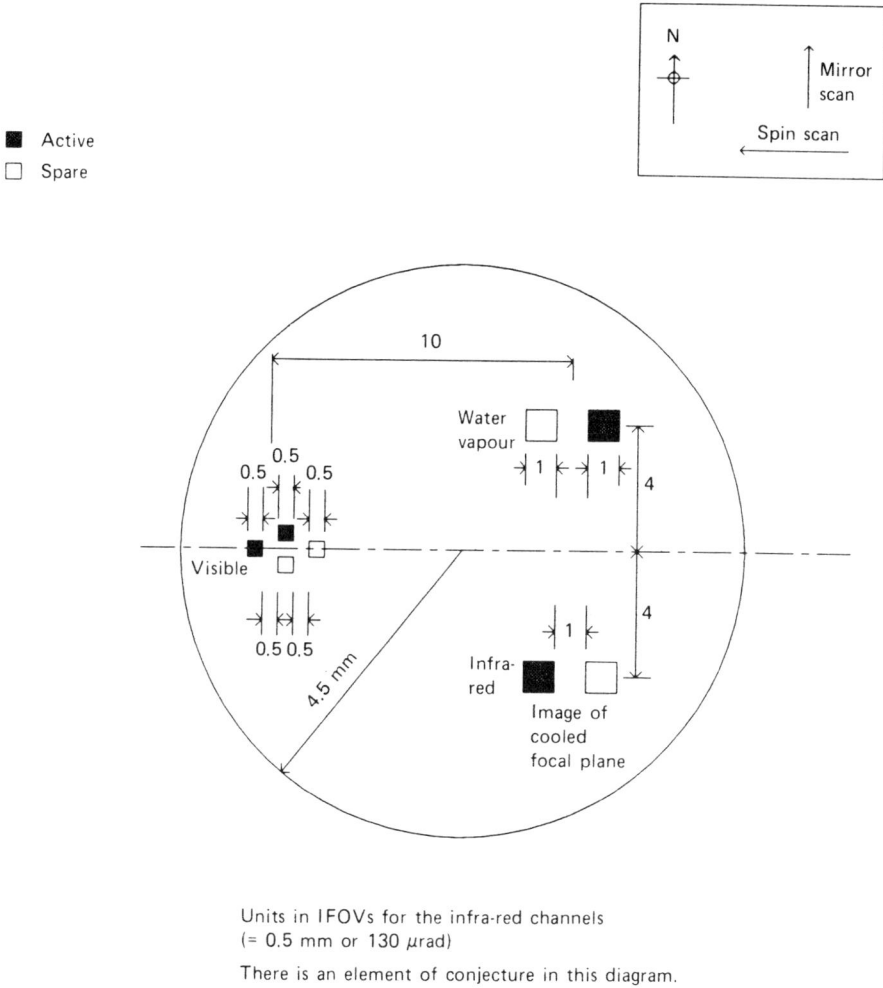

Units in IFOVs for the infra-red channels
(= 0.5 mm or 130 μrad)

There is an element of conjecture in this diagram.

Fig. 11.4 — The Meteosat Logical focal plane (from Meteosat 4).

much reduced radiance at the longer wavelengths by concentrating the available photons on to a smaller active area. This improves the signal/noise ratio (see Chapter 5), by reducing the volume of material in which the noise is generated. The seven-fold size reduction used in Meteosat, is somewhat greater than the norm for this operation. For Thematic Mapper, for example, the relay optics serve the same purpose by generating a size reduction of a factor of only two. Although the introduction of an extra lens of such power increases the aberration losses, and also the surface reflectance at the higher angles of incidence, a substantial net benefit is still obtained (Pick, 1989).

The two filters giving the short wave cut-off for the longer wavebands are multilayer interference filters deposited onto the two halves of a single disc of germanium. The disc can be removed during ground testing. This enables shorter-wavelength radiation to be used, and avoids the need for cooling for much qualitative development. At ambient temperatures, the sensitivity of HgCdTe detectors shifts towards shorter wavelengths, a factor which has to be taken into account during such testing. For critical tests, however, the cold of space was simulated using a black body, with the wall facing the aperture cooled with liquid nitrogen.

11.7   THE DETECTOR ELECTRONICS

There being no focal plane, as such, space is not at a premium as it is with many systems (but see also Chapter 12). The visible detectors are described as 'incorporated into the preamplifier housing', rather than the other way about as would better describe most systems.

Little information was made available on the Meteosat electronics. However, the infra-red electronics contain a feature that is unique among the instruments that we cover, in that the preamplifier stage comprises two amplifiers in parallel (Fig. 11.5). The high frequency components of the detector output are fed to an FET-fronted amplifer, for its low high-frequency noise characteristics. However, the FET amplifiers available at the time that the Meteostat electronics were designed were unable to handle the wide bandwidth, required by the high angular resolution of a satellite in geosynchronous orbit. FET amplifiers also tend to generate $1/f$ noise, which can degrade their performance at low frequencies. Therefore the low frequency components are dealt with by a conventional bipolar-fronted amplifier. Two separate drift compensation circuits are employed, the operation of which is described in more detail in Chapter 6.

Six-pole Bessel filters perform the anti-aliasing role; cutting off, for the infra-red bands, at 30 kHz and, for the visible band, at 60 kHz. The electronic gains can be altered, by ground command, by a factor of up to 18.5, in steps of 1.2. This is mainly to allow for the effects of any change in cool-patch temperature which may occur owing to seasonal variations or aging.

The electronic items are split among several boxes mounted on different parts of the structure.

Satellites in 'high' orbit are subjected to the full rigours of the solar flux, which can result in their picking up high voltages. This does not matter as long as all conducting parts of the structure are very well earthed. (Readers unfamiliar with this term will be relieved to learn than it does not imply the need for a connection to the ground.) Any deficiency in this regard, however, can lead to the build-up of voltage differences, which may become sufficient to generate arcing. Meteostat 1 suffered from this effect to a certain extent, in that discharges occasionally altered the readings stored in digital registers, and triggered spurious scan switchings or altered line identification readings. The offending items were easily reset from the ground, and so there was no significant impact on operations. Nevertheless, on later satellites, the earthing was improved.

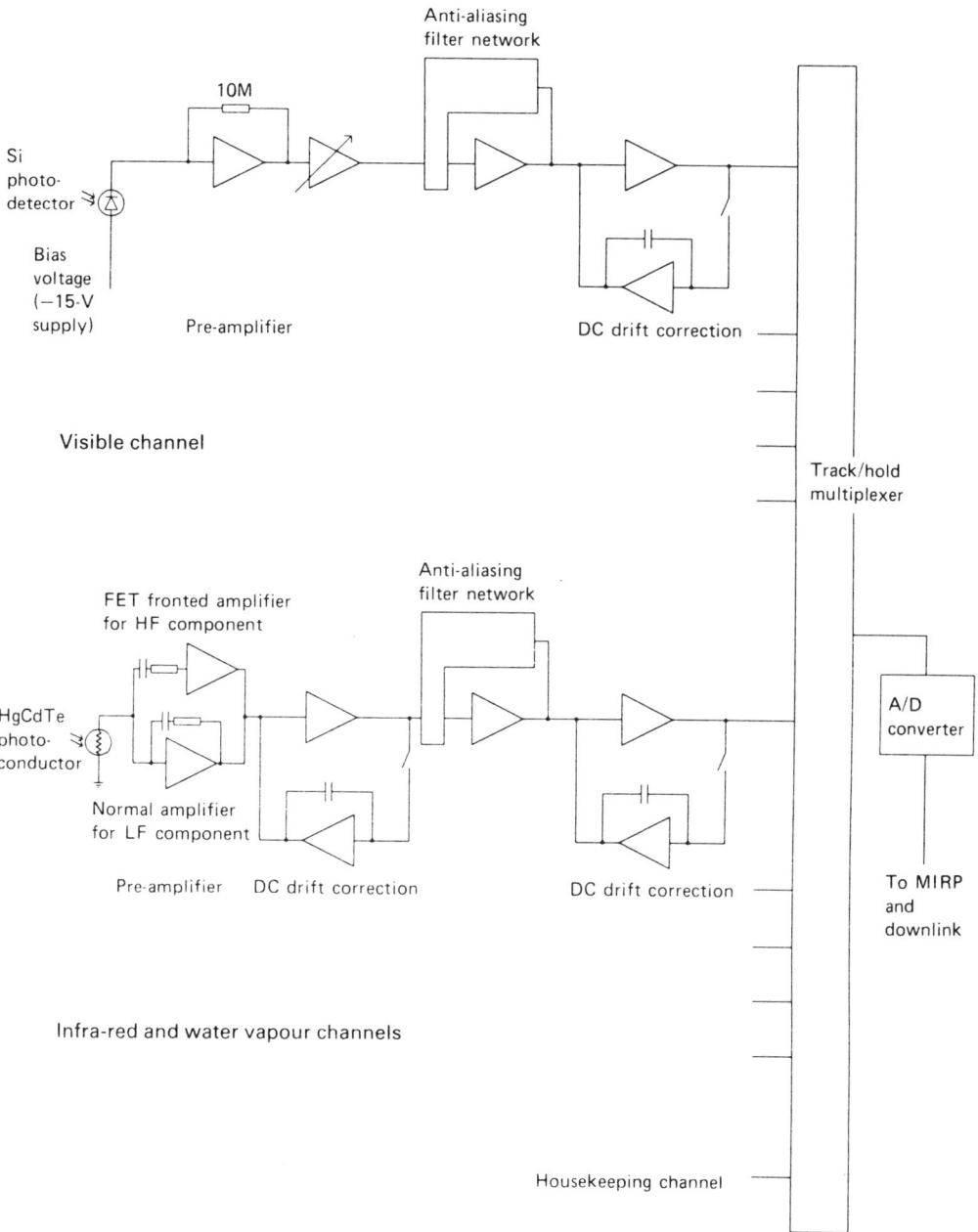

There is a degree of conjecture in this diagram

Fig. 11.5 — The Meteosat electronics.

11.8   THE DOWNLINK

In normal operation, a 30-ms scan is held in an on-board memory, and transmitted to ground over the remaining 570 ms of the spin cycle. In this 'stretched' mode the data rate is 166 kbit/s, which is within the capabilities of a wide range of low-budget receiving stations. However, should the memory fail, it is also possible to transmit in real time, or 'burst' mode at 2.7 Mbit/s for reception by Darmstadt only. The data are received at the Data Acquisition, Telemetry and Tracking Station (DATTS) at Darmstadt. The raw data stream is de-multiplexed, and has precise timing information added and then relayed to the dissemination station, also at Darmstadt.

In addition to its duty as an imaging instrument, Meteosat also carries a 'data collection system' for the relaying of environmental data form surface-based instruments and other satellites, such as GOES (GOES and the Japanese GMS satellites also perform a similar function). The sources of the surface-based data are a considerable number of 'data collection platforms' (DCPs). These may be installed on ships or aircraft, or on unmanned buoys, or they may be ground-based stations sited in inaccessible or hostile areas. Every half hour the DCPs are polled in sequence, on command from the central ground station at Darmstadt. The data are digested, packaged into a standard-format data stream and re-transmitted back to Meteosat, along with the processed image data from Meteosat, GOES and GMS.

Two output styles are provided, for the high-budget and the low-budget user. The primary data user station (PDUS) is typically a national forecasting service or major airport. It operates at 2400 bit/s, and can receive high-resolution imagery from Meteosat and GOES. It also receives regular DCP reports and the WEFAX data described in the next paragraph.

The secondary data user station (SDUS) is typically a university, yacht or flying club, or even a private individual. Such groups need a timely service for the online data, but are blessed with relatively small budgets. The SDUS data link carries the DCP reports and the WEFAX service (WEather FAX). WEFAX comprises lower-resolution processed imagery, and weather maps in facsimile format, as used in newspaper production and other intensive faxing operations.

11.9   CALIBRATION AND IN-FLIGHT CHECKS

Provision was initially made for the calibration of the thermal detectors only, because the water vapour channel was a late addition and no suitable source was available, at the time, for the visible waveband (Morgan, 1989). (See Chapter 14 for an account of the feedback system used on Thematic Mapper to obtain a constant output from incandescent lamps.) Two calibration mirrors were provided. The first reflected the detector front face back on to itself. Since its temperature was monitored at all times, this provided an excellent cold reference point. A small hole in this mirror also enabled sunlight to penetrate. The second mirror reflected the radiance from an internal black body on to the detector. The black-body temperature was maintained at approximately 30°C, and monitored to within 0.2°C. Both these calibrators bypassed the instrument optics, and were therefore supplemented by checking radiance values from Space, the Earth and the Moon. With a 360° East/West scanning, these bodies pass within view quite naturally, and without special steps needing to be taken.

From Meteosat 4, use of the detectors as calibration black bodies was replaced by a second unit. The two black bodies are maintained at 25 and 75°C. The remainder of the system is effectively unchanged. Also from Meteosat 4, a light-emitting diode system is installed to permit a relative calibration of the visible channel. The system is operable by ground command. Since the visible band is mainly required for position identification, relative calibration is all that is required.

An automatic calibration sequence takes place after each frame. The Sun (or Moon) calibration is carried out during the retrace period at the end of the scan. The black-body calibration is carried out during the stand-by period at the end of each frame.

11.10   DATA QUALITY

The spatial resolution of a geosynchronous imaging instrument falls off progressively as the angle subtended increases. Quantitative meteorological data may be derived out to about 55° of latitude or longitude relative to nadir. Qualitative interpretation is possible out to about 65°, and the ground disappears completely below the horizon at just over 80°.

11.10.1   Radiometric performance

The thermal channel is designed to measure the temperature of cold cloud-tops ($\sim$200 K), with a radiometric resolution (noise-equivalent temperature difference) of no more than 1.2°, and hot ground ($\sim$300 K) to within 0.4°. The visible channel has an analogue SNR of more than 400, which is much higher than demanded by the specification. This is a byproduct of its use of the same optics as the IR channels, at which the excitation levels are much smaller. Thus the noise content of the visible imagery is almost entirely dictated by the A/D conversion.

Within approximately a month of launch, the signal levels of both the infra-red bands began to fall off, with the thermal bands being the worst. The fall-off was attributed to moisture outgassing from the structure and settling, as ice crystals, on the optics. This is a problem that will become extremely familiar to the reader in due course. Any moisture (or other gasifiable contaminants) which might be trapped in porous parts of the structure will gradually escape into the vacuum of space. A proportion of the molecules will strike any adjacent cold surfaces and re-deposit as ices. Ideal candidates for suffering this effect are, unfortunately, the infra-red optical surfaces. Ice crystals absorb strongly in the region of the thermal band, and less strongly around the water vapour band, and this plays havoc with inter-band comparisons. The partial vapour pressure of water at 100 K is so low that re-evaporation in negligible. For other contaminants it is likely to be even lower. Unfortunately, if the contaminated surface is irradiated with ultraviolet from the Sun, then permanent damage can occur.

The solution is to warm up the cooler to drive off the contaminants. If, after recooling, some returns then the process must be repeated. In the case of Meteosat, both coolers were warmed up for a couple of days, which completely cured the problem. However moisture continued to escape, and de-contamination is now carried out approximately once a year. Péraldi (1978) makes the point that moisture penetrated the structure in spite of all the precautions that were taken at all stages of manufacture, testing and transport. In view of the importance of the decontamination procedure, later missions incorporate a fully duplicated heating system.

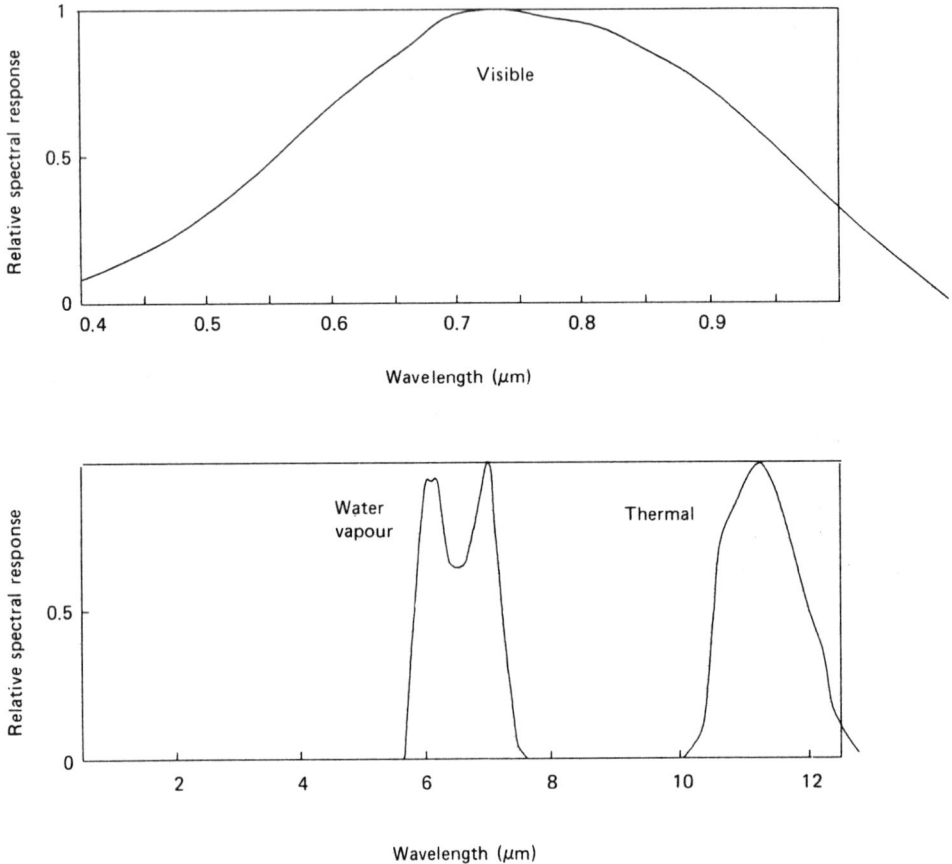

Fig. 11.6 — Meteosat spectral responses.

11.10.2 Spectral performance

It was noted that the initial imagery was exceptionally clear, by comparison with GOES, and free from the 'zenithal halo' produced by the effects of atmospheric scattering. This is attributed (Péraldi, 1989) to the spectral response of silicon diodes, which — unlike the photomultiplier tubes used in GOES — reduces towards the blue end of the band, which is the region of the spectrum most subject to scattering. Fig. 11.6 shows the responses for Meteosat. However equivalent data for GOES were not made available.

11.11   THE HOST SPACECRAFT

Meteosat is a spin-stabilized cylindrical body rotating at 100 r.p.m. The internal design is therefore also based on cylinders and discs, and is dynamically balanced to ensure that the completed vehicle rotates precisely about the design spin axis. These requirements do not, of course, lend themselves to the increasingly fashionable modular ('bus and payload') design approach. Meteosat, along with the other first generation geosynchronous satellites, is highly integrated. (See Chapter 15 for another example of an integrated design — in this case for totally different reasons.) The spin axis is aligned almost parallel with the spin axis of the Earth, but tilted slightly upwards towards the north. The alignment of the telescope is offset in the opposite direction, to provide a true vertical view.

Second generation satellites may be three-axis stabilized, as are all except the earliest Earth-resource satellites. However the stabilization problem is more severe for a high-altitude satellite because of the much greater pointing accuracies and stabilities required. So they also may not (Péraldi, 1989).

The main cylindrical body houses most of the satellite subsystems, including, of course, the main instrument. All the spare areas of its surface are covered with solar cells. Since, however, the apertures and other bespoken areas cannot be quite equally arranged around its circumference, the electrical output is somewhat uneven. In particular, a significant drop in output occurs when the telescope aperture passes in front of the Sun.

On the north-facing end of the cylinder are mounted the various antennae for communication with the ground. All spin-stabilized satellites face the problem of communicating with the (relatively) stationary ground. This is often side-stepped by attaching a 'despun' assembly to one end. This of course is a major technological challenge. With Meeteosat, the problem is neatly solved by careful antenna design. The antennae are bolted to the main satellite structure, and spin with it, thus eliminating the need for moving parts. The main S-band antenna can be seen clearly in Fig. 11.1(a). It comprises a ring of small dipoles mounted round the periphery of the antenna structure. The signal to be transmitted is fed to these slots in a phased manner, designed to transmit as from a single despun slot (hence the term 'electronically despun'). Other antennae are designed to achieve a similar effect. The early GOES also used an electronically despun antenna system, but Hughes went away from this, for the later GOES, in favour of conventional antenna design.

11.11.1   Attitude and orbit control

Like GOGS, Meteosat uses a combined thruster system for orbital and altitude control. The spacecraft has four main thruster motors and two smaller vernier motors, all hydrazine-driven. The fuel supply is appropriate for a five-year design life.

Two of the main motors are mounted with their thrust axes parallel to, but offset from, the satellite spin axes. They can be used for spin-axis precession manoeuvres and for inclination control. The two other main motors thrust through the centre of gravity. They are used for station-keeping and for active nutation control. The two vernier motors act in a plane offset from that of the radial thrusters, and are also offset from the centre of gravity. They act in opposition to each other to control spin rate and for active nutation damping. As with other vehicles (such as GOES) the latter operation is only required during the transfer orbit phase.

A degree of nutation (or wobble) is generated during scanning by the motion of the scanning system. As with other similar satellites, the motion is controlled by means of a 'passive' nutation damper. This comprises some damping mechanism designed to extract energy from any untoward oscillation. Such systems are not able to prevent a degree of nutation developing, and a 'rest' period of a few minutes is left between scans for the residual motion to die away.

As described in Chapter 9, GOES uses a viscous fluid for this purpose, while Meteosat employs a damped mass/spring system tuned to the nutation frequency. A tuned system can be made considerably more efficient than a viscous fluid device. On the other hand it is then more specific to a particular motion, and less effective as a general 'catch all' damper. It is perhaps not surprising that the dynamically simpler Meteosat is able to rely on such a system, whereas the dual-spinning GOES is safer with a less highly tuned device. It also appears that the $2\frac{1}{2}$ min rest period, allowed for Meteosat at the end of each scanning cycle, is much shorter than that required for GOES.

Attitude is sensed by four Earth sensors, operating in the 14–$16.6\,\mu$m region. Two of these scan the Earth's disc as the vehicle rotates and detect the ground/space transition. Two Sun slit sensors give a pulse each revolution, and two accelerometers monitor nutation. These data are used to activate the imaging scans and to control the electronic despinning algorithms for the antennae. The exercise of control over the satellite's attitude and orbital position is carried out through ground control.

For about six weeks around the Equinox periods, the satellite suffers eclipse for up to two hours either side of midnight (GMT). During the periods of eclipse, all the main functions, including imaging, are suspended to conserve power.

Whenever the ground antenna points directly at the Sun, the solar radiation has some tendency to swamp the wanted signal. A gap in reception occurs, lasting up to an hour or so, on certain channels only. Such an alignment occurs, at Darmstadt, for about a week at the winter side of each Equinox period. Other ground stations will be affected at different times, depending on their latitude and longitude.

11.11.2 Temperature control
The fact that a spacecraft is spinning simplifies the thermal control problem somewhat, because the incoming heat load is spread evenly over a large part of the surface. Of the end faces, one is occupied with the radiative cooler, while the other is largely taken up with antenna elements, which are relatively temperature-insensitive.

With Meteosat, as with GOES, temperature control is simply a case of paying careful attention to the average emittance and absorbance of the outer surface during the design phase, and then accepting the temperature balance that results. No active control, such as is used for example on AVHRR (Chapter 12), is provided. During normal operation, the balancing of radiated heat against that absorbed and generated internally, generates a relatively constant internal temperature of 25°C. During eclipse, the temperature drops slightly.

However the I.R. detectors are maintained at 90 K by a relatively conventional two-stage radiative cooler (the operation of these is described in more detail in Chapter 12). An outer sunshade (highly polished) protects the first stage of the cooler, which stabilizes at an intermediate temperature of about 140 K by the first stage radiator surface, which is coated with high-emissivity white paint. The heat radiated to the second stage cold patch is thus reduced substantially, enabling a second radiator to maintain it at a temperature of some 90 K. The second-stage radiator is a 12-cm^2 black-painted open-honeycomb plate, to which the relay optics assembly, incorporating the thermal band detectors, is attached. Fine control is, as usual, exercised by means of a small heater. Both stages are supported by low conductivity linkages of epoxy-impregnated glass wires. A heater is also fitted to the first stage. The heaters are capable of raising the cooler temperatures to about 20°C for in-flight outgassing purposes (see Section 11.10).

Although the design of the cooler is described above as conventional, it was in fact one of the first to be devised. It turned out to be by far the most difficult component to develop (Péraldi, 1978). The basic problem is that the radiating capability of a black body at 80 K is extremely small (at 0.2 mW/cm^2). A single detector, with a minimum of bias, dissipates a significant proportion of this, and so there is very little tolerance left for stray radiation or for conduction down the supporting structure. The first idea was to make the cooler large, in order to provide a good margin of safety. However, this overloaded the ground testing facilities, and made it impossible to obtain realistic temperatures. In addition, mathematical modelling had shown that increasing the size of the unit also increased the stray heat loads, and provided little improvement in performance. The size was therefore halved, with considerable savings in mass, and in the problems associated with its integration into the spacecraft. The performance of this second design was disappointing, however, because the first stage settled out at far too high a temperature. The reason turned out to be that some of the support tubes were absorbing excessive amounts of energy originating from the ambient-temperature interior of the spacecraft.

As part of a major design review, the supports were wrapped in aluminium foil, and the problem was solved. A temperature of 84 K was reached, which was acceptably close to the design goal of 80 K.

It was at this point that the additional dissipation of the water vapour channel had to be accommodated. The size of the cooler was increased again, by 28%, and improvements were made both in the thermal isolation of the second stage and in its radiating efficiency. In particular, the black-painted flat-plate radiating surface was replaced by black-painted honeycomb. In parallel with these developments in thermal performance, various structural problems were revealed and overcome during launch-stress, vibration and fatigue testing. The final design produced a flat-out temperature of 73 K, which is claimed to be the best performance ever achieved with this type of cooler. In-orbit tests, with the heater switched right off, confirmed the excellence of the design.

11.12 PRODUCTS

The incoming data stream arrives in pixel-interleaved format in the order that the detectors are scanned. Owing to the physical separation of the detector apertures on the logical focal plane (see Fig. 11.4), the bands are misregistered by significant (but accurately known) amounts. A number of radiometric, as well as other geometric, errors also exist in a raw data. In particular, and inevitably, areas away from nadir are grossly distorted. All data are immediately passed through a calibration, registration and geometric correction algorithm. The raw data are preserved on tape until this process has been successfully completed, when the processed data only are archived. When, rarely, the processing is not carried out successfully the raw data are archived instead. Geometric correction is applied in two stages. The first stage is automatic, and based on satellite position and attitude data, as well as on synchronizing pulses and other information concerning the condition of the instrument. The second stage is interactive and 'rubber sheets' the image to match visible 'ground control' points. Equipment is available to enable this to be done 'on line'. The **deformation** model thus derived is incorporated into the satellite and instrument control loops. It provides the basis for an accurate determination of attitude, the results of which determination are telemetered back to control the acquisition of the next image.

Control theory states that the amount of correction that may safely be applied to a system incorporating 'dead time' is small. However, the accumulation of error is also small so that little loss of accuracy need result.

Apart from the on-line production of imagery, the Meteostat ground station is also charged with the production of maps of:

— cloud-top height,
— cloud distribution,
— wind speed, derived from cloud-top movements,
— sea surface temperature,
— upper troposphere water distribution and (on an off-line basis) climatological data.

Because Meteosat is essentially an online tool, a great deal of automatic (and semi-automatic) checking and processing is carried out on a production-line basis. The result is a standard range of meteorological products which are then transmitted back to the satellite for speedy dissemination to users.

The first step in this process is to divide each image into segments of 32×32 infra-red pixels (or 160×160 km at the sub-satellite point), and this is the resolution at which meteorological products are derived. For each of these segments, therefore, the following are some of the quantities that are derived:

Wind velocity — or cloud motion vectors (CMV)
The attempt is made to estimate mean wind velocity, over a period of 60 min, for each of the segments., This is done by estimating the movement of clouds across three adjacent images. A spatial corelation is carried out between the first image of the threesome, and the second; and also between the second and the third. Whenever a good correlation is found between small patches on two images, but at a displacement commensurate with a realistic wind velocity, then the displacement found is reckoned to represent the mean wind velocity (speed and direction) over a 30-min period at that point. Two estimates are obtained at each point on the middle image, which are averaged to give the 60-min mean. The derived wind vectors are compared automatically with figures from forecasts, before dissemination, to eliminate unreasonable results.

Sea surface temperature
This is extremely important for forecasting temperature, humidity and rainfall in maritime regions of the world. It is derived from the thermal band data essentially by assuming the sea to be a black body. After correction for atmospheric absorption, the sea surface temperature may be derived directly from its radiance at $11 \, \mu$m. In fact, experimental curves are used to relate radiance to temperature, rather than placing reliance on Wien's law. The reader will recall comments made about the difficulty of obtaining reliable absorption figures for use with satellite-acquired imagery. However, two things are in the meteorologists' favour here. The first is that they have to hand the best possible network for obtaining data about the state of the atmosphere, and the second is that the absorption at $11 \, \mu$m represents a relatively small correction. As a result, it is normally possible to measure sea surface temperature, on a routine basis, to within $\pm 0.1°$C.

Upper tropospheric humidity
This is derived primarily from the water vapour band, and gives the mean humidity in the vertical layer from '700 to 300 hPa'. It is derived using vertical sounding techniques, which are beyond the scope of this book.

Precipitation index
This is based on the assumption that the colder the cloud top, the higher is the probability of rainfall occurring bneneath. The index gives an estimate of the accumulated precipitation over a five-day period.

Cloud-top height
This estimate assumes a correlation between cloud-top temperature and height. The cloud-top height is estimated at a resolution of 4×4 pixels. It is estimated between 3000 and 12 000 m, at a height resolution of 1500 m. The estimate is based mainly on the thermal band data, but with a correction applied based on the water vapour band.

Cloud analysis.
After carrying out the full automatic analysis, cloud amounts in each segment are estimated by inspection. Cloud-top temperatures for up to three cloud clusters are also quoted.

11.12.1   Quality control
All the above products are subjected to an online quality control procedure, in which the derived data are superimposed on the raw image and inspected by a trained meteorologist. Any data judged to be incompatible with known information may be deleted.

REFERENCES

ESA (1981) 'Introduction to the Meteosat system' ESA SP-1041.
ESA (1989) 'Meteosat, the Operational Programme', ESA brochure (in fact undated).

Morgan, J. (1989) Director Eumetsat, private communication.

NOAA (1987) 'The TIROS-N/NOAA A-G Satellite series' NOAA Tech. Mem 95 (reprinted 1987).

Péraldi, A. (1978) 'The Meteosat Radiometer' 15th European Space Symposium, Bremen June 1978.

Péraldi, A. (1989) 'Meteosat — a Historical Account' Matra. Also various corrections in draft.

Pick, D. R. (1989) Eumetsat, private communication.

Reynolds, M. (1989) ESTEC, private commnunication, 3/89.

12

AVHRR

Sources: **ITT (1982), NOAA (1987)**
Vetted by: staff of ITT, Fort Wayne

12.1   INTRODUCTION

The advanced very high resolution radiometer is one of several instruments flown on
the TIROS-N/NOAA series of weather satellites, which are operated by the
American National Oceanic and Atmospheric Administration (NOAA). Although
AVHRR data find their way into a variety of application areas, it is basically a
meteorological instrument. The philosophy behind it is geared to the provision of a
timely and dependable service to an international industry. As a result, reliance only
on tried and tested techniques is even more important for AVHRR, and the other
meteorological missions, than it is for the Earth-resource missions such as Landsat
and SPOT. In consequence, and as will become clear, designs are noticeably more
conservative.

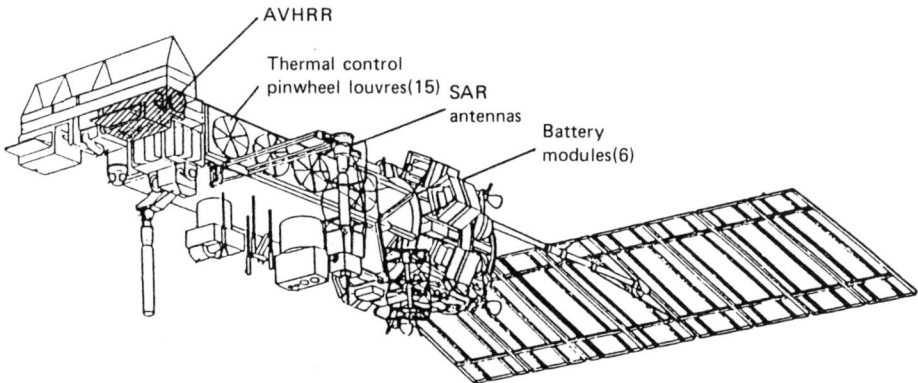

Fig. 12.1(a) — The NOAA Advanced TIROS-N satellite.
Courtesy: Aerospace/Optical Division ITT.

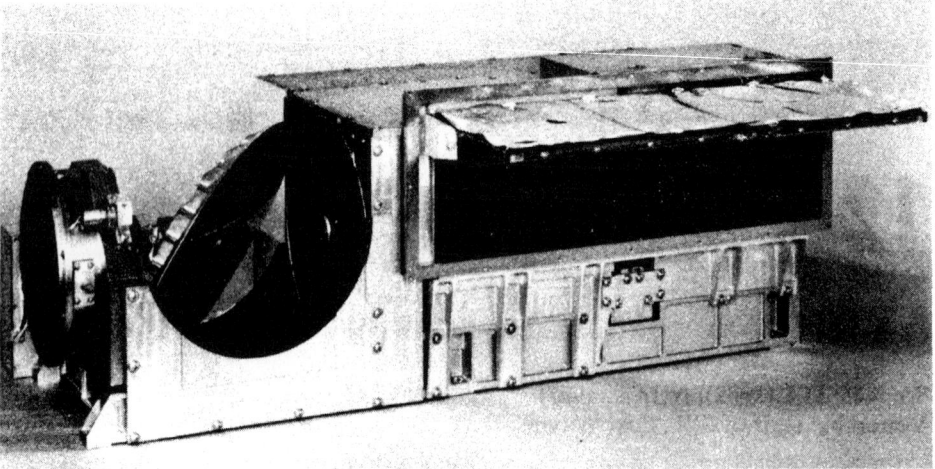

Fig. 12.1(b) — AVHRR. Courtesy: Aerospace/Optical Division ITT.

The instrument's name requires some explanation, since it will appear to readers in due course that AVHRR is neither particularly advanced nor of high spatial resolution. The simple answer to this conundrum is that it is not spatial resolution that is meant. The complete answer however lies partly in history, and partly in the nature of the meteorological requirement. AVHRR is the latest in a long line of instruments, each giving distinguished service to demanding customers, and stretching back to 1960. The original service was provided by a television-based system, using 'automatic (analogue) picture transmission', or APT. In 1972, the first radiometer-based system entered operational service in the form of the new very high resolution radiometer (VHRR). This instrument offered two bands, at a spatial resolution of just under 1 km at nadir. It was a true radiometer, in that its design and operation were both geared to the provision of quantitative radiance and temperature measurements. However analogue transmission was still used. Its radiometric resolution was poor by modern standards — though good for its day — and its imagery was noisy. In 1978, the first 'advanced' VHRR was launched, employing digital techniques throughout. The early AVHRRs offered four channels, at a slightly reduced spatial resolution, but at 10 bits (1024 grey levels) of radiometric resolution. At this time, the multispectral scanner was offering just 6 bits (64 grey levels); and so it could be argued that the NOAA series were serious measuring instruments when their Earth-observation compatriots were still little more than toys. To a certain extent, this difference still persists. The radiometric resolution of AVHRR is four times better than any of the Earth-observation instruments, one of which went up, as recently as 1988, with just 7-bit resolution (128 grey levels).

Two first-edition AVHRRs were launched between 1978 and June 1981, when the improved AVHRR/2 first flew. From then on, the two editions alternated until 1990, the time of writing. The main improvement introduced in AVHRR/2 was the provision of a second thermal band adjacent to the first. As will emerge, one of the channels (at 3.75 μm) operates at a very difficult part of the spectrum, and its imagery is plagued by severe noise problems. At the time of writing, the first AVHRR/3 is planned to be launched in 1993. It will incorporate a sixth channel, at 1.58–1.64 μm, to provide better discrimination between cloud and ice or snow.

As we have seen, the meteorological application demands high radiometric performance from AVHRR. Not only must small differences be resolvable but, for the infra-red channels at least, quantitative accuracy is demanded. This places great emphasis on calibration. Regular and accurate instrument calibration is required, together with continuous comparison with 'ground truth' data. The latter could of course be regarded as the ultimate right-through calibration of the entire system, such a vital feature of any well-conducted experiment.

By contrast, the spatial resolution requirements are relatively relaxed. Readers who have experienced the 'digital indigestion' that can be caused by using excessively detailed data will applaud this feature. Meteorological phenomena generally change with a relatively low spatial frequency, and a kilometre pixel size proved an acceptable compromise. It will become increasingly apparent that this relatively relaxed requirement colours almost every aspect of AVHRR's design. In particular it has permitted the use of a modular or 'optical bench' design, in which most components are bolted, separately, to a rigid structure.† This means that components can easily be changed or modified — or even added — at minimal cost or delay. It also facilitates adjustment of, for example, interband registration. We shall see, in due course, how small are the modifications required to introduce first one, and then two, additional bands. Perhaps the most striking use of these relaxed requirements is shared by the geosynchronous satellites GOES and Meteosat. It is the use of a continuously rotating scanning mirror, which spends most of its time observing deep space. This compares interestingly with the extraordinarily complex scanning system on Thematic Mapper, required to ensure that every possible millisecond of the scanning cycle contributes to the exposure of the detectors.

What meteorologists do require, of course, is frequent revisit capability. This is achieved by having AVHRR scan from horizon to horizon, and by flying two instruments, on separate polar-orbiting satellites, at any one time.

As with Landsat, the TIROS and NOAA satellites are identified by a letter suffix before launch, and by numbers once they are safely in orbit. They occupy a conventional Sun-synchronous orbit, although the constraints on its accuracy are also relatively relaxed. For example, a SPOT swath is about 120 km wide, and the SPOT satellite is required to retrace its published ground track pattern to within 5 km throughout the duration of the mission. Since the AVHRR swath extends from horizon to horizon (although the outer edges are perhaps of limited use), it matters little exactly where the NOAA satellites overfly from day to day. No fixed ground track, or orbital, pattern is published or adhered to. Instead, the NOAA satellites are injected into the most accurate possible orbit, and then left to fly free.

AVHRR is basically a thermal-band instrument, with additional visible bands (unlike Thematic Mapper, for example, which is the opposite). This means that AVHRR is as active during the night-time pass as it is during the day.

---

† It is the author's personal view that such a design philosophy would not have worked for Thematic Mapper or SPOT. Readers will have to decide for themselves whether they agree.

ITT (1982), together with Ames & Koczor (1989), constitute by far the most comprehensive account of an instrument that has come the author's way. In compiling this chapter, it has not been possible to do more than skim the surface of the report in the light of the author's preconceptions of what might interest the typical reader. Any reader who really wishes to know what makes AVHRR 'tick' is strongly recommended to obtain access to a copy.

12.2   THE ORBIT

The NOAA satellites are injected into conventional Sun-synchronous orbits, similar to those of most of the other instruments described, at altitudes between about 810 and 870 km. The NOAA orbit is illustrated in Fig. 12.2. As we have seen, the tight control exerted on most of the other satellites covered is not applied to AVHRR (NOAA, 1987). (The position is, nevertheless, known at all times, to an accuracy that enables the nadir point to be calculated to within 1 km (Pick, 1989).) Because of the large degree of overlap between adjacent images, and also because the principle phenomenon being monitored is unpredictable, the timing of passes is not critical. What remains critical, however, is that the solar panels receive their standard ration of sunlight, and that the radiative cooler surfaces be protected from sunlight or earthlight; and these are the principle reasons for constraining NOAA satellites to follow an approximately Sun-synchronous orbit. Sufficient propellant is supplied to provide for a degree of control adequate for these requirements, but beyond that the satellite is left to fly free. This means that the strictly repeatable ground track pattern which (as we have discussed) is so important a feature of Earth-observation satellite orbits is not featured with the NOAA spacecraft. Instead, future ground tracks are computed from a knowledge of those recently traced out.

Whenever reliability and launcher problems permit, two NOAA satellites are in orbit at any one time. One is arranged to have a morning descending node, and the other an afternoon ascending node. Thus every part of the globe can be imaged, to a reasonable resolution, four times a day. The orbital periods are arranged to be about a minute apart, with a suitable inclination adjustment to retain nominal Sun-synchronism. This ensures that their orbital cycles are out of step, and means that they are normally orbiting over different parts of the globe at any one time.

12.3   SCANNING

The scanning system on AVHRR utilizes a 45° scan mirror, rotating continuously at 360 r.p.m., or one revolution in 167 ms. Since (at a nominal altitude at 830 km) the Earth only subtends an angle of 124° at the instrument, this might seem a somewhat profligate use of available scanning time. However, as we have already discussed, it is made possible by the large pixel size appropriate to meteorological imagery, and eliminates the mechanical problems of oscillating mirrors and the field-of-view problems of pushbroom systems. The latter were, in any case, not available when AVHRR was designed. Calibration was also a consideration. A true radiometer requires this to be frequent, accurate and utterly dependable. The reader is invited to compare the simple system described in Section 12.9 with its equivalent for Thematic Mapper, described in Chapter 14. The 'dead' part of the scanning cycle is also used

Table 12.1

Spacecraft
Mass — satellite (kg)	1030	Launched	1979, 1981, 1983, 1984
— AVHRR (kg)	29		9.86, 9.88
		Target	Cloud etc.

Orbit — average values
Altitude @ Equ. (km)	850	Orbits/day	14.15
Semi-major axis (km)	7220	Orbits/cycle	n/a
Eccentricity (%)	0	Days/cycle	n/a
Period (min)	101.8	Shift/orb (°)	25.4
Inclination (°)	98.8	(km) @ Equ.	2830
Descending node	Morning	Gnd Track spacing (°)	n/a
Ascending node	Afternoon		
Ground speed (km/s)	6.5	(km) @ Equ.	n/a

Telescope
Type	Afocal confocal paraboloid		
Focal length (m)	0.461/0.132	80% blur c. d. (μrad)	—
Aperture (m)	0.203	F.P. dia (mm)	1.02
f-number	2.27/0.65	Field-of-view (°)	±.55
1y mirror dia. (m)	0.203	IFOV (μrad)	1310
2y mirror dia. (m)	0.0254	(μm @ F.P.)	605†
Total obscuration (%)	6.25	(m on ground)	1082† (@ nadir)
Clear area (m²)	0.030	EIFOV (m on ground)	—
Exit pupil	—	Inter-band reg. (pix)	0.016

Scanning
Method	360° rotating mirror		
Rate (revs/s)	6	Mirror swing (°)	360
(radians/s)	37.70		
Period (ms)	—		
Active scan time (ms)	—		
Efficiency (%)	100×(110/360)×(9.25)=11		
Sampling interval (ms)	0.025 (across scan)		
Integrating time (ms)	0.009		

Channel details (nominal)

Band‡	Wavebands (μm)	Detectors	Dynamic range (W/m²/sr/μm)	Noise equiv. reflectance (% F.S.)	MTF @ pix rate across trk
1	0.58–0.68	Si	— — —	>3:1§	0.3
2	0.725–1.0	Si	— — —	>3:1§	0.3
3	3.55–3.93	InSb	— — —	≥0.12°C¶	0.3
4	10.3–11.3	HgCdTe	— — —	≥12°C¶	0.3
5	11.5–12.5	HgCdTe	— — —	≥0.13°C¶	—

Electronics
Consumption (kW)	0.05
Tape recorder?	Yes (some products only)
Compression	Reduced resolution (some products only)

Downlink
Data rate (Mbit/s)	—
(GHz)	S-band/VHF

Data
Word length (bits)	10
Grey levels 1024	
Image width (pix)	2048
Image width (km)	2500

† For bands 1 and 2, actual; for others, effective.
‡ Some authorities reverse the numbering of Bands 3 and 4.
§ In fact, SNR at 0.5% albedo.
¶ NEΔT at 300 K.

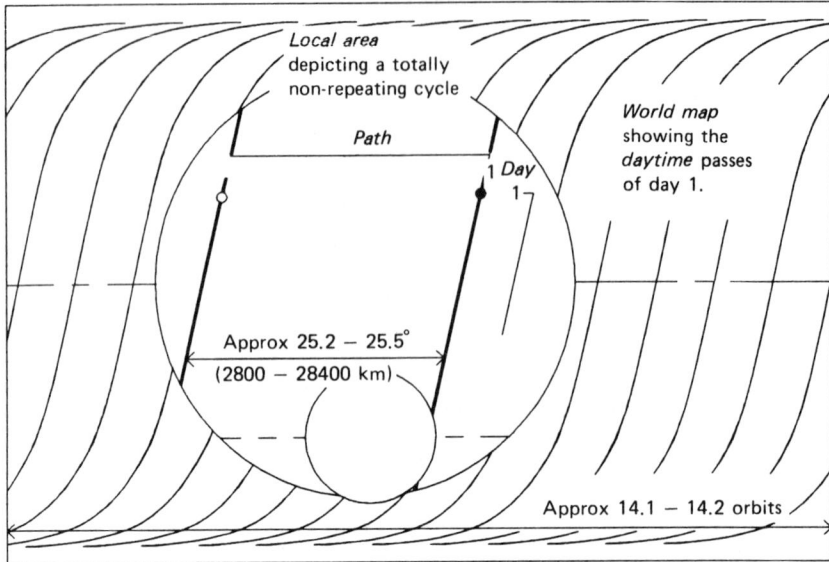

AVHRR orbits are Sun-synchronous, but,
in contrast to Landsat, SPOT, etc., they
are not quantized into exactly *n* orbits in
d days

Fig. 12.2 — Ground-track pattern for NOAA series.

for the transmission of housekeeping data. The mirror is made from the ubiquitous beryllium, with a waffle construction for lightness. It is coated with electroless nickel, polished and then overcoated with aluminium. The back of the mirror is coated with gold.

The scanning sequence starts with the receipt of a synchronization pulse obtained from a pick-off on the mirror drive shaft. This occurs when the mirror is pointing about 80° to the west of nadir, and the instrument is viewing deep space. Over the next 25°, calibration data are collected, as described further in Section 12.9. At 62° from nadir, the line of sight crosses the western horizon. Acquisition starts at 55°. A total of 2048 samples (pixels) are collected, over an angle of 110°. Seven degrees later, the line of sight leaves the eastern horizon and re-enters deep space. Part of the next 100° or so are used for the transmission of housekeeping data. Ten samples are taken while the instrument is viewing the infra-red calibration target (described in section 12.9), which is positioned around the zenith line of sight.

An auxiliary timing system is provided to maintain operations in the event that the signal from the position pick-off should be lost. The assumption is that the spinning speed of the mirror is extremely accurate, and that drift off synchronism is slow. An auxiliary trigger signal is generated by the timing clock, and checked by reference to the signal from Channel 4. Channel 4's output is at its lowest when the instrument is viewing deep space, and will therefore drop as the line of sight emerges from the cold target. This is very close to the point at which the position pick-off should have operated. (It will also drop at the end of 'Earth view' and so, if the error should have become large, a more complex situation is created. A procedure for dealing with this is also provided.) Two longish samples are taken, at what should be either side of the cold target viewing period. If the timing is correct then, over 16 scans, the two samples should average out to the same value. If they do not, then the timing is adjusted on the basis that the higher sample is picking up some cold-target readings. Correction is only made every 128 scans, to allow the electronics to settle out between adjustments.

The pixel size is defined as 1.3 mrad, or 1.08 km at nadir. This figure, together with mirror rotational speed, was chosen to ensure that the satellite moves forward by exactly one pixel length for each revolution. Off nadir, the pixels get progressively elongated. Note, however, that, unlike a photograph, the imagery is basically stretched in the cross-track direction only (see also Chapter 16). For certain products, the imagery is transmitted raw; for others, however, a crude correction algorithm is applied before transmission.

The across-track sampling interval is 25 μs.

Interline 'jitter', or timing irregularity, is specified to be less than 17 μs between scan lines. The procedure for measuring this irregularity, and for correcting it, is discussed in Section 12.7.

12.4 THE TELESCOPE

The AVHRR telescope is an afocal confocal paraboloid design, and is shown in Fig. 12.3. The telescope is simpler and cheaper than the Ritchey Chretien system favoured by NASA for the Landsat series, but it is markedly inferior in its off-axis performance. This is not a problem for AVHRR, however, because, as we shall shortly see, the relaxed spatial resolution requirements permit each band's single detector to be placed exactly at the focal point.

The telescope is possibly the area that benefits most from the relaxed spatial reauirements of AVHRR. In spite of the demanding specification for radiometric accuracy, there is ample flux density 'left over', as it were, to enable a number of design simplifications to be employed. First, it will be observed that AVHRR's aperture, at 0.04 m^2, is a quarter of that of Thematic Mapper, and just over a third of that of SPOT. Second, it has been possible to split the incoming beam into five (four with AVHRR/1) and thus enable each band to observe the same point on the ground. Compare this with the situation faced by the designers of Earth-resources instruments. The SPOT HRVs, with the gains obtained by the use of a pushbroom system, do use a single incoming beam split into three (although at a price in terms of radiometric performance). However each of Thematic Mapper's 100 detectors needs to exploit the full radiometric flux available, and must occupy a separate position on the logical focal plane. This means that each observes a different point on the ground at any given instant, and makes the reconstruction of well-registered imagery, as demanded by the customer, no trivial task.

The third design simplification is that the scanning rate can be made high enough to allow a single detector (per band) to scan the entire image. This is almost unique.

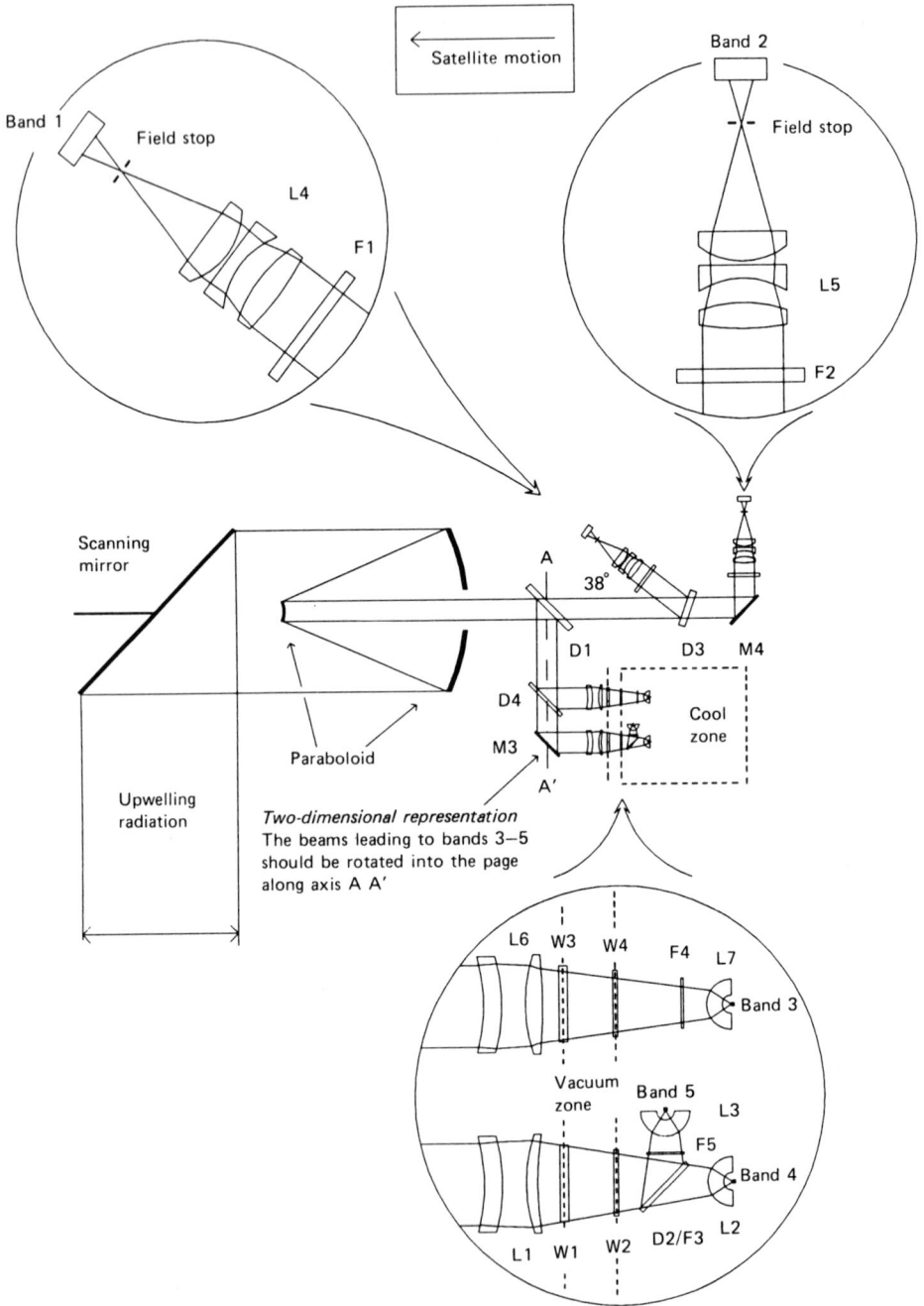

Fig. 12.3 — The AVHRR telescope.

(Compare, for example, Thematic Mapper, which requires 16, or SPOT, which employs 6000.) It means that each detector can be placed exactly at the logical focal point, and renders the instrument relatively insensitive to the aberrations to which optical systems are subject.

Finally, as we have already seen, the 'zigzag' motion of the scanning mirror can be replaced by continuous rotation, thus eliminating, at a stroke, one of the main design headaches of non-meteorological non-pushbroom instruments such as MSS and Thematic Mapper.

The **afocal** design feature means that the second mirror re-collimates the incoming beam, having concentrated it by a factor of 8 (the diameter ratio of the primary and secondary mirrors). This feature enables the filters and other optical elements to be small and spaced out on the 'optical bench'. It also helps to reduce the size of the windows into the cooler, and thereby to reduce its heat load. Focussing lenses are, of course, required, but each deals with but a single nominal wavelength and needs relatively little correction. Finally the design means that some, at least, of the interference filters can be arranged to receive strictly normal radiation which, as is discussed in Chapter 4, is desirable for optimum performance. The physical aperture of the instrument being fixed, the 'speed' of the instrument is dictated by the focal lengths of the individual paths (to be described in the next section). They emerge as f/2.27 for the two reflective bands and f/0.65 for the thermal bands.

Although the dispersed arrangements of the remaining components might make it logical to discuss them here, for consistency with other chapters they will be discussed in the next section.

Problems arose with the choice of coatings for the scan mirror. For polarization compensation, it was necessary to use aluminium coatings, both for the scan mirror and for M4 (see next section). However, because overcoated aluminium mirrors have a dip in sensitivity in the region of Band 2, it was planned to use silver coatings for the main telescope mirrors. Unfortunately, no reasonably priced silvering process was found which would pass the humidity test. In the event, therefore, they also were aluminium-coated. A proprietary 'enhanced' coating was used, which significantly reduces the degree of polarization produced.

12.5 THE FOCAL PLANE AND BEAM SPLITTING

The AVHRR optics have already been well described as an 'optical bench' design. As implied in Fig. 12.3, discrete components are bolted separately to a rigid structure. This design makes heavy demands on the stability of the structure, and would probably not be possible on a high-resolution instrument such as Thematic Mapper or SPOT.† Its advantage, however, is that developments and enhancements are particularly easy to implement.

As with all these instruments, it is necessary to distinguish between the 'logical' and physical focal planes. As seen from the telescope aperture, all the detectors are co-registered on the axis. This is the logical focal point, as depicted by the rather uninformative Fig. 12.4. Physically however, as shown in Fig. 12.3, the detectors are well separated, the radiation being directed to them by a system of beam splitters and

---

† As mentioned earlier, this is the author's personal view only.

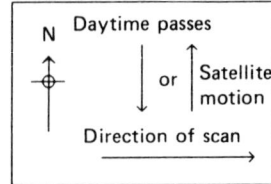

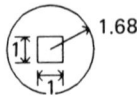

The diagram shows the 'logical' focal plane,
and illustrates the area over which the
detectors may be moved without compromising
the MTF performance.

The beam-splitting arrangements are shown
in Fig. 12.3.

Units in IFOVs (=605 μm or 1310 μrad)

Fig. 12.4 — The AVHRR focal plane.

mirrors. The three thermal-band detectors are housed in an evacuated box which is cooled by a radiative cooler.

The single light beam that leaves the telescope first meets a simple dichroic mirror D1, comprising a thin gold layer evaporated on to glass. Its function is to provide high transmission (70–75%) to the reflective wavelengths (Bands 1 and 2), and high reflection (82%) to the longer thermal wavelengths (Bands 3 to 5). Because of the spectral gap between these two regimes, the effect is moderately easy to achieve. This mirror introduces a degree of polarization into the transmitted beam, and so an additional folding mirror, M4, is added to Band 2 to compensate for this. (The neutral density beam splitter D3, about to be described, serves the same purpose for Band 1.)

The reflective wavelengths next meet a second beam splitting mirror, D3. On AVHRR/1, Bands 1 and 2 overlapped, and so D3 is a traditional half-silvered mirror, providing 34% transmittance for Band 2 and 22% reflectance for Band 1. (In fact it comprises a thin layer of inconel evaporated onto glass.) To minimize polarization effects, the angle of reflection is made as close as possible to 180°. The design of the filters, F1 and F2, is described in the next section. The lenses for the reflective bands are made interchangeable, for reasons of overall design simplicity. This means that each must handle almost the full visible spectral range, and that a degree of chromatic aberration control is required. As we shall be discussing, a somewhat extended field of view was in fact specified for all channels. To achieve these two effects required the use of a lens triplet, as indicated in the figure.

The thermal-band energy is separated, by dichroic mirror D4, into the medium wavelength Band 3 and the far infra-red Bands 4 and 5. Mirror M3 then deflects the far infra-red, so that all the thermal bands are directed into the cool zone. The cool patch is housed in a vacuum chamber to facilitate ground testing. Four clear windows, W1–W4, enable the radiation to penetrate the box. The windows are in fact band-pass filters in their own right. Their function includes keeping thermal radiation from the surrounding structure out of the box. The windows for Band 3 are of sapphire, which excludes virtually all energy below (of longer wavelength than) 6 μm. Those for Bands 4 and 5 are of zinc-selenide (ZnSe), and incorporate an additional filter, passing wavelengths between 9.6 and 13.2 μm. As is discussed in Section 12.11, a considerable 'spurious' heat load still gets through. All the windows are anti-reflection-coated, using a single layer of magnesium fluoride ($MgFl_2$). The outer windows are relatively thick, because they have to withstand atmospheric pressure during bench testing. Bands 4 and 5 are separated by a more sophisticated dichroic mirror (D2/F3), which doubles as the band-pass filter for Band 4. Filter F5 generates the long-wave cut-off for Band 5. The other filters, F1, F2 and F4, are designed as complete band-pass filters. The focussing lenses for the thermal bands are all similar in design, although, unlike the reflective bands, each is individually tailored. A lens doublet is used, made from germanium for Band 3, and from zinc selenide (ZnSe) for Bands 4 and 5. The lenses are placed outside the cool zone, partly to facilitate adjustment, but also to generate a degree of convergence in the beams before they enter the cool chamber. This enables the size of the windows to be reduced, to keep out more stray radiation.

Whereas with most instruments, except MSS, the Airy disc/blur circle is matched to the detector aperture, we have seen that with AVHRR it is deliberately made larger. This is to enable the position of the detectors to be adjusted, to optimize interband registration, without compromising the instrument performance.

The telescope barrel is made of Invar.

12.6 THE DETECTORS

As with other wide-band instruments (e.g. Thematic Mapper) a substantial part of the detector problem is generated by the disparity in excitation levels between the various bands. Fig. 2.1 shows how the level of excitation varies with wavelength. The reflective bands, 1 and 2, sense the solar energy reflected back from the scenery beneath, and so the flux received is dependent both on the solar radiation reaching the surface (or cloud tops) and on its albedo. (It also depends on the transmittance of the atmosphere, which is of course variable.) The net spectral flux reaching the telescope is relatively generous at these wavelengths, particularly so in the case of AVHRR because of its large IFOV footprint. In common with almost all the other instruments covered, AVHRR uses silicon diodes for these two bands, run (as always) at ambient temperatures.

The thermal bands, 4 and 5, sense directly the black-body radiation from the scene surface. Even by day, the solar irradiation is negligible at these wavelengths. The net flux available, per unit bandwidth, is about 50 times less than is enjoyed by the reflective channels (there are so many factors to be taken into account that this is a very crude estimate). Although wider bandwidths are normally used at these wavelengths, there is still invariably a need for the detectors to be cooled to, in general, some 90–95 K to avoid self-generated noise swamping the desired signal. In fact the AVHRR detectors are run at a slightly higher temperature than normal, namely 105 K. The detectors used are mercury–cadmium–telluride (HgCdTe) photoconductors. As we discussed in Chapter 5, HgCdTe is an alloy whose proportions may be adjusted to match the response to the waveband of interest.

Channel 3 is situated in the crossover zone, where both the reflected solar radiance and direct thermal emission are significant. It is basically intended as a night-time channel, although successful use by day has been reported. As the figure shows, therefore, Channel 3 has to cope with a flux per unit bandwidth nearly 10 times lower again, than the traditional thermal-band region. As we shall see in due course, more sophisticated preamplification arrangements are required for this channel than for any other — although they are considerably simpler than those of, for example, Thematic Mapper. Even so, as we shall see in a later section, compliance with the performance specification is considerably more precarious, and noise levels are high enough to be troublesome to users. The detector chosen in an indium antimonide (InSb) photodiode, operated in photovoltaic mode. However, a small reverse bias, of around -30 mV, is used because it improves this particular detector's noise characteristics.

The effective detector aperture, at 605 μm, is relatively large. The actual size of the two visible-band detectors is even larger. Unlike more integrated designs, the detector field stops are not built into the detector housing, but are situated at the focal point of the telescope, with the detectors themselves mounted 3700 μm behind. Tests indicate that they are 99% efficient, in that 99% of the rays passing through the field stops are collected by the detector's active region.

The infra-red channels are more conventional. The beams are further concentrated by use of a field lens to enable the active area of the detector to be made as small as possible. In the case of AVHRR, a $3\frac{1}{2}$-fold reduction is obtained, enabling the actual apertures to be made 173 μm square. The Band 3 sensing element is deliberately made some 7% longer in the electrode to electrode direction, to compensate for the end effects resulting from the presence of the electrodes. Uniquely, at least in this book, the fields lenses are of 'aplanatic' design. It can be seen from Fig. 12.3 that the rays always pass radially through an aplanatic lens; which therefore introduces no additional coma or spherical aberration. Its magnification effect is strictly equal to the refractive index of the material used. The AVHRR field lenses are made from germanium, with a refractive index of about 4. The detector field stops are built into the housing in the traditional manner. The field lenses are hermatically bonded to the detector cavities to provide both sealing and accurate alignment. The cavities are filled with inert gas. (Contrast this with the decision, for IRS-1, to install a bursting disc into each telescope to allow the contained gas to escape once in orbit.)

The optical coatings and the beam splitters were all carefully chosen, or designed, to have flat responses over the spectral bands that they affected. However it was not possible to provide a flat detector response for Band 1. The Channel 1 bandwidth for the original AVHRR/1 instruments was relatively wide, from 0.55 to 0.90 μm. As

Fig. 12.6 shows, this is considerably wider than the peak response of a typical silicon detector. This meant that a flat response could only be achieved by deliberately attenuating parts of the passband, which would have reduced the signal/noise ratio to an unacceptable degree.† Therefore the filters were designed with flat tops, and the slope imposed by the detector response was accepted. For AVHRR/3, it is planned to arrange a flatter top for both visible-band channels (Ames & Koczor, 1989). With later AVHRR/1 instruments, Band 1 was changed to 0.58–0.68 μm — a modification which was of course retained for AVHRR/2.

The reflective-band optical filters use the bulk absorption properties of 'Schott' glass for the short-wave cut-off, and multilayer interference techniques at the long-wave end. The interference layers are sandwiched between the Schott filter and a 2-cm-thick glass substrate upon which they are evaporated. The edges are epoxy-sealed to exclude moisture. While the Schott filter provides total blocking to short wavelengths, the long-wave blocking of the interference filters contains gaps. However the gaps occur at wavelengths to which the detectors are not responsive, and so no additional blocking is required.

12.7 THE DETECTOR ELECTRONICS AND TAPE RECORDERS

The NOAA electronics is a central function built into the equipment support module (ESM), which collects raw data direct from each instrument, including AVHRR. The AVHRR electronics comprises the, by now familiar, signal processing chain for each channel, together with a single A/D converter. A simplified diagram is presented in Fig. 12.5. The design is based, as far as possible, on space-proven techniques and components. The voltage and current ratings of components are reduced from the published figure to maximize life.

The pre-amplifiers for the two visible channels are identical, except for individual final adjustments to match each amplifier to the characteristics of its particular detector. The design is considerably less exotic than is used in, for example, Thematic Mapper. As discussed in Chapter 14, TM uses a JFET mounted, together with the 10^9-Ω feedback resistor, on the same substrate as the detector. These precautions are necessary to obtain a sufficient signal to noise ratio from the small ground pixels and the very short sampling times. By contrast, the relatively relaxed regime aboard AVHRR permits the visible channels to use off-the-shelf operational amplifiers and (as we shall see) conventional printed circuit techniques even for the detectors. As shown in Fig. 12.5, however, a 'differential' detector is used, in which a second diode is provided which does not receive any illumination. Its output is fed to the non-inverting input of the amplifier in order to compensate for the effect of temperature variations. This feature is not met in any of the other instruments described. The bias voltage for the detector is obtained directly from the −15-V stabilized supply. The detectors are both mounted on one printed circuit board, which also incorporates the two pre-amplifiers. The board is shielded in copper, and is hard-wired, rather than employing connectors.

All three infra-red channels use discrete-component front ends. Channel 3

---

† The reader may find this a little surprising, particularly after he has read on a few pages. However, by the time he has read on a few chapters, he will find that sloping tops are extremely common.

Anti-aliasing
filter network,
4-pole Butterworth

4M

Si
photo-
detector

Pre-amplifier

Post-amplifier

Bias
voltage
(−15-V
supply)

Calibration   DC drift correction
input

Channels 1 and 2

Anti-aliasing
filter network,
4-pole Butterworth

Track/hold
multiplexer

16M

+15 V

24K   24K

0.1μ   4M

Post-amplifier

1.2K

1.2K   20K   100K

−15 V

InAs
photo-
voltaic
detector

Calibration   DC drift correction
input

Channel 3

A/D
converter

Anti-aliasing
filter network,
4-pole Butterworth

−15 V

3K   35K

3.3K   3.3K

10K

To MIRP
and
downlink

Post-amplifier

HgCdTe
photo-
conductor

10K

Calibration   DC drift correction
input

+15 V

constant
current
source

Note
To emphasize the similarity
between the two circuits,
this pre-amplifier stage has
been drawn with the positive
and negative rails inverted

Channels 4 and 5

Housekeeping channel

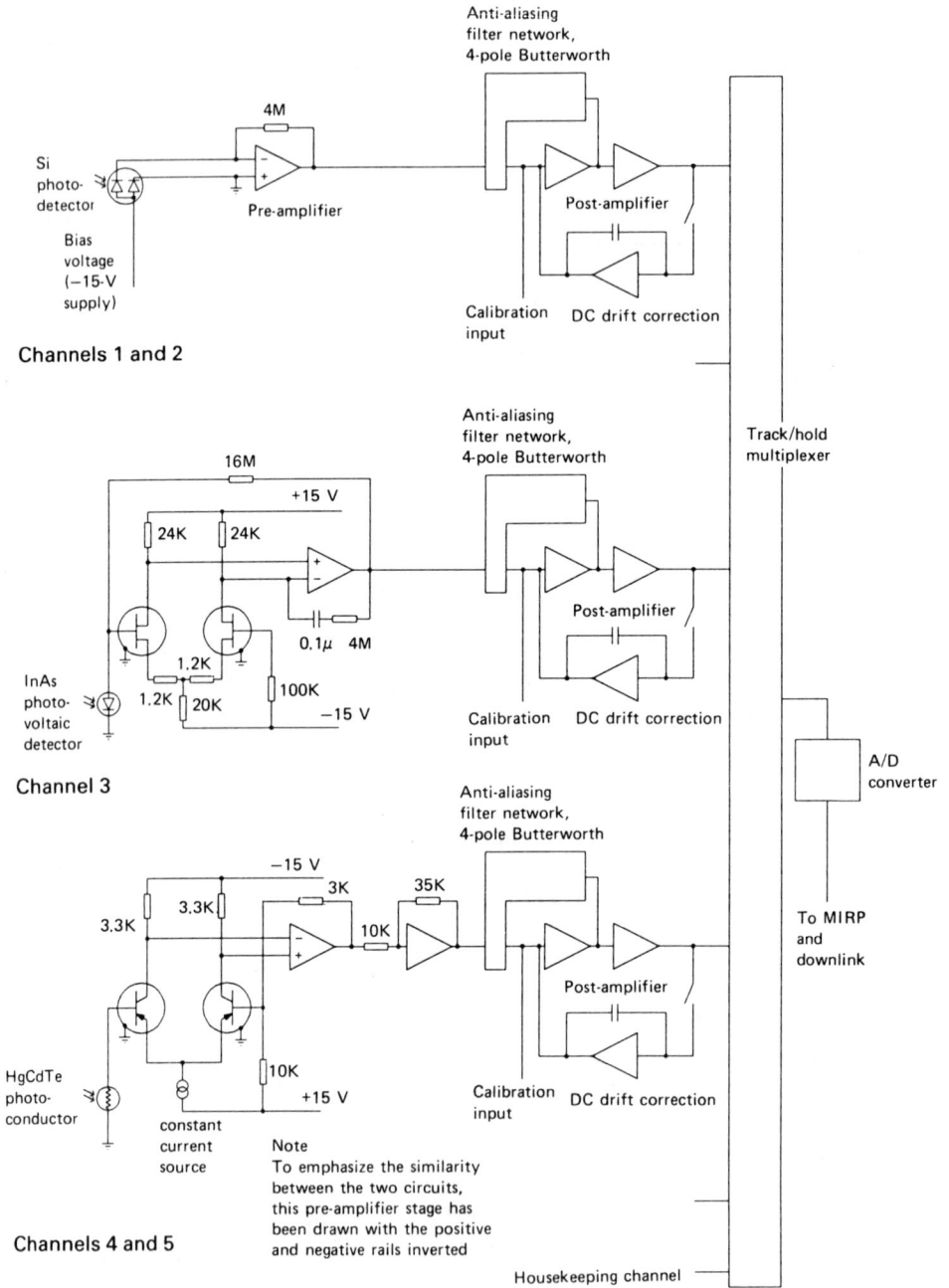

Fig. 12.5 — The AVHRR electronics.

employs a relatively conventional current-driven circuit, using FETs for their high input impedance and low noise. Channels 4 and 5 use a similar circuit except that, because of the low impedance of the HgCdTe detectors, conventional bipolar transistors are employed. Operational amplifiers provide further gain. The infra-red pre-amplifiers are housed in metal cans. The Channel 3 pre-amplifier is mounted on the outside of the electronics module, those for Channels 4 and 5 are mounted on the outside of the cool box; i.e. as close as possible to the detectors, while remaining accessible.

The supplies for the pre-amplifiers are given an additional degree of stabilization, over and above that provided for the rest of the electronics.

The remaining analogue functions are built into 'post-amplifiers', which are identical for all channels, again apart from an element of individual final adjustment. Each comprises a four-pole 'transitional Butterworth Thomson' anti-aliasing filter, rolling off at 15 kHz. This is a compromise between the standard Butterworth, which provides excellent characteristics in the pass band but which tends to overshoot when faced with a transient, and the linear phase filter, which provides excellent transient response but is poor at pass-band frequencies close to cut-off. The filter is followed by further amplification. The post-amplifiers, track and hold units and the A/D converter are all housed in the electronics package. A single connector is employed, to enable the electronics package to be easily removed.

An additional function of the post-amplifier is to provide dynamic correction for DC offsets and drift. The basic technique is described in more detail in Chapter 6. Before the start of each scan, while the detectors are viewing deep space (see section 12.3), their outputs are effectively forced to zero by shorting the bias voltages to zero. During this period, the output of the signal chain should theoretically also be zero. In practice, however, it is not, owing to lack of perfection in the dark current correction, and also to 'offsets' in the various amplifiers. This 'error' signal is corrected out by applying a classic control system technique. During the period when the input is clamped to zero, the error signal is inverted and fed back to the input of the post-amplifier (or, in the case of Channels 4 and 5, to that of the final stage of the pre-amplifier). Feedback is through the integrating circuit shown in Fig. 12.5. This means that the error signal will charge up the integrator's capacitor, and thus increase the inverted voltage fed back to the input. In accordance with feedback theory, this system can only settle down when an error signal is zero, in other words when the voltage across the capacitor generates a feedback signal equal and opposite to the net effect of all the spurious DC voltages. Before the start of the scan, the switch shown is opened. The circuit is designed so that the capacitor holds its charge, without significant voltage drop, for the duration of the scan. It therefore corrects for whatever DC errors existed when the switch was opened. If the errors change but slowly, which they do, then the result is almost perfect correction at all times.

A single six-channel, 10-bit, A/D converter incorporates individual track/hold circuits. It processes the five imaging channels, with one channel for multiplexed housekeeping data. All five optical channels are sampled simultaneously every 25 μs. Of this time, 16 μs is required for the convert phase, leaving 9 μs for tracking the incoming signal.

After digitization, the 10-bit data from the five AVHRR detectors are passed straight to the manipulated information rate processor (MIRP), installed in the

ESM. Here the signal is buffered, to smooth out the highly irregular data flow that is produced as a result of the low duty-cycle scanning method. It is then multiplexed into the common data stream, which incorporates data from the five other instruments carried as well as the spacecraft-wide housekeeping data. It is passed to the multiple data recorders described below, and to the high-speed downlink channel. A copy of the AVHRR data is also compressed, as described in the next section, and passed to the low-speed APT transmitter.

As discussed earlier, interline irregularity is specified to be less than 17 μs between scan lines. The task of measuring the irregularity, and for correcting it, falls to the MIRP. The irregularity is measured by comparing the timing of the mirror-position synchronization pulse with the synchronization pulse output by an internal clock, which is assumed to be stable. If the difference between the two synchronization pulses exceeds ±256 μs for two successive measurements, or 511 μs for one measurement, then a 'rephasing' is necessary. The rephasing takes place at the end of the next 'HRPT major frame' (three scan lines), and results in up to one scan line being lost.

Several digital tape recorders are installed. Their function is to enable a number of different derived products to be generated onboard and stored for high-speed transmission to selected ground stations. Details of these products are given in section 12.15.

12.8   THE DOWNLINK AND DATA COMPRESSION

The equipment support module contains several transmitters, at different frequencies and data rates, as appropriate to the different instruments and user types. Some transmit the complete data stream, others transmit it in compressed form. Compression in this instance implies permanent degradation of the data. Those appropriate to AVHRR are:

— An S-band transmitter for the real-time transmission of the complete AVHRR output, including housekeeping data, to any suitably equipped ground stations.
— Two more S-band transmitters for playback transmission from the onboard digital tape recorders. If required, all three S-band transmitters can be used for this purpose.
— A VHF transmitter, intended to be received by even modestly equipped users, and transmitting the reduced-resolution APT data of the area local to the receiver.

12.9   CALIBRATION AND IN-FLIGHT CHECKS

Because of its use as a true measuring instrument, online calibration is a particularly important feature of the AVHRR system. A two-point calibration of the infra-red channels takes place every scan. As we saw earlier, the electronics are adjusted to zero shortly before the start of each scan, while the instrument is viewing deep space. At the same time readings are taken of any residual DC output. In this way, dark current and long-term drift are either eliminated or calibrated out.

As the line of sight passes zenith, the detectors are exposed to a black body of known temperature, and further readings are taken. The temperature of the calibration source varies between 283 and 293 K (Pick, 1989), and does not therefore represent the top of the dynamic range. It does, however, fix the most important part of the scale. The visible channels do not, of course, receive any benefit from this reading, and therefore effectively receive a zero-point calibration only. Hence they are really only suited to relative measurements. The reason for this is that no simple and reliable source was available when the system was designed (Pick, 1989).

The key to the accuracy of this calibration is, of course, the performance of the black body. ITT (1982) contains a detailed account of the design of the black body, and of the steps taken to check its performance. Also covered are the arrangements for measuring its temperature to the required accuracy.

A comprehensive 1024-point linearity check is also carried out for all channels at a rate of one point per scan. The calibration source for this is a voltage divider chain energized by a standard voltage.

During parts of the night-time pass, scattered sunlight can reach the calibration source, past the edges of the scanning mirror. Studies are described in ITT (1982) showing that the energy reaching the target would be insufficient to affect its temperature. However, under the worst possible conditions, enough sunlight could be reflected back to the Channel 3 detector to increase its reading by up to 0.65°C. (For Channels 4 and 5, the predicted increase was 0.002°C.) It was concluded that Channel 3 calibration data taken in the danger zone should be disregarded.

It is possible, should it be considered necessary, to check the performance of all parts of the system downstream of the detectors, including the onboard processing and the downlink. This is done by inserting a known test pattern of pseudo-random imagery into the MIRP.

12.10 DATA QUALITY (AND QUALITY CHECKING)

Many aspects of predicted data quality have been subjected to intensive scrutiny, as recorded in ITT (1982). The procedure is covered in reasonable detail in Chapter 5, and only the briefest overview is included here.

12.10.1 Radiometric performance

ITT (1982) offers a comprehensive analysis of the sensitivity requirements for the various bands. For Bands 1 and 2, the spectral flux reaching the detectors is reasonably generous, and the object of the analysis is to demonstrate that the predicted noise levels are low compared with the minimum specified flux. The minimum flux is specified in terms of the threshold surface albedo of 0.5% reflectance, and of a standard spectral flux, incident upon the top of the atmosphere, of about 400 $W/m^2/\mu m$ (the figure is very slightly different for the two bands). A surprisingly large source of losses is the optics. It emerges, from an analysis of the optical performance of the various elements, that the overall transmission efficiency is around 5%. Much of the greatest single loss comes, of course, from the half-silvered mirror separating the two visible bands. In addition, however, each ray has to run the gauntlet of about seven other elements, each of which has the transmittance averaging out at, say, 0.85. A 50% allowance for dirt accumulation during service in space is included in this competition. From these figures a 'minimum' spectral flux, at the detector, of about 1.5×10^{-9} W is computed. It must be noted, of course, that what is measured is not actually the albedo, but the intensity of the reflected/scattered radiation. It will be a measure, largely of the albedo, but also of the transparency of the atmosphere and (marginally) of the incident solar intensity. It must also be noted that, if the electronics gains are set such that 100% albedo represents full house on the digitizer, then 0.5% albedo is still five grey levels above the discrimination threshold of the electronics.

The probable signal/noise ratio is also computed. The principle noise sources considered are (see Chapter 5) the shot noise due to the current through the detector and the Johnson noise generated in the pre-amplifier feedback resistor. From knowledge of the characteristics of the noise-producing mechanisms, a noise equivalent power of about 2×10^{-11} W is deduced, leading to an NEP of around 35:1. (There is a small difference between the figures for the two channels. Those quoted are an approximate mean.) This is a handsome margin, and indicates that, under the conditions used for the computation, the radiometric performance of the complete system is dictated largely by the resolution of the 10-bit A/D converter.

The flux available in Bands 4 and 5 is a good deal less generous, and the margin of performance is much lower than for the visible bands. As a result, it is not possible simply to prove that the level of the instrument noise is negligible. We have to demonstrate directly that adequate radiometric resolution is provided in the temperature range of maximum interest. This is normally done by computing the noise equivalent temperature difference (NEΔT), at a standard source temperature of 300 K (27°C). If we imagine the noise being superimposed on the required signal, then it will induce fluctuations in the measured output, leading to uncertainty in the reading. The NEΔT, therefore, is a measure of the radiometric resolution of the instrument at this relatively high point on the range. It says nothing about the dynamic range of the instrument, or the minimum temperature that can be measured. The computation is based on the slope of the black-body radiance curve at this temperature and these wavelengths, and also on the detectivity and other characteristics of the optical and detecting chain. The transmittance of the AVHRR optics, at about 35%, is much greater at these wavelengths. This is largely because it has been possible to do the beam splitting entirely by dichroics, whose transmittance and reflectance can be made relatively high. The principle noise sources considered are the $1/f$ noise from the detector, and a small additional contribution from the pre-amplifier. The report concludes that the specified NEΔT figures are achievable, but gives no predictions.

Band 3 is the most critical of the five bands, because of the reduced radiance of a black body at this wavelength. At such levels, noise external to the detector chain become significant, and has to be included in the analysis. The principle external noise source is the 'background photon flux' (the random fluctuations in the arrival of 'background' photons from sundry nefarious sources — see Section 5.1). It is of particular importance down the scarp (high energy) slope of the black-body curve, not only is the energy flux falling, but the photons are getting more energetic, and so fewer are needed to carry a given flux. Thus the 'quantum efficiency' of the detector assumes an importance in the analysis. (A minimum figure of 75% is quoted, although 75% of what is not specified.) The analysis evaluates the detector current appropriate to each of these noise sources, of which the largest turns out to be the background photon flux. The total noise current is computed to be 2×10^{-11} A r.m.s. (This compares with the required signal current, which emerges as 7×10^9 A.) Next, the specified NEΔT (0.12 K) is converted to a detector current, and emerges as 3.6×10^{-11} A. From this it is deduced that the specification is met with a signal/noise ratio of 1.85, and that the effective NEΔT, at 300 K, will be 0.065 K. Finally it is computed that, for Band 3, a single digitizer step is about equal to the estimated noise level.

How any of the infra-red channels perform, when viewing lower temperature scenery, is not predicted.

12.10.2 Spectral performance
The results of the pre-launch spectral evaluations are shown in Fig. 12.6.

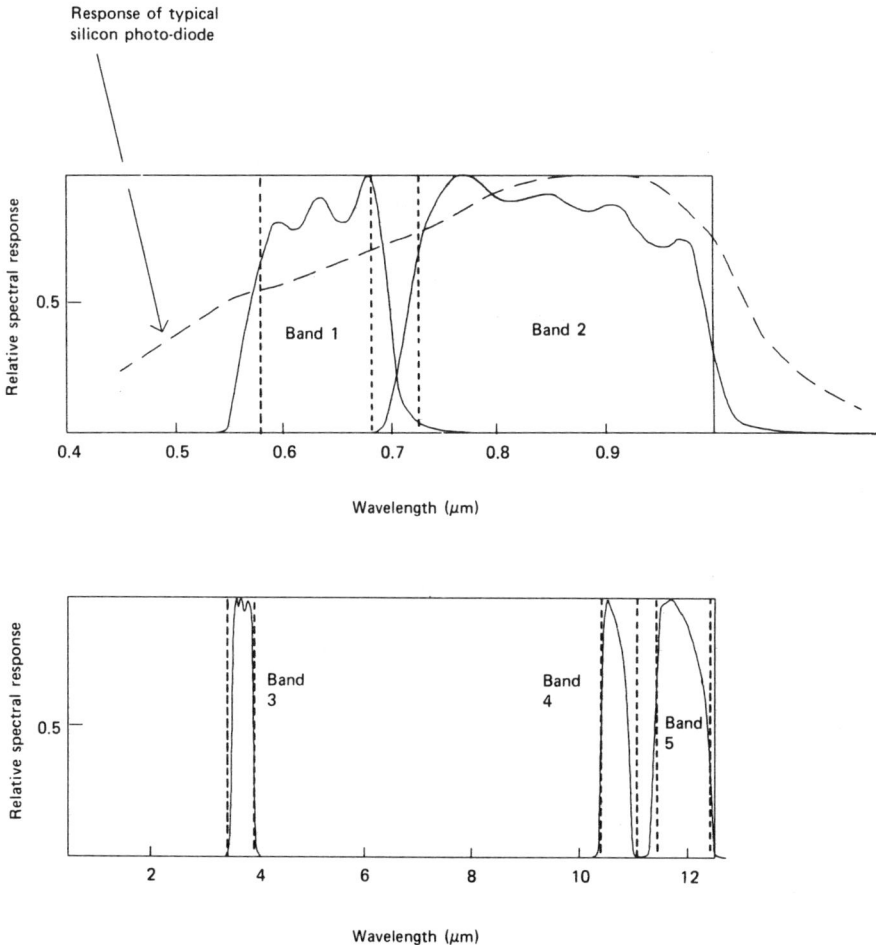

Fig. 12.6 — AVHRR spectral responses.

12.10.3 Spatial frequency response

The spatial specification for AVHRR is defined in terms of the usual black and white bars, and requires that the MTF at one bar per IFOV (half a cycle per IFOV) be at least 0.3. This is a relatively relaxed specification. In direct consequence the anti-aliasing filter is reasonably conservatively rated. At 14.5 kHz, its 3-dB down point is 0.36 IFOV. As Fig. 12.7 shows, the filter is in fact the main factor contributing to the

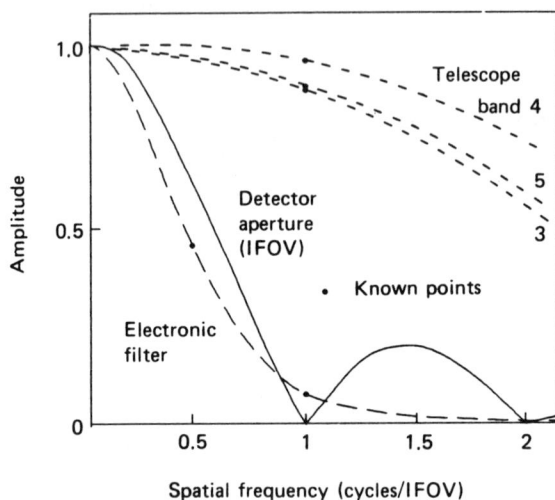

Component MTFs — IR bands

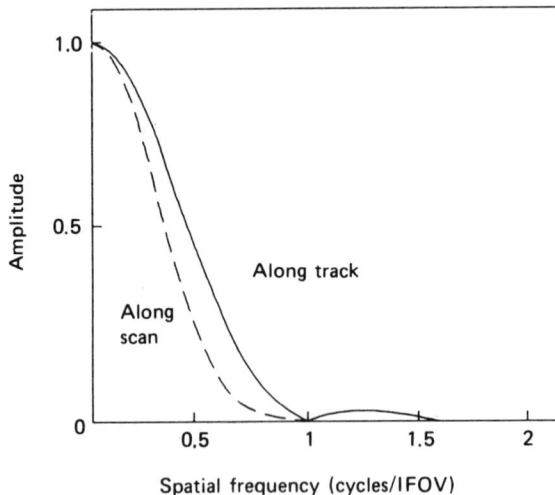

Net MTFs — schematic only — IR bands

Fig. 12.7 — AVHRR spatial responses.

net along-scan MTF. It follows that the along-track MTF must be noticeably higher, as suggested in the lower graph of the figure. The lower graph is in fact illustrative only, and has no data to support it.

As has already been discussed, the telescope MTFs were required to be considerably higher than simple logic might dictate. This was to provide a degree of freedom in the precise positioning of the individual detectors, to enable the tight inter-band registration specification to be met. In spite of the fact that, as Fig. 12.7 shows, the telescope optics contribute little to the MTF limits, ITT (1982) records a comprehensive analysis spatial performance of the various channels. The analyses are based on a sinewave excitation (see Chapter 5) for a discussion of the meaning of the term). The results have had to be modified somewhat to suit the book's standard presentation.

It is interesting to compare the figure with the equivalent for Thematic Mapper (Fig. 14.7). Again, in the higher frequency bands, the telescope MTFs are high, although on this occasion the reason is that the same optics must also encompass a thermal band. Because the MTF specification is tighter for Thematic Mapper, less margin was available for aliasing elimination, and the filter cuts off at a higher frequency than the detector aperture stops. It follows that there is a greater danger of spurious small features showing up on Thematic Mapper than on AVHRR. (However, it may well be that genuine small features have a greater contrast in the reflective bands than they do in the thermal bands.)

12.11 THE HOST SPACECRAFT

The TIROS-N/NOAA satellite is a large vehicle, the bulk of which is taken up with the equipment support module, which provides housekeeping and central facilities (Fig. 12.1). The sides and top of this module are fitted with a total of 12 thermal control pinwheel louvres. Those in shadow can be opened selectively to control the radiation of heat away to space. The instrument platform is a relatively small appendage, attached to the end remote from the solar array. Instruments are bolted to the underside of the instrument platform. Instruments are thus shielded from sunlight by the body of the satellite. However, they must make their own provisions, if necessary, for earthlight protection. A sunshade is provided to protect the upper side of the instrument platform.

In accordance with stated policy, the NOAA satellite that carries AVHRR is based largely on components developed for the Defense Meteorological Satellite Program. The contractor for both series is RCA. The satellite is designed to support a number of instruments, of which AVHRR is just one. Others carried include the Tiros operational vertical sounder (TOVS), the high resolution infrared radiation sounder (HIRS/2), the strtatospheric sounding unit (SSU), the microwave sounding unit (MSU) and the international search and rescue receiver (SAR).

12.11.1 Attitude control

All the three-axis-stabilized satellites that we cover use basically the same method of attitude control. Since this is the first time that we meet the problem, we will discuss it in slightly more general terms than might otherwise be appropriate.

The function of attitude control is to keep the satellite oriented correctly in relation to the Earth. In principle, an object in Space will maintain a constant orientation in relation to the star background. In practice, however, rotating parts within the satellite, such as the solar panels, will impart angular momentum to the structure when their rotation is initiated, and again whenever their speed is adjusted. There are a number of residual forces acting on the satellite: for example, atmospheric drag, solar radiation pressure, particle bombardment and interaction with the Earth's magnetic field. Unless these forces are precisely symmetrical they will cause a slow drift of attitude.

An Earth-observation instrument needs to keep its field of view pointing permanently towards the ground. To achieve this, a satellite in sun-sunchronous orbit must spin slowly about its pitch axis, at a rate of one revolution per orbit. It must also spin more slowly about its roll axis, to maintain its orientation in line with the orbital plane. About the yaw axis, there should be no motion. Tidal effects might engineer this situation over millions of years, as they have with the Moon. However, for an artificial satellite, a more immediate solution is required.

Roll and pitch attitude may conveniently be sensed by infra-red horizon sensors. The sensors are aligned so that, when the attitude is correct, they observe the edge of the Earth's disc. Any mis-orientation results in their observing either the apparent cold of deep space or the relative warmth of the Earth's disc. Yaw attitude cannot, however, be monitored in this simple way. Short-term monitoring of yaw angle is normally by the use of a 'rate gyro', which senses any movement away from the nominal orientation. Gyros, however, measure angular movement, not position. Estimates of position have to be obtained by integration, and this gets increasingly unreliable as the time over which the signal is integrated is increased. Zero frequency, or steady-state, feedback must therefore be obtained from another source. This is normally done by the occasional observation of a point of Space whose orientation at the time is either fixed or can be accurately computed. Many systems sense the position of the Sun, at carefully controlled times, such as its emergence from eclipse. Others use prominent stars.

It might be thought that the attitude control system now has all the information that it needs, and if infinitely slow response were acceptable this would indeed be the case. However the way in which systems respond to stimuli is the main reason why control is difficult. 'Lags' and other dynamic effects invariably require to be taken into account, either by modelling or by having their effects measured. Measurement is safest, and 'derivative sensing' is a classic feature of advanced control systems. Rate gyros are therefore commonly used in the control of roll and pitch, as well as for that of yaw.

The NOAA satellites use four horizon sensors operating in the 15-μm region of the thermal band, and designed to operate over an altitude range of 740 to 926 km. Long-term yaw monitoring is through a single Sun sensor, which senses the edge of the Sun's disc. The relative position of the Sun at the time of the observation is computed from ephemeris data and from the outputs from the various attitude readings. (Contrast this with the SPOT method, whereby readings of the edge of the Sun's disc are taken only at the moment that the satellite emerges from eclipse (night-time), thus eliminating variation in Sun position and the need for much complex computation.)

Having measured an error in satellite attitude, it is next necessary to correct it.

All the low level satellites that we cover use **reaction wheels** for this purpose. Rotating the wheel will rotate the satellite in the opposite direction. Three such wheels mounted orthogonally, will enable any required attitude correction to be generated. In the absence of external forces, the attitude control system will return the yaw wheel to rest, at the end of a correction. The pitch and roll wheels should be returned to constant speeds, to generate a constant rotation, as already discussed. In practice, as we have seen, not inconsiderable external forces do exist. As a result, the reaction wheels do not return to rest; and the speed can eventually build up to an uncomfortable degree. When this happens, the excess angular momentum is 'dumped to the Earth's magnetic field' by the use of **magnetic torquers**. These are coils wound round the body of the satellite. When energized, they generate a torque by reaction against the Earth's field. The torque produced is small (for a reasonable current) and dependent on the local magnitude and direction of the Earth's field. It is therefore ill-suited to a direct role in attitude control (although some of the satellites that we cover appear to use it). However the generation of a small torque, even if not too well controlled, can cause the fast-response attitude control system gradually to slow a speeding reaction wheel — while maintaining the satellite attitude under perfect control.

Many satellites (including the NOAA series) install an additional, skewed, reaction wheel as a spare. Rotating it generates rotation about all three axes. However, if any two wheels are still working then they can be used to counter the components of the torque that are not required.

Some satellites (though not the NOAA series) use **momentum wheels**. These rotate continuously to lend gyroscopic stability to a spacecraft. A term such as **biased momentum** is often used to describe this usage; because the system as a whole (including the spinning wheels) possesses a non-zero angular momentum. A fixed momentum wheel can have its speed adjusted to act also as an attitude control actuator, as described above. A fixed-speed wheel can be swung through a small arc, to generate a precession force on the spacecraft. It is then also called a 'control moment gyro'.

Satellites in geosynchronous orbit require a large supply of propellant for station keeping. It is not uncommon for 50% of the initial on-station mass of such a satellite to be propellant. The mass of fuel required for attitude control is relatively insignificant in comparison; and such satellites often avoid the additional complexity of a reaction wheel system by using the station-keeping system for attitude control also.

For completeness, it might be mentioned that some more recent satellites use **solar sails** to control attitude. In the case of the Eurostar satellite, flaps are designed into the **solar panels**. These may be moved to vary the pressure exerted by the solar radiation. Moving the flaps in opposition generates the required torque.

12.11.2 Temperature control

The temperature of the body of the AVHRR instrument is maintained at 15°C (288 K) ±1°, by the use of controllable louvres in front of a radiating surface.

Further cooling, of the thermal-band detectors only, is provided by a two-stage radiative cooler built into the down-sun face. As we have discussed before, the temperature of any object is dictated by the balance of heat transmitted to and from it. Space being effectively extremely cold, a good radiating surface will get moderately cold if it sees nothing but deep space, and if it is insulated from all internal sources of heat (except that generated by an array of infra-red sensors). The latter is a conventional, though difficult, design problem. The sheltering of the 'cold patch' from sunlight and earthlight is the task of the two-stage cooler. On a satellite in Sun-synchronous orbit, the patch can be protected from sunlight by mounting it on the face that is in permanent shadow. But, in general, a screen is necessary to keep earthlight away. However the screen will inevitably 'see' both the patch and the warm earth. If its temperature is allowed to rise, then it will itself radiate to the patch. Therefore the screen must also be cooled, by a separate radiating surface, which is out of the line of sight of the patch.

AVHRR's cooler is the large rectangular aperture, just visible in Fig. 12.1(b). The flap 'above' it is the earthscreen already mentioned. It is lowered, during orbital acquisition, to protect the cooler from direct sunlight. The first stage radiator of the cooler is the blackened ribbed structure visible within the cooler aperture. It is not in fact deeply embedded, as might appear from the picture, and would see very nearly a hemisphere (2π sr) of deep space, were it not for the earthshield. The earthshield is thermally bonded to the radiator, so that the heat that it gains from the earthlight can be dissipated. The heat balance, in normal operation, ensures that the radiator and the earthshield maintain a temperature of about 170 K. The radiating area is 356 cm^2 (55.2 in^2); and, in normal operation, it radiates 1.6 W.

The second stage is the cool 'patch' to which the infra-red detectors are thermally bonded. It is also the external face of the evacuated box described in earlier sections, and is visible in Fig. 12.1(b), occupying the upper middle third of the radiating surface. (Note that less than half the area of the patch is visible in the picture.) The patch is insulated from the main radiating surface, which cannot radiate significantly to it because both are in the same plane. A thermal balance is struck between the heat loads introduced by the rest of the instrument, together with that received from the partly cooled earthshield, and the radiating capabilities of the patch external surface. This thermal balance leads to a 'flat-out' patch temperature, under optimum conditions, of about 95 K. However, as is conventional with these systems, there is no mechanism for controlling the radiator's output. Instead, the load is adjusted, by means of a heater, to enable a constant temperature of 105 K to be maintained as the external conditions vary. The radiating area of the patch is 145 cm^2 (22.4 in^2) and, at 105 K, it radiates 97 mW. Should the radiating surfaces degrade, perhaps through contamination from material outgassed from the structure, to the extent that 105 K cannot be maintained, then provision is made for restabilizing the temperature at 108 K with a slight loss in radiometric performance.

ITT (1982) provides a comprehensive thermal analysis of the behaviour of the AVHRR cooler. From it emerges the interesting fact that most of the heat input to the patch arrives through the structure, and that it is by improving the insulation that additional capacity is gained whenever a channel is added. The second largest heat load is radiation, from the surrounding components, arriving via the optical windows. As we discussed in Section 12.5, this is one of the reasons why the lenses were positioned outside the box. By arranging a degree of convergence on the infra-red beams, the windows can be made significantly smaller. They also, as we also discussed earlier, take the form of filters, to exclude as much of this load as possible.

To put these matters into perspective, however, it must be reported that 35% of the heat load on the patch was expected to be generated by the heater.

12.12   APPLICATIONS

The bands were chosen to suit investigations in the areas of hydrology, oceanography and meteorology. Bands 1 and 2 are designed to discern clouds, land/water boundaries, snow and ice extent and indication of melt inception. The thermal bands are designed for the measurement of cloud distribution and the measurement of surface temperature (cloud or ground/water). The latest addition (Band 5) is particularly useful for eliminating contamination of surface temperature measurements by atmospheric water vapour.

The new Band 6 (1.6 μm), which will arrive with AVHRR/3, is intended to assist in snow/cloud discrimination.

12.13   PRODUCTS

Four standard products are generated in real time by the satellite electronics, and transmitted separately. These are:

High-resolution picture transmission (HRPT)

This is direct transmission of the complete output of AVHRR including telemetry data, via S-band, in real time. It can be received by any ground station within line-of-sight range of the satellite at the time. The data have variable resolution, being 1 km at nadir gradually deteriorating towards the horizon.

Automatic picture transmission (APT)

This is a reduced-resolution product transmitted at VHF frequencies, also in real time, to any user within line-of-sight range. One scan line in three is transmitted, of Bands 2 and 4 only (i.e. one visible and one thermal). The data are averaged along each scan line at a variable rate which roughly corrects for the increasing distortion as the viewing angle increases. Thus the spatial resolution of the APT image is roughly constant. The pixel size is around 4 km, in the across-track and along-track directions. To achieve this effect, the APT image is split up into zones, as shown in Table 12.2. Each zone is derived by averaging the number of raw pixels shown

Table 12.2

Pixel number	Compression ratio	
0–121	1 : 1	Western horizon
122–184	1.5 : 1	
185–268	2 : 1	
269–378	3 : 1	
379–533	4 : 1	nadir
534–642	3 : 1	
643–726	2 : 1	
727–790	1.5 : 1	
791–900	1 : 1	Eastern horizon

derived by averaging the number of raw pixels shown in the compression ratio column. From these figures, it can be seen an APT image line is approx 910 pixels long, and the nadir point is approximately pixel number 455.

Global area coverage

This is a low-resolution product, collected from an entire orbit and stored on a digital tape recorder for transmission to the central ground station at Wallops Island, USA. The product uses one scan line in four, and averages four adjacent pixels to produce one data point. Its resolution is therefore variable, as with the HRPT, but unlike APT.

Local area coverage

This is essentially 10 minutes' worth of HRPT data, recorded over any part of the orbit and transmitted to the central ground station.

An internally generated time code is multiplexed into all digital output formats. Satellite time is kept as close as possible to Greenwich Mean Time, but will not normally be identical to it.

REFERENCES

Ames, A. J. & Koczor, R. J. (1989) 'Imaging components of the 5 channel AVHRR/2', ITT Aerospace/Optical Division. Unpublished report, American Society for Photogrammetry and Remote Sensing annual conference (April, 1989).

Foote, R. & Draper, L. T. (1980) 'TIROS-N Advanced Very High Resolution Radiometer (AVHRR)' *Coastal and Marine Applications of Remote Sensing* Remote Sensing Society (UK).

ITT (1982) 'Advanced Very High Resolution Radiometer: Technical Description' ITT A/O division, Indiana, USA, September 1982.

NOAA (1987) 'The TIROS-N/NOAA A-G Satellite Series' NOAA Tech. Men. 95 (reprinted 1987).

Pick, D. R. (1989) Earnetsat, private communications.

13

The Indian INSAT series

Consultant: **Dr S. V. Kibe**†
Sources: ITT (1977), INSAT (1988)

13.1 INTRODUCTION

The Indian National Satellite (INSAT) is a multi-role satellite, covering not only meteorological Earth observation, but also nationwide telecommunications plus radio and television transmission. Because of its multi-purpose mission, the INSAT satellites are three-axis-stabilized, using sensitive attitude detectors and inertia wheels.

The INSAT system is a joint venture between the Indian Department of Space, the Department of Telecommunications, the India Meteorological Department. All India Radio and the government television networking agency. As with the later IRS project (Chapter 18), INSAT is very much a pragmatic response to India's particular needs. It is basically as good as it needs to be to meet realistic objectives, and designed and built strictly to a budget.

The INSAT-1 system has been in operation since 1983. It involves the deployment of two, or sometimes three, identical geosynchronous satellites, serving different roles, but each available as a spare for the other(s). The primary satellite provides the complete range of services, and is stationed over 74°E, a point on the Equator just west of the axis of the sub-continent. The second is stationed at 93.5°E and, in addition to its role as a spare, augments the communications capacity. The demand for communications capacity has grown considerably faster than anticipated, and when a third satellite is available, it is stationed over 83°E.

The INSAT-1 satellites were built by the Ford Aerospace Corporation of the United States to an Indian specification. In addition to being three-axis-stabilized, the satellites feature an asymmetrical solar array to avoid obstructing the view of deep space seen by the radiative cooler. The asymmetric solar radiation pressure that

† Deputy Director, INSAT project office.

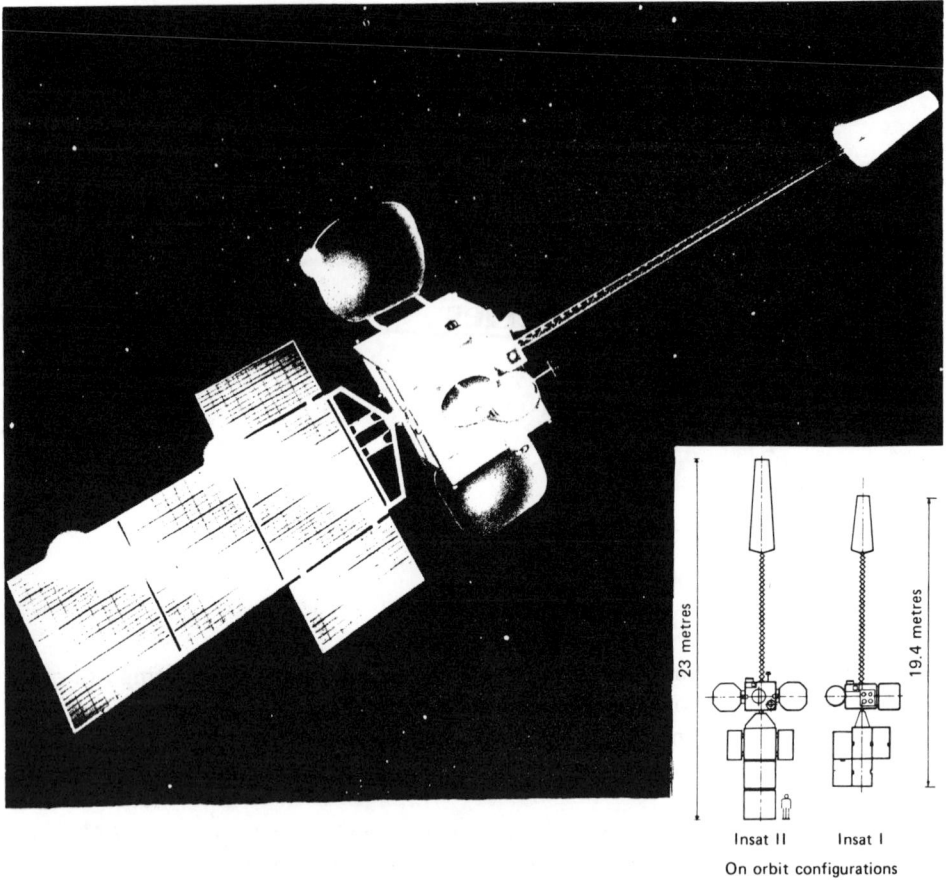

Fig. 13.1(a) — The INSAT satellites. © ISRO.

it generates is balanced by a solar sail, another feature that we have not met before. The satellites are designed for a minimum life of seven years. INSAT-1A was launched by a Delta rocket in April 1982. However, the solar sail failed to deploy, and a series of other mishaps led to total fuel depletion. The second satellite, INSAT-1B, was launched by the U.S. Space Shuttle in August 1983. At the time of writing, it is working well, and playing the primary role. INSAT-1C was launched by Ariane in July 1988, but a partial power failure has led to a loss of half its payload capacity. It lost Earth-lock in November 1989, and is now unusable. The replacement for INSAT-1B, INSAT-1D, was due for launch by a Delta rocket in June 1989. However, the launch was postponed by an accident on the launch pad. It was eventually launched in June 1990.

Of principle interest to us, of course, is the VHRR earth imaging instrument. This is basically a development of ATS-6 GVHRR flown by NASA in 1974. GVHRR is not covered elsewhere in this book. It is a version of the forerunner of AVHRR,

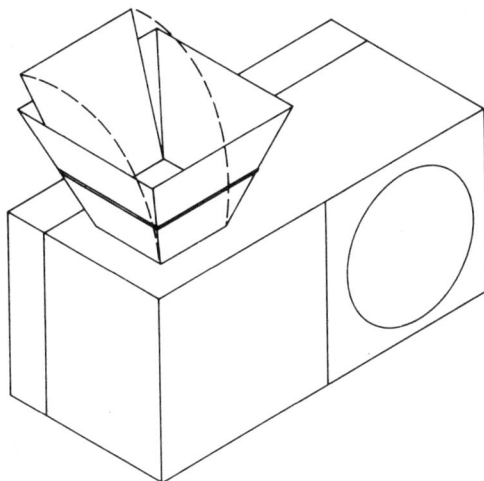

Fig. 13.1(b) — The INSAT VHRR radiometer. © ISRO.

adapted for operation from geosynchronous orbit. The design of INSAT's VHRR also draws on experience gained in the creation of AVHRR. All four instruments were built by ITT Fort Wayne. All but AVHRR provide two moderately broad bands, one visible and one thermal. The main difference between the low- and the high-altitude versions lies in the scanning mechanism. As we have already seen, the low altitude instruments employ the somewhat wasteful but rugged spin-scan system, whereas the geosynchronous versions use a double-gimballed zigzag mechanism. (The ATS-6 also used a high-speed mechanical beam chopper, which served the double purpose of beam splitting and of stabilizing the DC electronics. **Chopper stabilization** was once a regular feature of earthbound DC amplifiers.)

The foreign procured (and launched) INSAT-1 spacecraft will be gradually replaced, during the nineties, by the indigenously developed INSAT-2. This will eventually be launched by the Indian Geo-stationary Launch Vehicle. INSAT-2 spacecraft will be about 50% heavier than INSAT-1, to incorporate increased communications capability, and better resolution for the VHRR. The two initial INSAT-2 spacecraft will be test vehicles, which will co-exist with INSAT-1 satellites. If they work, however, which is of course confidently expected, then their capabilities will be put to full operational use. INSAT-2A is scheduled for launch, by Ariane, in the last quarter of 1990, with INSAT-2B following a year later.

The rapidly increasing demands on all the facilities provided by the INSAT system demand a satellite that is far too large for any realistic indigenously designed launch vehicle. This limitation will continue to force the payload to be split up into several satellites. However, trade-off studies still point to the superiority of the multi-purpose vehicle. The current concept is therefore for a total of four nominally identical spacecraft: two co-located at the prime station of 74°E, with one at 93.5°E and a spare located halfway between the two.

Table 13.1

Spacecraft			
Mass — satellite (kg)	650	Launched	4.82, 8.83, 7.88
— VHRR (kg)	40		
		Target	Cloud sea surface temp.

Orbit			
Altitude @ Equ. (km)	35 786	Orbits/day	1 (per sidereal day)
Semi-major axis (km)	42 164	Orbits/cycle	1
Eccentricity (%)	—	Days/cycle	1
Period (min)	1436	Shift/orb (°)	0
Inclination (°)	0	(km) @ Equ.	0
Descending node	n/a	Gnd Track spacing (°)	0
Ground speed (km/s)	0	(km) @ Equ.	0

Telescope			
Type	Afocal paraboloid+refractive		
Focal length (m)	—	80% blur c. d. (μrad)	—
Aperture (m)	0.203	F.P. dia (mm)	—
f-number	0.81	Field-of-view (°)	20
1y mirror dia. (m)	—	IFOV (μrad)	76.8/307
2y mirror dia. (m)	—	(μm @ F.P.)	51/204
Total obscuration (%)	—	(m on gnd @ nadir)	2750/11000
Clear area (m^2)	—	EIFOV (m on ground)	—
		Interband reg. (pix)	0.25 1.12, pixel

Scanning			
Method	Double zigzag		
Rate (radians/s)	0.175	Mirror swing (°)	20×20
Scan repeat time (ms)	—		
Time for 1 line (ms)	1200		
Efficiency (%)	—		
Sampling interval (ms)	—		
Dwell time (μs)	880		

Channel details (nominal)

Band	Wavebands (μm)	Detectors	Dynamic range (W/m^2/sr/μm)	Noise equiv. reflectance (% F.S.)	MTF @ pix rate across trk
1	0.55–0.75	Si	0—100% albedo	17*	
2	10.5–12.5	HgCdTe		.11*	

Electronics	
Consumption (kW)	0.052
Tape recorder?	No
Compression	No

Downlink	
Data rate (Mbit/s)	0.4
(GHz)	4

Data	
Word length (bits)	6/8
Grey levels	64/256
Image width (pix)	4548
Image width (km)	n/a

*SNR at 2.5÷albedo.
**NEDT at 300 K.

13.2   THE ORBIT

The orbit is a conventional geosynchronous orbit, with the satellites stationed at several positions local to the Indian subcontinent, as discussed earlier. Fig. 13.2 shows the view from all three stations specified for the INSAT satellites.

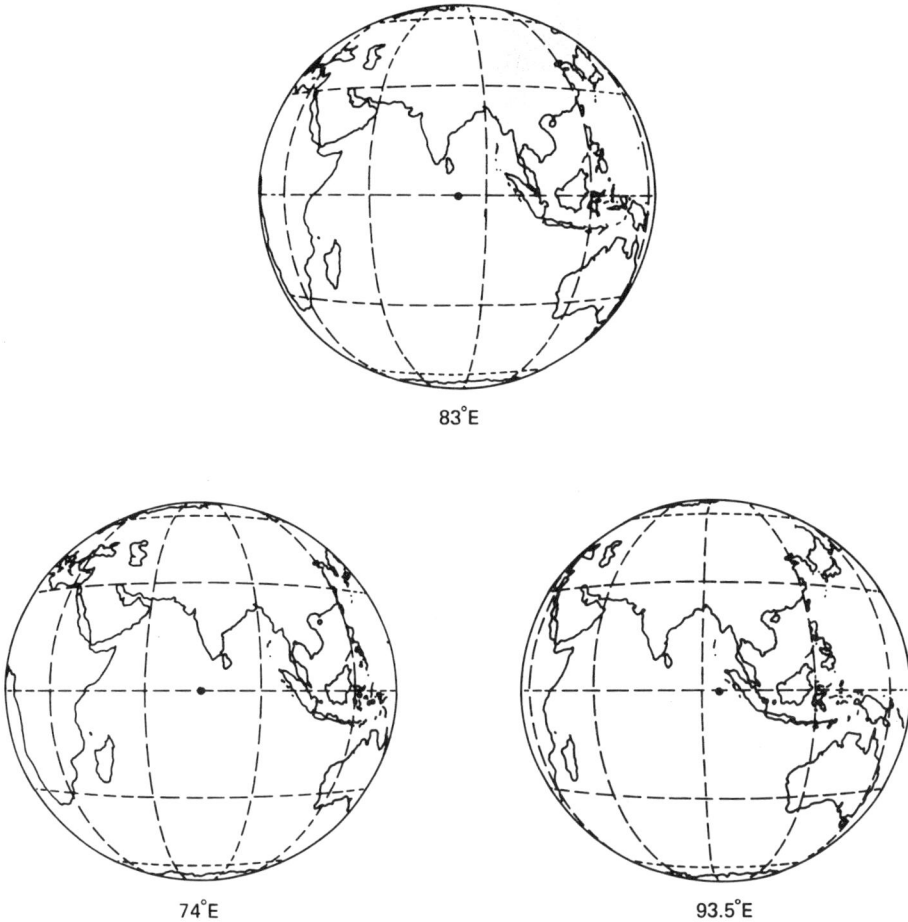

83°E

74°E   93.5°E

Fig. 13.2 — Ground coverage from the three standard INSAT stations.

13.3   SCANNING

The orientation of the satellite being fixed in relation to the Earth, a two-dimensional scanning system is required. The mirror is therefore double-gimballed, and two straightforward servo systems generate a fast zigzag scan east/west and a slow, *incremented*, zigzag scan north/south. Thus there is no high-speed flyback nor is any

potential scanning time wasted. End stops are used to provide positive location of the
nominal starting point for each scan.

An acquisition starts from a corner of the image, and the 'fast scan servo' swings
the mirror steadily east or west at 10°/s (0.175 rad/s) for 20° (0.35 rad). The mirror
velocity is then carefully reversed, over about 0.2 s, during which period the 'slow-
scan servo' increments the assembly northwards (or southwards) through 1'
(0.29 mrad). During this period also, the detectors obtain a view of deep space.
When a frame is completed, the servos move the aiming point to view the calibration
patch, which is mounted on a sidewall of the mirror housing. In addition to scanning a
full frame, the system can also be switched to 'sector' mode in which north/south
scanning is restricted to a selected 5° band.

13.4   THE TELESCOPE

The INSAT telescope is of similar design to that of AVHRR, namely an afocal
confocal paraboloid, as shown in Fig. 13.3. The afocal feature means that the
principle telescope elements do not focus the incoming rays. Instead they produce a
concentrated collimated beam of diameter some $2\frac{1}{2}$ cm. The beam is passed through a
beam splitter, and thence to individual optics for each of the two channels. The fact
that there are only two is part of the reason why liberties may be taken with other
aspects of the instrument's design.

The mirrors are made of ULE fused silica, and the supports are of Invar. An
'athermalization' study was carried out to identify parts of the optical system that
were the most sensitive to thermal changes. These were then modified or redesigned
to reduce their sensitivity. From experience of previous designs, ITT were confident
that no defocusing or registration changes were likely during the operational life of
the instrument.

As in other chapters, the remaining components are considered in the next
section.

13.5   THE FOCAL PLANE

The first element that the rays meet after leaving the telescope is a dichroic beam
splitter. This takes the form of a 'half-silvered' mirror, in fact coated with gold, which
transmits 75% of the Band 1 radiation and reflects 92% of that of Band 2. The visible
radiation passes through an optical filter, and hence to the focussing optics and a total
of eight silicon detectors. The assembly comprises four active detectors and four
spares, all built into the same package. The optical speed of this beam, in the camera
sense, is f8. The thermal radiation is fed to focussing optics and through two windows
into the cool chamber and on to a pair of HgCdTe detectors. Again, one is a spare
and is built into the same package. One of the windows doubles as a band-pass filter.
The speed of the reflected beam is about f3.6, and is increased to f0.9 by the
inevitable field lens.

The 10 optical elements have an average transmission efficiency of about 0.9,
leading to an overall optical transmission efficiency of 0.41.

The detector active areas are all the same, at 51 μm square. This size was chosen
because it is small enough to generate a realistic noise contribution, yet large enough

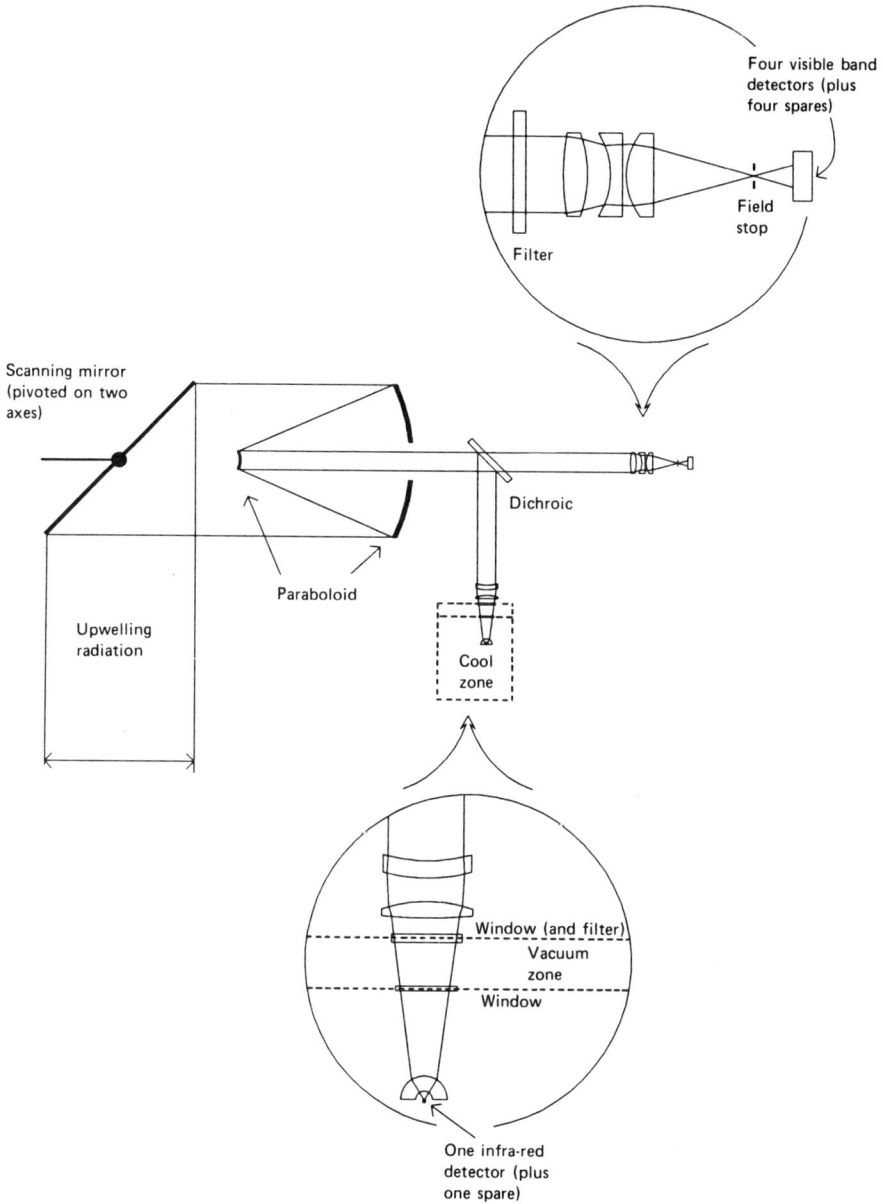

Fig. 13.3 — The INSAT VHRR telescope.

to fabricate. However, an aplanatic field lens (see Chapter 12), made of germanium, is mounted onto the aperture housing of the infra-red detector package. It provides a magnification of 4, and brings the effective infra-red detector size up to 204 μm. The active and spare detector IFOVs are separated by a distance equal to the infra-red IFOV, as shown in Fig. 13.4.

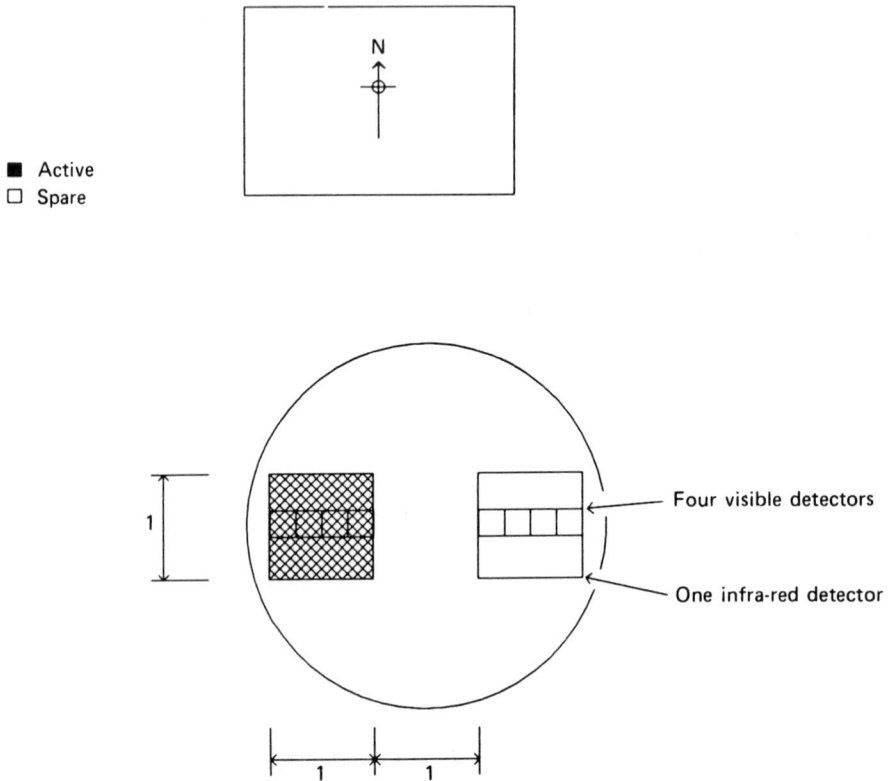

There is an element of conjecture in this diagram

The beam-splitting arrangements are shown in Fig. 13.3

Units in infra-red-band IFOVs (= 204 μm or 307 η μrad)

Fig. 13.4 — The VHRR logical focal plane.

13.6 THE DETECTORS

As we have just discussed, the visible band uses a total of eight silicon detectors (for operational, and four available as spares) covering the band 0.55–0.75 μm. As is normal for these wavebands, they are run at ground-ambient temperatures.

The thermal band covers the normal range of 10.5–12.5 μm, and employs a dual package of HgCdTe photoconductive elements (one active and one spare). They are optimized for a working temperature of 105 K, which is somewhat higher than normal. The temperature was chosen to minimize contamination from moisture escaping from the constructional materials — a problem from which most instruments appear to suffer.

13.7   THE DETECTOR ELECTRONICS

The VHRR electronics are shown in Fig. 13.5. Each channel, including the spares, is provided with an independent electronics chain. The chains are similar to those of AVHRR, with the difference that the Channel 1 amplifiers parallel that of AVHRR's Channel 3.

The silicon photodetectors are driven by a bias supply. They operate at ground ambient, as do their associated electronics. The pre-amplifiers are each a differential pair of FETs, to provide the very high input impedance needed for this type of detector. Further amplification is provided by an operational amplifier before the signal passes through the anti-aliasing filter, shortly to be described.

The infra-red channels use low-impedance HgCdTe detectors which, for that reason, are less susceptible to noise pickup and the effects of stray capacitance. Therefore exotic pre-amplifier designs are not required. Furthermore, for the same reason, the pre-amplifier components do not need to be cooled to cold-patch temperatures, and are mounted remotely. However the HgCdTe detectors exhibit a significant internal resistance, and a constant current bias supply is used to obtain a high impedance source.

DC offset is corrected by the traditional method, described in more detail in Chapter 6. For a short-period at the end of each scan, the detectors view deep space. During this period their output should be zero. However, dark current and amplifier drift will normally combine to generate a finite signal even under these conditions. To counteract this signal, a feedback loop injects an additional signal into the input of the post amplifier. This signal is opposite in sign to that of the detector, and is manipulated by the control loop until the amplifier output is zero. At that point it exactly cancels out the effect of dark current and any other DC offset present. The input voltage required to produce this cancellation is stored in a capacitor. Before the start of the next scan the switch shown in the figure is opened, and the output of the control loop is frozen for the duration of the scan. Thus all pre-existing zero errors are cancelled, leaving only that small amount of drift that occurs during the short period of a single scan.

The anti-aliasing filters are four-pole 'transitional Butterworth Thomson', similar to those used on AVHRR.

Each channel is provided with its own track and hold circuit, so that each may be sampled simultaneously one per IFOV. The single A/D converter is similar to that used in AVHRR, and it encodes the digital data stream into 10-bit words. House-keeping data are multiplexed into the data stream at the end of each scan. A digital buffer is used to enable a smooth flow of data to be presented to the downlink.

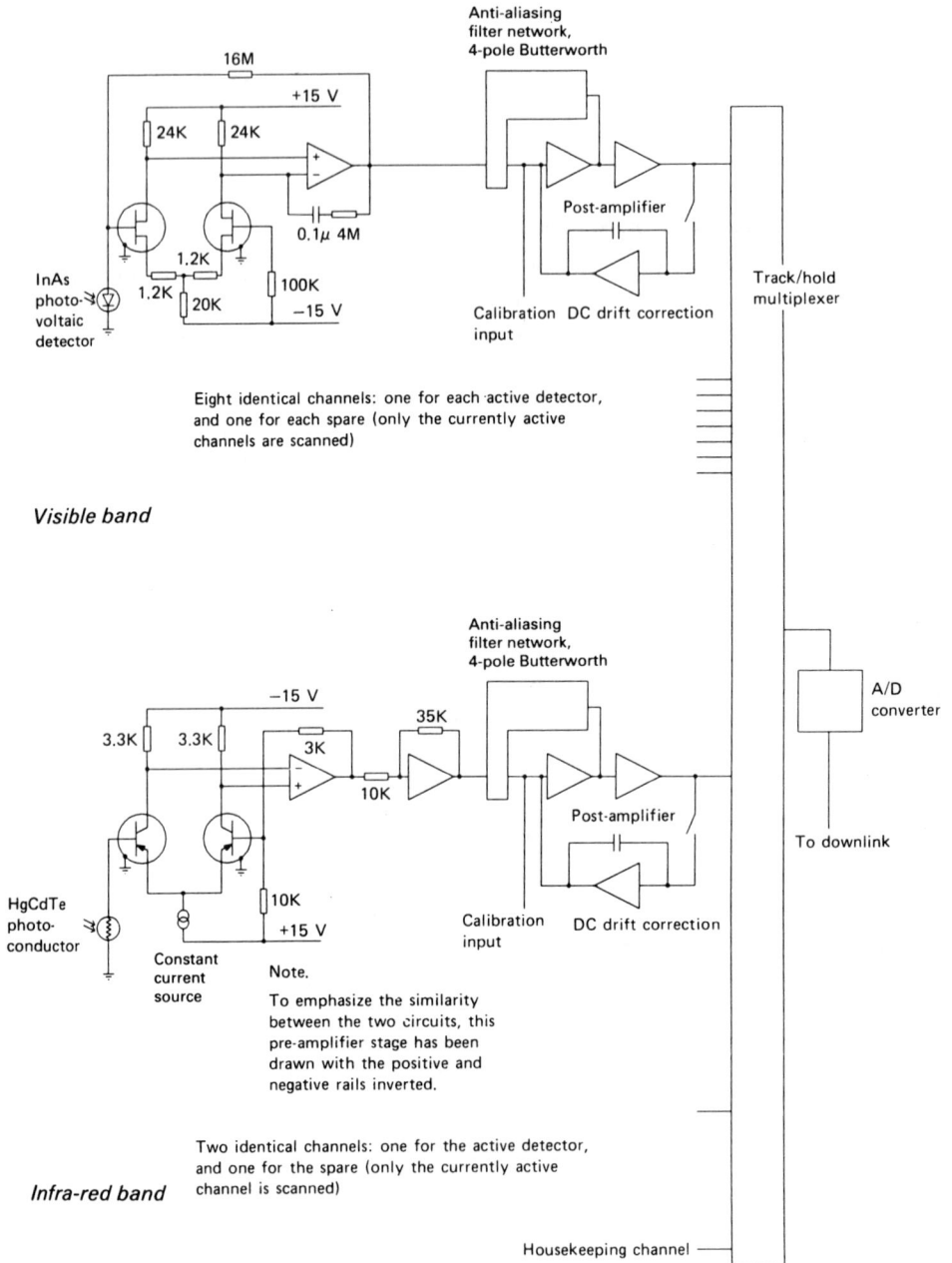

Fig. 13.5 — The VHRR electronics.

13.8 THE DOWNLINK

From the VHRR output, the 400-kbits/s pulse code modulated signal passes to the output multiplexer. Here it is multiplexed with the housekeeping data and transmitted down a 4-Ghz C-band link to the master control facility at Hassan near Bangalore.

13.9 CALIBRATION AND IN-FLIGHT CHECKS

The detectors view deep space once per line; and a black body once per frame.

The black body is in fact a black honeycomb target, larger than the mirror, and mounted into the side of the mirror housing. As discussed earlier, the mirror slews to view the target at the end of each frame. Its temperature is monitored to within $\pm 0.1°C$ by five platinum resistance thermometers. To the extent that it represents a genuine black body, this device provides an absolute one-step calibration of the thermal band chain.

In common with many other instruments however, no provision is made for absolute calibration of the visible channels. This is because no technique was available that was capable of generating the high degree of confidence required. To provide a *relative* calibration capability, such as installed on many Earth-observation instruments, was not thought to be justified for a meteorological mission.

At the end of each line, the detectors obtain a view of deep space. This provides an effective zero input to enable the electronics to be reset to zero output.

Non-linearities are calibrated out, in the traditional manner, by providing a sequence of equispaced voltages. One voltage from the sequence is injected into the analogue data stream at the beginning of each scan line.

13.10 DATA QUALITY (AND QUALITY CHECKING)

13.10.1 Radiometric performance

ITT (1977) provides a comprehensive analysis of the expected performance of each channel, carried out for the purposes of securing the order. No information is available on actual performance. However, the INSAT project office have expressed themselves fully satisfied with the performance of the instrument.

13.10.2 Spatial frequency response (see Fig. 13.6)

The telescope MTF was predicted to be about 68% at a spatial frequency of 0.5 cycles/IFOV for both bands. This implies that the increased IFOV of the infra-red band almost exactly compensates for the degraded resolution at these longer wavelengths.

The filters are designed to have cut-off frequencies (3-dB-down points) that match this spatial frequency almost exactly. For the visible band this emerges as 2.276 kHz, whereas for the IR band it is 569 Hz.

13.11 THE HOST SPACECRAFT

INSAT is a three-axis-stabilized geosynchronous satellite, operating over India. The same propulsion system provides power for both the orbital transfer manoeuvre and

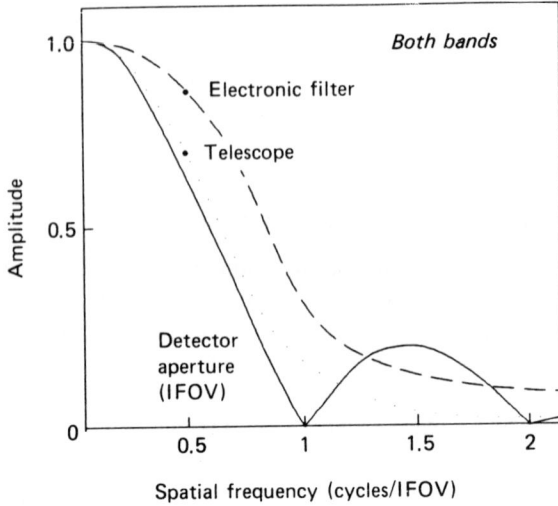

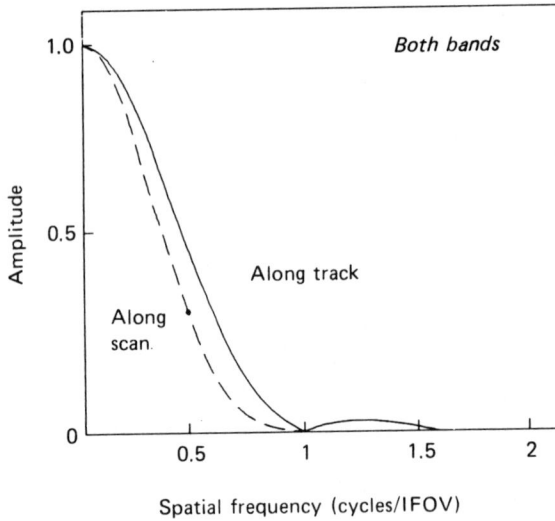

Fig. 13.6 — The VHRR spatial responses.

station and attitude maintenance. Transfer is made in three-axis-stabilized mode, to avoid sunlight reaching the cooler.

The total deployed length of INSAT-1 is about 19.4 m. Its start-of-life on-station mass is 650 kg.

Its solar panels generate 1200 W at start of life and about 930 W at the end.

13.11.1 Attitude control

The attitude control system provides 'biased momentum three-axis stabilization', using two skew pitch momentum wheels. A third, yaw, reaction wheel is available as a back-up. A magnetic torquer, with a current coil placed round the periphery of the satellite body, provides fine control. A microprocessor-based attitude control system employs duplicated infra-red scanning Earth sensors to sense attitude.

13.11.2 Temperature control

The two-stage radiative cooler is similar to that used on AVHRR. However, it did not prove possible to keep the solar panels clear of the cooler's field of view, and so extra shielding was added to keep out radiation reflected therefrom. In addition the shutter used to protect the cooler during launch can be closed during later decontamination cycles.

Silver Teflon is used, in place of white thermal paint, on the radiating surfaces of the shields and cooler housing. This material has been shown to maintain its performance better over long periods.

The use of multilayer insulation, a common feature of satellite design, is abandoned on VHRR in favour of gold-plated metallic shields. Readers will be aware of the problems experienced on most instruments with moisture escaping from the spacecraft structure and settling on the cold optics. It is obviously particularly difficult to keep moisture out of multilayer materials, and so this is a natural change, provided that adequate thermal performance can still be obtained.

The outer casing of the cooler is maintained at 0°C. The first stage proper operates at 156 K, and radiates 0.66 W. The second stage is maintained at 105 K and radiates 47.3 mW. As has already been discussed, this temperature is slightly higher than is usual. The taking of this liberty is made possible by the relatively easy INSAT specification. A higher operating temperature reduces the tendency of the optics to collect contamination. This operating temperature is some 7° higher than the flat-out capability of the cooler, and temperature is controlled in the conventional manner by the use of heaters. Although the performance of the cooler was expected to degrade in flight, this margin was felt to be more than adequate. In any case some at least of the cooler performance loss may be expected to be recovered when the temperature is raised to decontaminate the optics.

The cooler incorporates three heaters, two on the second stage and one on the first. One of the second stage heaters is a low power device and is used to maintain the patch at its working temperature of 105 K. A platinum resistance thermometer, on the patch, drives a proportional control loop. The controller adjusts the heater current to maintain the patch temperature constant. The remaining heaters are both 'area heaters' capable of raising the entire cooler assembly to about 290 K for outgassing and decontamination purposes. During testing, the heaters were also used to simulate expected heat loads from solar and Earth radiation. Temperature regulators, also using platinum resistance thermometers, are incorporated to prevent any possibility of accidental overheating. Thus the second stage is in fact provided with both a wide-range and a close-range temperature controller, to cover the two different regimes.

The avoidance of contamination of the optical elements is regarded as particularly important, because there are indications that its effects are not completely reversed by heating. Part of the function of the windows into the cooler (see Section 13.4) is to eliminate outgassing paths past the cold optics. The volumes within the cooler can outgas only directly to space. 'Cold traps' are built into areas where external contamination could enter the cooler. The first stage window, which is also the infra-red filter, is maintained at a nominal 10°C above its mounting temperature. It is protected by a **cold trap**, comprising a small patch maintained at the mounting temperature, to collect contaminates that get that far. This temperature difference should provide an order of magnitude difference in the contaminant vapour pressures around the two items. The optical elements on the second stage operate slightly above the temperature of the main cold patch, and are similarly protected by a trap maintained at patch temperature. The outer window is mounted on the cooler housing, and is heated to ground ambient (20°C) to eliminate totally any risk of contamination.

The cooler was maintained at ground ambient temperatures, by means of the heaters, for the first few weeks after launch, to allow outgassed contaminants to dissipate.

A similar thermometer monitors the temperature of the calibration target.

13.12 SPACECRAFT OPERATION

In view of its multi-function role, the operation of INSAT is the joint responsibility of the Departments of Space, Telecommunications and Meteorology, together with All India Radio and the government television networking agency. However, the running of the spacecraft operation is delegated to the Department of Space. The satellites are operated from the INSAT master control facility at Hassan in Karnataka State. This is towards the tip of the sub-continent, roughly on a level with Madras and Bangalore. The facility incorporates the complete range of ground station capabilities. However specialist monitoring, online, is also carried out at the Indian Space Research Organisation (ISRO) Satellite Centre at Bangalore.

Meteorological data are relayed direct to the Meteorological Data Utilization Centre in Delhi, from where the raw and processed data are relayed by ground link to a number of large earth stations dotted around the sub-continent. The raw data can also be received by any small user equipped with suitable inexpensive equipment. In addition a complex network of other links, involving all the satellites available at any one time, close the communications links and disseminate the radio and television signals. Improved telecommunications, particularly to and from the less accessible areas, has been one of the major benefits brought about by the INSAT mission. The unifying effect of improved broadcast reception is also not without significance.

On the whole the INSAT project emerges as a major contribution to many aspects of Indian development. Its meteorological role is possibly far from being its most important function.

13.13 APPLICATIONS

The meteorological component of the INSAT system provides round-the-clock, half-hourly images of the hemisphere. From these — also half-hourly — synoptic images are generated, on the ground, featuring severe weather, cyclones, sea-surface and cloud-top temperatures etc. over the entire sub-continent as well as adjoining land and sea areas. These are transmitted back to the satellite, and broadcast. Explicit storm and disaster warnings are also disseminated some 12–23 hours in advance of what was possible in pre-satellite days.

As is usual in these matters, INSAT also relays data from unattended remote platforms.

13.14 PRODUCTS

Incoming imagery is processed to produce the usual range of products.

Cloud-cover images are produced at least once every three hours, and more frequently if weather conditions demand it. By day, both visible and infra-red images are produced. Night-time imagery is infra-red band only. These images are particularly useful for identifying weather systems developing over the oceans, where no other observations are available.

Cloud motion vectors are derived from consecutive images by observing the movement of identifiable cloud structures. They are derived by a fully automated pattern-matching technique.

The ability to estimate sea surface temperature from a single infra-red band is inevitably limited. However, experimental derivations have been compared with equivalent data derived by the USA, and have indicated an accuracy to within 1°C. Attempts have been made to generate sea surface temperatures on an operational basis, with encouraging results.

It has been found possible to produce precipitation figures, from the three-hourly INSAT data, which are accurate when averaged over large areas ($2\frac{1}{2}°$ of latitude/longitude square) and over several days. This figure is, again, particularly valuable over oceans, where traditional readings are not available. Another derivation of interest is the 'outgoing longwave radiation'. This is derived over $2\frac{1}{2}°$ squares, as before, and averaged over a month. It is expected that these two measures together will provide a valuable input to medium-range weather prediction models.

REFERENCES

INSAT (1988) 'The Indian National Satellite system' INSAT co-ordination committee, 1988.

ITT (1977) 'INSAT VHRR' proposal for the design and manufacture of the INSAT imaging instrument. ITT Aerospace/Optical Div. Fort Wayne, Indiana, USA.

14

Thematic Mapper

Consultant: **Dr Robert Murphy†**
Principle sources: Hughes (1984), NOAA (1984)

14.1 INTRODUCTION

The first Thematic mapper (TM) was flown on Landsat 4 in July 1982, and the second

Fig. 14.1(a) — Landsat 4/5. Courtesy: NASA.

on Landsat 5 in March 1984. During the early years, TM and MSS (Chapter 8), were operated and managed by the Goddard Space Flight Center of NASA, with data disseminated from the EROS Data Center, South Dakota. Imagery was made

† Chief, Land processes branch, NASA.

Scan mirror
assembly

Electronics boards

Scan mirror

Multiplexer

Redundant shutter

Radiative cooler

Secondary mirror
assembly

Sun shade

Secondary mirror

Primary mirror

Focal plane array

Along-track
direction

Alignment and
focal assembly

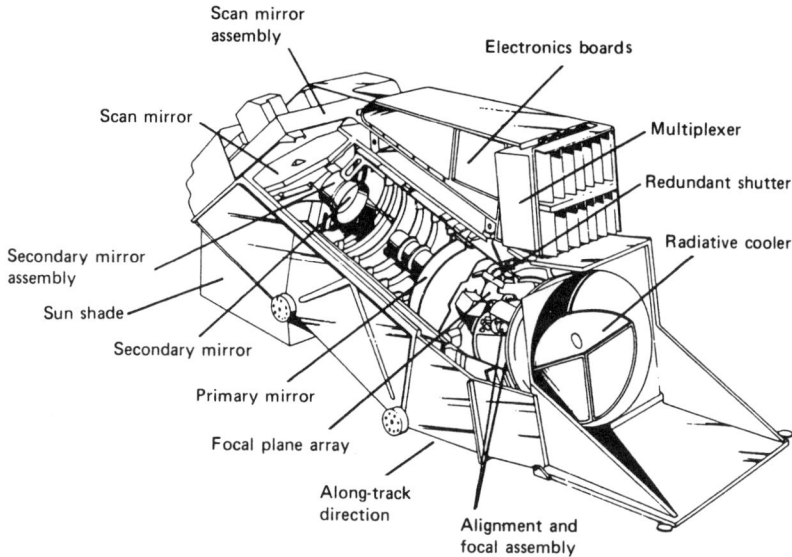

Fig. 14.1(b) — Thematic Mapper. Courtesy: NASA.

available virtually for the cost of running the ground segment. However, in 1985 a company was set up to exploit the commercial potential of MSS and TM data, and to run the Landsat system. Since about 1987, Eosat Inc of Maryland USA has been the sole outlet for new Landsat data. However, at the time of writing, prices are still based on what it is thought that the market will bear, and Eosat are still in receipt of substantial subsidies.

The TM project could be said to have started as early as January 1970, when a Hughes Aircraft Company report to the Goddard Space Flight Center suggested an advanced seven-band scanner with an IFOV of 37 m. The suggestion was taken up. Breadboard work on the instrument was started in June 1974. The Hughes Aircraft company was awarded the contract, and embarked on the development programme in the spring of 1977. This was about 18 months before AVHRR (Chapter 12) first flew.

One of the things that most strikes the chronicler of these events is the massive amount of attention devoted to TM as compared with the other instruments covered. The instrument was specified with the utmost care, and was tested with extreme thoroughness. Hughes (1984), for example, describes the problem of producing sufficiently high quality equipment for testing the instrument as rivalling that of producing the instrument itself. The topic was also seized upon by the scientific community with an avidity that filled countless specialist journals to bursting point.

TM represents such a major advance in so many areas of technology, that it is difficult to give a succinct summary. However, the principle feature of the instrument which singles it out from all others in its class is its combination of good spatial resolution and wide spectral range. As we have already seen in Chapter 3, such a combination demands a high degree of freedom from chromatic aberration. This in turn requires a fully reflecting telescope system. At the state of the art at the time, detector technology did not permit the use of a large-array 'pushbroom' system such as is a feature of the later SPOT (Chapter 16) (Murphy, 1989†). Therefore active scanning is required which must be very well done if acceptable ground cover and inter-band registration are to be combined with good reliability.

The chief characteristics of TM are given in Table 14.1 The instrument covers six reflective bands from 0.45 to 2.35 μm, plus one thermal band at 11 μm. The pixel size is 30 m for the reflective and 120 m for the thermal band. The selection was based on the need to provide at least three bands in ranges suitable for a number of different applications which, in the nature of things, tended not to coincide. In the event priority was given to agricultural applications, followed by those of geology and hydrology (Murphy, 1989). The bands chosen also had to avoid regions of atmospheric absorption. The bands were required to be narrow, both to assist in the avoidance of absorption bands, and also to improve the effectiveness of classification techniques. The green and red bands were made narrower than their MSS equivalents to improve their sensitivity to agricultural phenomena (Freden & Gordon, 1983).

Because of the late addition of an extra mid infra-red band, the bands are not numbered in an entirely logical order. The thermal band is numbered Band 6, while the late-added 2.1-μm band, is numbered 7.

TM basically comprises a Ritchey Chretian telescope preceded by a scanning mirror which follows a symmetrical zigzag motion. Acquisition takes place on both sweeps, and an ingenious rotating periscope device deflects the incoming beam and eliminates the overlaps and gaps that would otherwise occur at the ends of the scan. The focal plane is occupied by an assembly of 100 individual detectors, comprised of six assemblies of 16 detectors each; and one assembly of four lower-resolution detectors in the thermal band. For space reasons, detectors are offset by amounts equivalent to a whole number of pixels. Thus interband registration is assured without resampling (although a degree of resampling may be required; see Section 14.5). The pixel numbering has to be rearranged to bring the bands back into line. The detectors covering the four shorter wavebands are mounted at the 'prime' focal plane itself. The longer-wavelength detectors are mounted at a second focal plane positioned on the active surface of a radiative cooler. The cooled focal plane is imaged onto the prime focal plane by means of relay optics.

A word of warning is appropriate before the reader embarks upon this very important chapter. The organizations behind most of the instruments covered are relatively compact, to the extent that the fountainhead of all knowledge is to be found at one or two closely related sites. It is reasonably easy therefore for the chronicler to encapsulate the available material and have his text vetted, updated and given an unofficial 'seal of approval' by said fountainhead.

However, the effort devoted to TM was and is so vast, and so diffuse, that it has

---

† In Chapter 3, it is stated that reflecting systems are not suitable for wide field applications such as SPOT or a hypothetical pushbroom TM. This is the view of Matra, who have built (or been involved in) three of the intruments described in this book. NASA however do not necessarily agree. The author has been advised (Murphy, 1989) that reflecting systems can 'easily handle 5° field-of-view' although problems are generated with focal plane curvature.

Table 14.1

Spacecraft			
Mass — satellite (kg)	—	Launched	6.82
— TM (kg)	244		3.84
		Target	Land

Orbit			
Altitude @ Equ. (km)	705	Orbits/day	14.5625 $(14\frac{9}{16})$
Semi-major axis (km)	7083	Orbits/cycle	233
Eccentricity (%)	—	Days/cycle	16
Period (min)	98.9	Shift/orb (°)	24.73
Inclination (°)	98.2	(km) @ Equ.	2752
Descending node	9:45	Gnd track spacing (°)	1.545
Ground speed (km/s)	6.79	(km) @ Equ.	172

Telescope			
Type	Ritchey Chretien		
Focal length (m)	2.44/1.22	80% blur c. d. (μrad)	5
Aperture (m)	0.406	F.P. dia (mm)	19.5
f-number	6/3	Field-of-view (°)	±0.27
1y mirror dia. (m)	0.4115	IFOV (μrad)	42.5[†]
2y mirror dia. (m)	0.157	(μm @ F.P.)	103[†]
Total obscuration (%)	44.8	(m on gnd @ nadir)	30[†]
Clear area (m^2)	0.1056	EIFOV (m on ground)	32–36[†]
		Interband reg. (pix)	±0.2 or better

Scanning			
Method	Corrected zigzag		
Rate (double scans/s)	7.00	Mirror swing (°)	±3.85
(rad/s)	4.42		
Period (ms)	142.9		
Active scan time (ms)	2×60.743=121.486		
Efficiency (%)	85		
Sampling interval (ms)	0.0013		
Integrating time (ms)	0.0096		

Channel details (nominal)

Band	Wavebands (μm)	Detectors	Dynamic range[‡] (W/m^2/sr/μm)	Noise equiv. reflectance (% F.S.)	MTF @ pix rate trk/scan
1	0.45–0.52	Si	−1.5–152	0.8	0.46/0.34
2	0.52–0.60	Si	−2.8–297	0.5	0.46/0.34
3	0.63–0.69	Si	−1.2–204	0.5	0.46/0.34
4	0.76–0.90	Si	−1.5–206	0.5	0.46/0.34
5	1.55–1.75	InSb	−0.4–27.2	1.0	0.42/0.30
7	2.08–2.35	InSb	−0.2–14.4	2.4	0.42/0.30
6	10.4–12.5	HgCdTe	200–304 K	0.5 K[§]	0.48/0.35

Electronics		
Consumption (kW)	0.335	
Tape recorder?	No	
Compression	No	

Downlink		
Data rate (Mbit/s)	85	
(GHz)	8.2 (X-band)	

Data		
Word length (bits)	8	
Grey levels	256	
Image width (pix)	6000	
Image width (km)	185	

[†]For Band 6, multiply by 4.

[‡]These are the figures generated by the 'TIPS' processing system, after Jan. 1984, from 0 and 255 digital counts respectively (NASA, 1984).

[§]NEΔT at unspecified temperature.

not proved possible within a reasonable timescale to achieve anything like the same degree of certainty in the initial encapsulation. The attempt to impose a reasonable degree of brevity, lucidity and order into this chapter inevitably meant resorting to many interpretations and simplifications, and it has by no means always been possible to check these out. In some instances material inserted at the behest of one referee raised objections from another.

This chapter is offered, therefore, as a general introduction to TM, as a chronicle of its unique features, and as a comparison with a number of equally interesting (if simpler) other instruments. The student who requires an impeccable understanding of every aspect of this remarkable instrument is strongly advised to refer to the original references.

14.2 THE ORBIT

The orbit of the TM-carrying Landsats is unusually low, compared both with earlier vehicles of the series, and also with the norm for Sun-synchronous missions. A lower altitude was chosen mainly to facilitate access from the Shuttle for servicing. Although this strategy has never actually been implemented, a repair mission was considered in 1986 (Murphy 1989). The plan was abandoned, however, because it proved too costly for the relatively small performance degradation being experienced. However a lower orbit also eases the problems of achieving the resolution requirements of TM. The price of course is reduced orbital life or, more to the point, frequent altitude boosting.

The swath width of TM is the same as that of MSS, and so the number of orbits per cycle is similar (Chapter 1). The cycle chosen for Landsats 4/5 was 233 orbits in 16 days. This gives rise to k (orbits/day) of $14\frac{9}{16}$ or 14.56, an inclination of 98.22°, and an altitude of 705 km at the Equator. As discussed in Chapter 1, the ground-track pattern for a Sun-synchronous orbit is dictated by the fractional part of k. For the early Landsat missions, k was $13\frac{17}{18}$, which meant that tomorrow's orbits were displaced by 17 ground tracks to the east of today's, or one ground track to the west. For Landsats 4/5, however, tomorrow's orbits are displaced by nine tracks to the east, or seven to the west, so that they split almost equally the gaps left by today's. The complete pattern is shown in Fig. 14.2.

The altitude varies slightly with latitude, on account of the oblateness of the Earth. The scanning parameters of TM were optimized for an altitude of 712 km, which is the nominal altitude at 40°N.

14.3 SCANNING

As may already be apparent from a study of earlier chapters, the scanning method employed is both dictated by, and has a direct bearing on, the design of other parts of the instrument.

The TM scanning system is described as a 'hybrid' between the single-detector 'whiskbroom' approach, employed by meteorological instruments such as AVHRR, and the 'pushbroom' approach popularized by SPOT. The hybrid approach employs a small assembly of detectors, over which the image is whisked from side to side by means of a scanning mirror. Because more than one detector is scanned on each

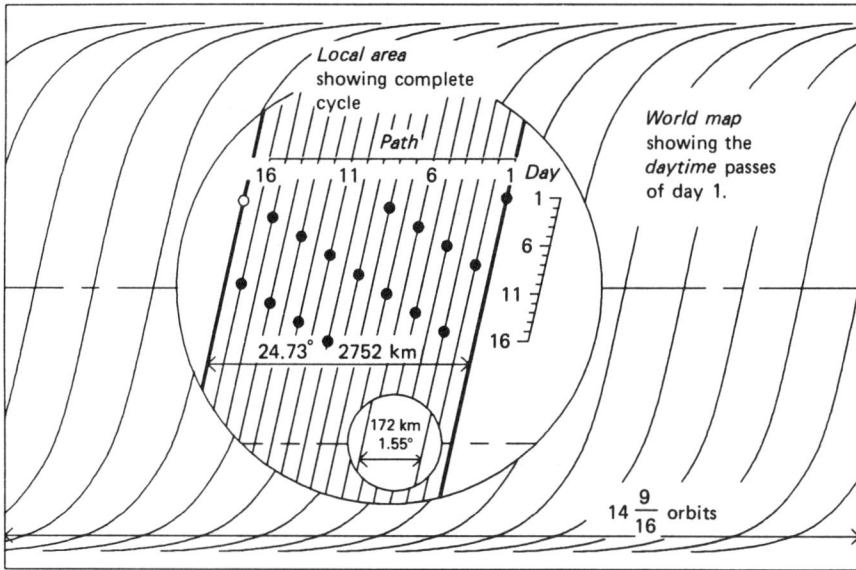

Fig. 14.2 — Ground-track pattern for Landsats 4/5.

stroke, the speed of the scanning motion can be slowed down in proportion, thus increasing the exposure and reducing noise.

High-speed flyback is not practicable in satellite-borne instrument because of the large momentum reversal generated. As is discussed in Chapter 8, the MSS scanning mirror returns at scanning speed, thus losing half the available exposure time. This time is saved, on TM, by using both the forward and reverse strokes of the mirror for active scanning. The total period of a complete double scan is 142.3 ms; the active period of each scan is 60.17 μs.

Thematic mapper employs 16 detectors in each of the reflective bands, which leads to a 32-fold improvement in exposure over a simple whiskbroom concept, and to an eight-fold improvement over MSS (Chapter 8). If other design features remained the same (which of course they do not) then this should lead to a directly comparable improvement in signal to noise ratio. However the thermal band (Band 6) employs four detectors to cover the same strip width. This is because the weaker signals available at thermal wavelengths, and the lower efficiency of thermal detectors, means that a larger exposure is required (Murphy, 1989). The IFOV for Band 6 is therefore four times larger than that for the reflective bands.

The symmetrical zigzag scanning motion, in which both sweeps are active, would naturally result in overlaps and gaps in coverage at the ends of each scan. To overcome this problem a 'scan line corrector' is introduced between the telescope secondary mirror and the focal plane. The scan line corrector consists of a rotating periscope device, as shown schematically in Fig. 14.3. Its operation is best understood by fashioning a simple instrument out of cardboard and two plane mirrors. The experiment is to observe an object, through the periscope, from a close distance. If the instrument is rotated slightly about the vertical axis, with the head remaining still, then the operation of the scan line corrector becomes clear. The image is translated linearly, in the direction of rotation, through a distance related to the separation of the mirrors and the angle of twist. Note that there is no rotation of the image. The emerging ray remains parallel with the incoming ray. (There is a slight defocussing effect due to movement of the focal point, but the effect is well within limits.) At the start of a scan, the TM scan corrector is angled such that an area slightly ahead of the nadir point is imaged onto the focal plane. As the scan progresses, the device rotates to translate the line of sight backwards, at precisely the rate that the satellite is moving forwards. At the end of the scan, the periscope executes a fly-back, to be correctly oriented for the start of the next scan. The motion required of the corrector is the same for forward and backward scans. Before this particular design was chosen, trade-off studies were carried out featuring both simpler and more complex designs.

Data acquisition, the operation of the scan line corrector, and the other scan-related features are all triggered by position sensors on the scanning mirror assembly, and are thus synchronized with it. Because of the importance of this synchronization, a back-up system obtains its information from pick-offs in each end-of-stroke bumper assembly.

It will be observed from Fig. 14.3 that the TM instrument is not mounted parallel to the ground, and that the incoming radiation is deflected through 110° instead of the more usual 90°. This is to reduce the size and weight of the scanning mirror. It is easy to see that the closer the deflection angle approaches to 180°, the smaller and lighter may the mirror be, and therefore the lower its moment of inertia. However, as the scanning mirror oscillates then the entire instrument, being unsupported in space, must oscillate (at a much lower amplitude) in antiphase with it. This has the effect of shortening the scan lines and reducing the swath width. The effect can be corrected for by increasing the mirror swing. Now, reducing the moment of inertia also lowers the impact on the buffer stops at the end of each swing and reduces the vibration transmitted to the rest of the structure. On the other hand, increasing the deflection angle increases the incidence of stray light being reflected into the telescope. Larger baffles are required round the secondary mirror, which increase its effective diameter and hence its obscuration It was found that the use of a 70° deflection angle gave a useful reduction in moment of inertia, without a significant increase in obscuration.

The scanning mirror is elliptical. It is made from pure beryllium, flat on both faces but hollowed out with an eggcrate structure. It was found, however, that bare beryllium could not be polished to a sufficiently high degree to avoid excessive scattering, so the faces were plated with 0.005 in (0.13 mm) of nickel. (Both faces were plated to equalize thermal stresses.)

The original design for the mirror comprised a beryllium eggcrate structure with a beryllium plate brazed to either side. Unfortunately, thermally induced bimetallic

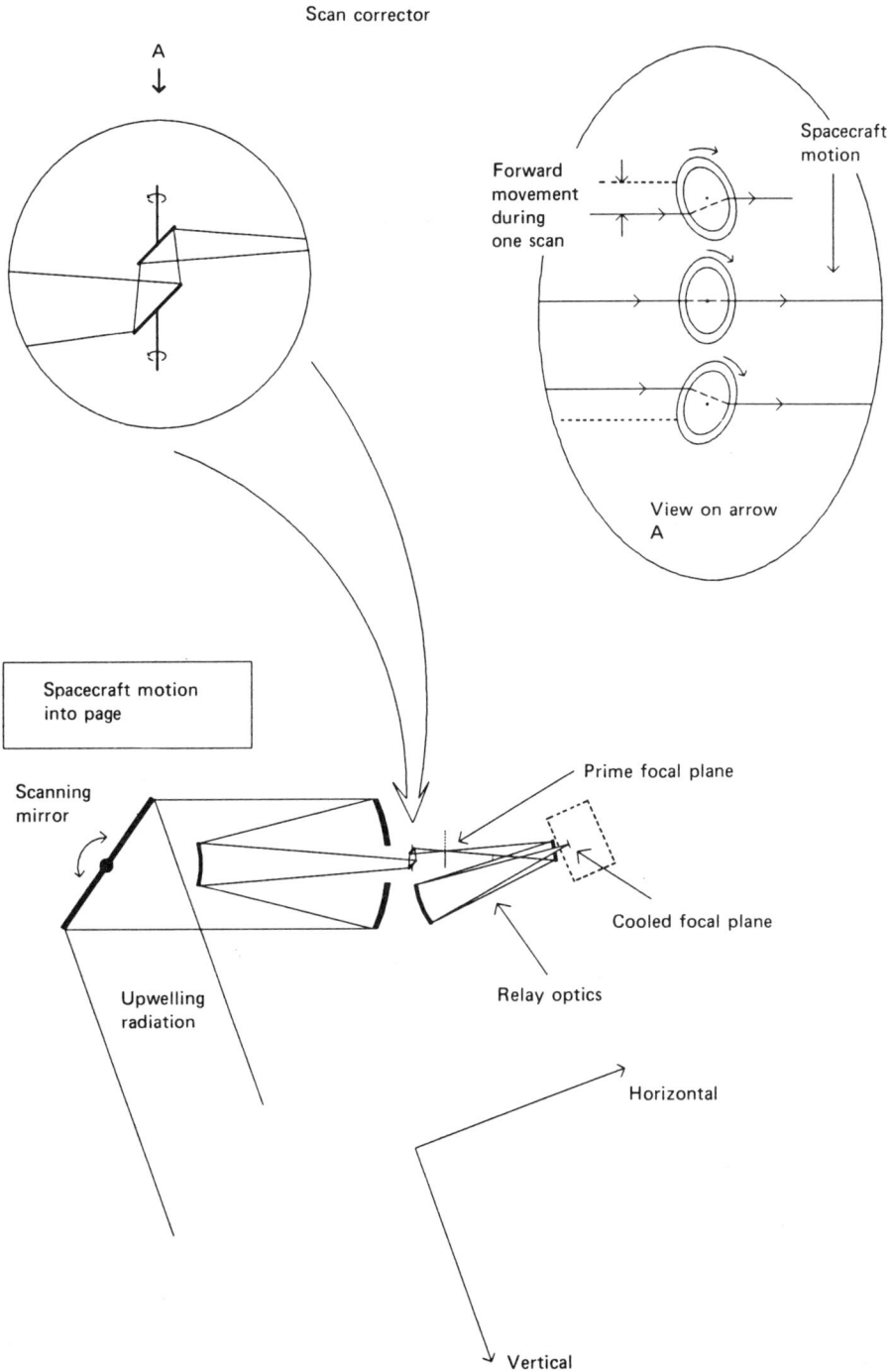

Fig. 14.3 — The TM telescope.

stresses distorted the mirror face in between the stiffening bars. The loss in flatness was small, but it was sufficient to put the optical performance of the mirror outside tolerance. A second design involved machining the eggcrate pattern into the back of two thicker beryllium sheets, which were the brazed together. This new design brought the degree of surface distortion well within tolerance.

Two particular design steps were taken to ensure that the scan velocity was maintained constant and equal for both directions of scan. First, the mirror is kicked into motion at the start of a scan and then left to swing freely. Second, magnets are used to counteract the forces in the flex pivot springs, thus leading to an almost friction-free swing (Engel & Weinstein 1983). Hughes (1984) gives a detailed table of the trade-offs that led to the chosen design for the drive for the scan mirror drive. Other problems associated with the design of the scan mirror assembly are also described.

14.4 THE TELESCOPE

Both reflecting and refracting designs received initial consideration for the TM telescope. However, refracting designs ruled themselves out early because of the wide spectral range demanded of TM. The only materials transparent over the range from 0.45 to 12.5 μm are halogen salts, which are hygroscopic. This problem might have been overcome by stringent environmental control during all ground handling phases. However, the level of chromatic aberration could not have been reduced to an acceptable level, except by the use of so many lens elements that transmission losses would have been excessive (Hughes, 1984) (the IRS telescopes use eight lenses to cover the bandwidth of a single band).

Of the wide range of reflective configurations available, a two-mirror layout was selected as being the optimum compromise between bulk and light loss due to the existence of multiple reflective surfaces. The Ritchey Chretien telescope represents the ultimate in two-mirror design, and employs a hyperbolic profile for both. This design provides a greater degree of control over off-axis quality, as is discussed in more detail in Chapter 3. This is important for any but a single detector per band instrument such as AVHRR. It provides an image that is more than adequate for the requirements, being really excellent on the axis, as well as extremely good up to about ±1° away from it (Reulet, 1988). (It is indicative of the subjective nature of many of these 'rules of thumb' that Chipaux (1990) quotes the field of view of the Ritchey Chretien telescope as ±0.4°.) The layout of the telescope is shown in Fig. 14.3. Its effective focal length is 244 cm, and the aperture is 41 cm (f/6).

The focussing mirrors are made from ultra-low-expansion (titanium silicate) glass. The smaller mirrors were made from solid glass blanks. The main mirror was hollowed out in a similar manner to the scan mirror, though with much greater difficulty on account of the material in use. The mirror was made from three components, all fabricated separately. These were the two face sheets, the eggcrate structure, and an outer ring with mounting lugs. The parts were then fused together, eventually successfully, through successive steps of heat sealing and 'slumping', and then annealed. Finally the correct profile was ground and polished into one face. It was found virtually impossible to avoid a degree of edge damage to the blanks during subsequent handling and finishing. Each chip was carefully inspected, both for surface distortion optically and for potentially damaging cracks, and then sealed with black paint. Finally, the mirror surfaces were overcoated with silver to give a high reflectance over the full spectral range of TM (Engel & Weinstein 1983).

The telescope structure was fabricated from a graphite epoxy composite material, which is stiff, light, and has a very low thermal expansion which matches that of the ULE glass. A feature of this material, however, is that it is somewhat hygroscopic, which dictated strict humidity control during all ground operations. Problems were experienced in mounting the main mirror rigidly to the telescope structure without introducing distortions. At least one mirror was destroyed by being bonded in too firmly for removal in one piece. Success was eventually achieved by inserting the mirror *after* the telescope structure had been bonded to the instrument mainframe.

A particular problem, related in Hughes (1984), was the difficulty in locating the actual focal plane in order to position the detector arrays correctly. Special measurement techniques had to be developed to achieve the high accuracy required.

As is discussed in the next section, the detectors for the three longer-wavelength bands require to be cooled, which means that they cannot be located with the short-wavelength detectors. To provide two separate focal planes, a relay optical system is used. Hughes (1984) describes the trade-off study carried out to optimize the compromise between performance and practicality. The chosen design is illustrated in Fig. 14.3. It comprises a plane folding mirror, and a concave focussing mirror. The mirrors are coated with aluminium, because of its high reflectivity in the infra-red. The two-mirror design gives an inferior performance to the four-mirror design originally proposed, but it is perfectly adequate and much easier to make. The relay has a magnification of 0.5, to allow the detector active areas to be reduced. As is discussed in Section 14.10, this is important from the point of view of noise reduction.

In view of the accurate interband registration demanded of TM, the need for two physical focal planes posed severe construction problems. The situation was of course made worse by the fact that the cooled assembly had to be cycled over substantial temperature differences many times during development and testing, culminating in the trauma of launch and the onset of weightlessness. A means of extremely accurate adjustment was required. Again, Hughes (1984) describes the results of a trade-off study. The solution adopted was to use thermally stable materials throughout, with a back-up adjustment mechanism operable by ground control. Three piezoceramic 'inchworm' devices are incorporated into the mountings of the relay spherical mirror. These change their dimensions by very small amounts when an electric field is applied, and (appear to) retain the new dimension when the field is removed. Differential application of such fields enables the alignment to be adjusted, while equal movements of all three provide a means of focus adjustment. At the time of writing of Hughes (1984), no such adjustments had been attempted or required. Part of the reason for eschewing such adjustments is the difficulty of assessing the effects thereof (Murphy, 1989). In-flight focussing checks represent a problem that has beset other operators also. Chapter 16 describes a strategy for comparing the spatial spectrum of an image with the probable spectrum of the scene imaged. It is less than perfectly straightforward.

The structural design of the telescope, as well as the overall instrument, posed considerable technical problems. Many of these are considered at length in Hughes (1984).

14.5   THE FOCAL PLANE

In this book, we define the 'logical' focal plane as that which is seen from the telescope. This is by contrast with the 'physical' focal planes of which TM has two, and AVHRR five or six. TM's logical focal plane is shown in Fig. 14.4, and comprises the prime focal plane together with the image on it of the cooled focal plane.

The reasons for gaps in, and between, the arrays are discussed in the next section. Note however that each array of 16 detectors is designed as two rows of eight; with a gap of exactly one IFOV between each detector. The rows are staggered, so that the detectors in the second row sense in the gaps of the first. The rows themselves are displaced by 2.5 IFOVs, as shown in the figure. For the thermal band (Band 6), the four detectors are laid out in a similar manner. For all but Band 6, the swath is scanned past the 16 detectors in 60.74 ms. Each is sampled every 9.61 μs which is equivalent to some 6300 samples (pixels) per line. The digitization is arranged so that the even detectors are sampled half a sample time after the odd detectors. This means that, during the 'forward' scan, the even detectors effectively 'catch up' half an IFOV, so that their samples are just two IFOVs behind those of the odd detectors (in the scan direction). During the 'reverse' scan, the even samples emerge effectively three IFOVs ahead. Thus a considerable amount of unscrambling is required on the ground, which some have called 'resampling'. For Band 6, the procedure is the same, though modified to allow for the larger detector size and coarser spacing.

In fact, true *resampling* involves the computation of new radiance values for a pattern of artificial pixels as, for example, in geometric correction. The term therefore has connotations of degradation of data quality, and in particular of an increase in the incidence of 'mixels', or pixels covering a mixed background. As long as unchanged radiance values are simply ascribed to a different position then no such danger exists.

Having said the above, second order effects are still possible. These can give rise to a need for genuine resampling, particularly at the ends of scan lines. Causes include attitude and altitude errors, and irregular motion of the scanning mirror. In extreme circumstances these errors can generate gaps and overlaps at the ends of scan lines, which are removed by procedures described in NASA (1983).

The active areas of each detector assembly are mounted at the centre of a moderately large IC package, and so the groups of detectors sensing the different bands have to be separated by a minimum of 25 IFOVs. Again, the separation is strictly an exact number of IFOVs, so that accurate registration can be assured without resampling. There remains a time difference between the acquisition of different bands which might have significance in certain specialized applications.

As discussed in the previous section, the relay optics used to image the cooled 'focal plane' on to the prime focal plane are designed to give a magnification of 0.5. This serves the same purposes as the 'field' lens(es) which are a feature of many thermal-band detector systems. It concentrates the radiation on to a smaller area of active material, which helps to improve the signal/noise ratio (see Chapter 5). The

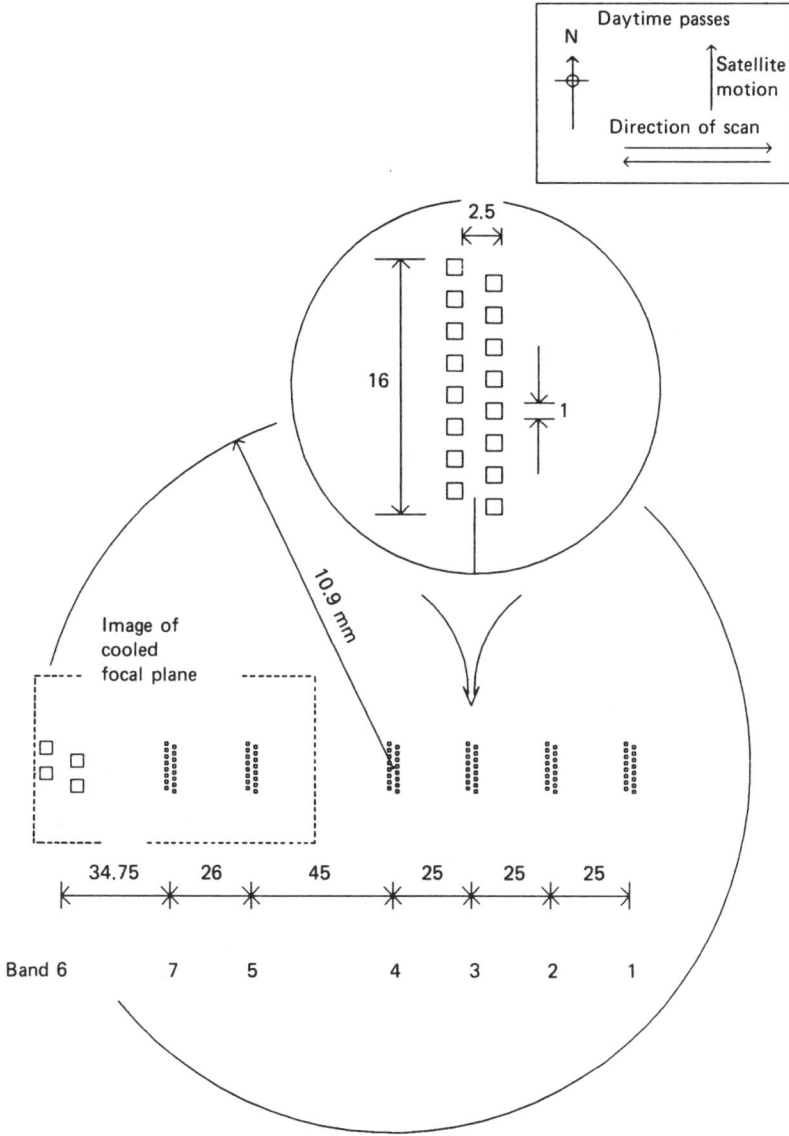

Units in IFOVs (= 0.103 mm or 42.5 µrad)

Fig. 14.4 — The TM (logical) focal plane.

active areas of Bands 5 and 7 are in fact 3% greater than exactly half-size, although the inter-band spacings are exact (Markham, 1984).

The cooled focal plane is housed in a dewar, and cooled by a radiative cooler (see

Section 14.11). Two windows, an ambient window and a dewar window, are used to allow light into the dewar.

The spectral filters are mounted directly in front of each detector bank. A single sheet of filter material covers each 16 channel detector array, thus maximizing the uniformity of spectral response of each detector. The spectral response of the filters is discussed in Section 14.10.

The interband registration obtained on TM is 0.1 pixel. This is remarkably high by the standards of some of the later systems. On the other hand it is probable that interband registration is one of the areas which benefits from the basic design philosophy of TM.

14.6   THE DETECTORS

A number of different detector configurations was investigated before the final choice was made. Fibre optics, as used on MSS, were ruled out because of their transmission loss. This in turn ruled out discrete detectors, because they could not be packed close enough together to fit on to the focal plane. CCDs looked promising, but were not yet sufficiently developed to be seriously considered. (They have, of course, since become the almost universal solution for the shorter wavelengths.) This left monolithic detector arrays, with various possible combinations of input circuitry (first-stage amplification). To minimize noise pick-up, each individual detector requires its first-stage of amplification to be mounted as close as possible to the active area (see Section 14.7 and Chapter 6). The effective size of each detector is thus too large to enable them to be butted up directly against each other. As discussed in the previous section, the spaces between the detector apertures are accurately matched to their IFOV, to facilitate scanning without resampling. The solution adopted was to fabricate multi-element arrays, with field effect transistors (FETs) fabricated directly behind them on the same chip. The rest of the pre-amplifier circuitry was placed remotely. This arrangement enabled the leads carrying the raw (totally unamplified) signal to be kept extremely short, while minimizing the physical size of the chip, and enabling the detectors for adjacent bands to be packed reasonably close together.

As part of the initial design exercise, 16–element arrays had been specified for all the reflective bands, with a four element array for the thermal band (Band 6). Silicon diodes were chosen for the uncooled bands (1–4), although some compromise in material was required to enable the photodiodes and FETs to be produced on the same substrate. The cooled reflective bands use indium antimonide (InSb), while Band 6 uses a four-element mercury–cadmium–telluride array (HgCdTe). All are used in 'current' mode.

Hughes (1984) describes the problems met in installing the detector assemblies on to the focal plane assemblies without distortion or breakage.

The detectors inevitably gradually deteriorate in use, and the A/D conversion process is adjustable to maintain a sensibly constant calibration.

14.7   THE DETECTOR ELECTRONICS

The silicon diode detectors for Bands 1- 4, and the indium antimonide (InSb) units for Bands 5 and 7, are all high-impedance devices. **Transimpedance** amplifiers are used for the first stage of amplification. These are simply amplifiers driven by a current source rather than the more usual voltage source. The difference may well lie only in the absence of a resistor in the input circuit. They are fronted by junction field effect transistors (JFETs) (see Chapter 6). These, and the feedback resistors, are mounted on substrates immediately behind the focal plane (prime or cooled as appropriate). The rest of the components for Bands 1–4 are mounted in the electronics bay, along with the rest of the electronics. The remaining pre-amplifier components for Bands 5–7 are housed on the ambient stage of the radiative cooler, and are therefore partly cooled. The remainder of the electronics are housed in the ambient-temperature electronics bay. The Band 6 detectors are low-impedance devices, of approximately 50 Ω, and employ conventional bipolar pre-amplifiers (Engel & Weinstein, 1983b).

A simplified diagram of the main channel electronics is given in Fig. 14.5. The transimpedance pre-amplifiers require very large feedback resistors in order to optimise the noise levels from the different potential sources (see Chapter 5). Bands 1–4 are provided with 10^9-Ω resistors, while those for Band 5 and 7 are only slightly smaller, at 2×10^8 Ω. The use of such high impedances invariably introduces problems with pick-up and to minimize this, the feedback resistors are mounted on the same substrate as the detectors and the JFETs. Stray capacitances remained a problem, and would have limited the response to about 10 kHz. However, being uncontrolled, such effects are particularly difficult to compensate for, and therefore a 'swamping' capacitor was introduced in parallel with the feedback resistor. This provides a smooth and controlled roll-off above about 3 kHz, which can be counteracted by suitable high-pass filter networks in the next stage of amplification (the postamplifiers). The low-impedance detectors for the thermal band (Band 6) presented no such problems, and use flat-response postamplifiers.

Inter-detector cross-talk is prevented by means of an aluminium shield between each pair of channels. At the end of each scan, all the detectors view an area of zero irradiance. This enables the amplifier outputs to be 'DC restored', to combat the effects of the detector dark currents, as described in Chapter 6. The process involves forcing the output of one amplifier in each channel to zero by the application of an equal and opposite input signal. The additional input signal then compensates exactly for the dark current existing at the start of the scan. It is held constant for the duration of that scan, so that only (very small) changes in dark current remain to contaminate the wanted signal.

The specification for the electronic filters was, as is always the case with filters, self-conflicting. It required a sharp cut-off in the 'stop' band, above the Nyquist frequency (see Chapter 6), to minimize aliasing. But it also demanded minimal attenuation — and phase shift — in the pass band. In addition, overshoot had to be virtually eliminated; the noise levels introduced had to be low; and it had to be possible to produce 100 copies, all accurately matched. The design eventually chosen was a three-pole 'Goldberg' filter, which is a modified Butterworth filter designed by the Goddard Space Flight Center. The filters are built into the postamplifiers and cut off at 52 kHz.

Preliminary noise analysis suggested that the Band 4 channels might require 'compression' amplifiers to bring them within specification. A compression amplifier

Not Band 6   Band 6: four channels only

Supplies
filter   Positive supply

R1

C1   R1

JFET   Three-pole
                                        'Goldberg' filter

Frequency boosting
filter

Track and
hold buffer

To A/D
converter

Detector

Pre-amplifier   Post-amplifier

Bias supply

Dark current
compensation
signal

16-channel
track and
hold
multiplexer

Focal plane
assembly
(cooled for
bands 5–7)   Electronics
                    bay

		Bands 1–4	Bands 5&7	Band 6
R1	feedback	10^9	2×10^8	low
C1	swamping	.03 pF	>.03 pF	–

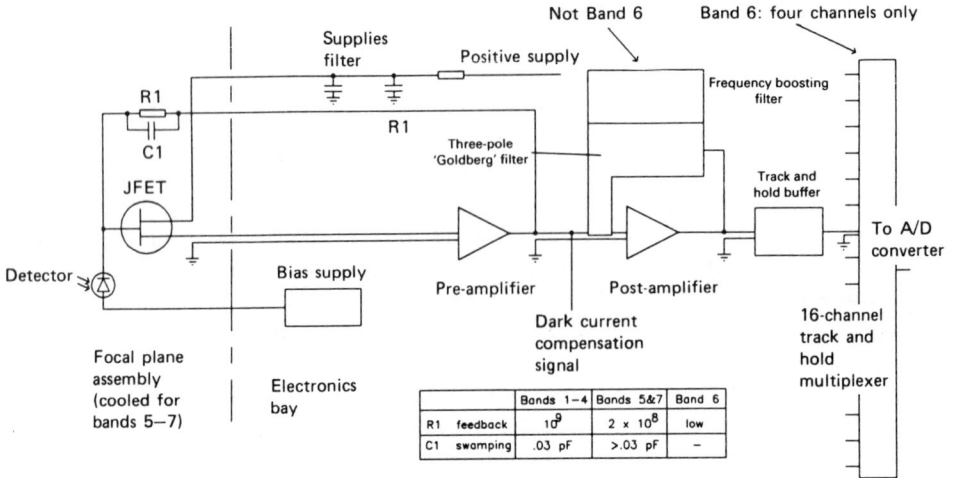

A typical channel

There is a degree of conjecture in this diagram

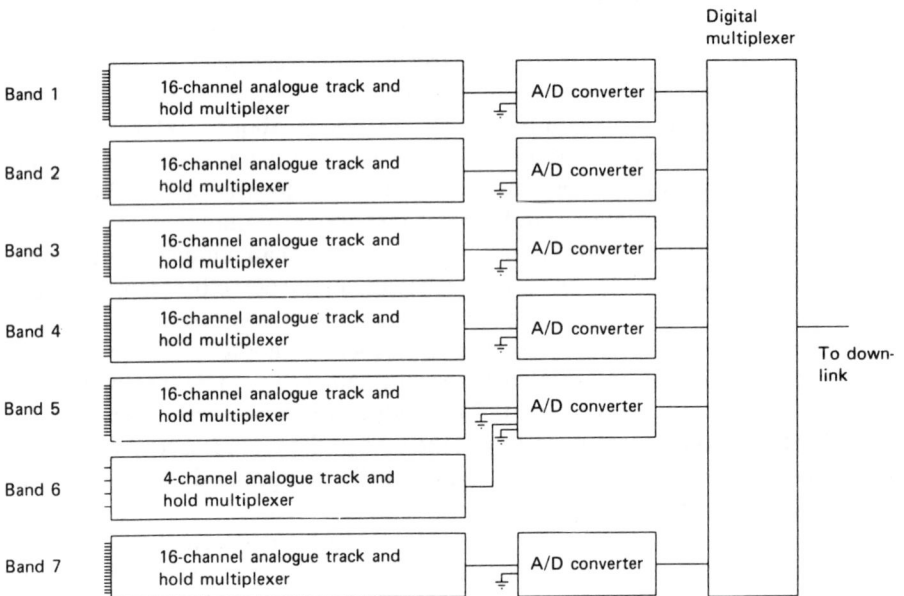

Digital
multiplexer

Band 1 | 16-channel analogue track and hold multiplexer | A/D converter

Band 2 | 16-channel analogue track and hold multiplexer | A/D converter

Band 3 | 16-channel analogue track and hold multiplexer | A/D converter

Band 4 | 16-channel analogue track and hold multiplexer | A/D converter

Band 5 | 16-channel analogue track and hold multiplexer | A/D converter

Band 6 | 4-channel analogue track and hold multiplexer

Band 7 | 16-channel analogue track and hold multiplexer | A/D converter

To down-
link

Fig. 14.5 — TM electronics.

has a gain that varies with amplitude. Low signals are amplified more than high signals, so that the dynamic range of the digitizer is used more efficiently. The use of such an amplifier complicates the data reconstitution process, and there was no doubt some relief when it eventually proved to be unnecessary.

After analogue amplification, and filtering, the signals from the 100 detectors pass to the multiplexer unit for sampling and digitization.

The sampling rate is often thought to be necessarily synchronized to the IFOV or pixel size. In fact it is advantageous to sample as frequently as possible, consistent with the need to economize on data rates and on ground storage capacities. The MSS detectors, for example, were sampled 1.4 times per IFOV, which maximized the resolution obtainable from MSS's relatively large pixels, while introducing minimal aliasing noise. Sampling once per pixel is in fact the minimum that can be tolerated. When given the standard square-wave test, the performance of such a system is poor. Either the low-pass filter must be 'slugged' to degrade the resolution quite seriously, or it allows through a wholly unacceptable degree of aliasing noise. However, an analysis of realistic imagery showed that the high frequency content of actual landscapes is relatively small, and that little aliasing noise is actually introduced. Because the quantity of data produced by TM was already very large, once-per-IFOV sampling was adopted.

Two possible methods of sampling are discussed in Hughes (1984). The first is 'integrate and dump', in which the output from each detector is (after amplification) used to charge up a capacitor. The voltage across the capacitor is read at regular intervals by the A/D converter, after which the capacitor is instantly discharged. Each sample thus represents the average illumination of the detector in the period since the previous sample. This algorithm incorporates an element of low-pass filtering which is a fixed function of the sampling rate. It is also relatively inflexible, from the point of view of digitization, because interrogation has to take place at precisely the correct moment for all detectors.

The second method of sampling, and that chosen, is 'track and hold'. In this method a 'track and hold' amplifier follows the fluctuating output from the detector, until it receives a sampling trigger signal from the controlling clock. The output of the amplifier is then frozen until it has been digitized, when it is freed to follow the detector signal once more. This method obtains the precise reading of the detector at the moment of sampling, irrespective of whether it truly represents the overall intensity level of that particular ground patch. However, this is only important if the scenery is changing rapidly, and the instrument is approaching its limits of response. By giving the designer complete control over the frequency response of the anti-aliasing filters, track and hold may in fact win out. It also permits economy in the use of digitizers, because the output of each amplifier can remain frozen until its turn to be digitized comes round.

After the usual trade-off studies, it was decided to use a total of six A/D converters, one per band with Bands 5 and 6 sharing. This permits a degree of graceful degradation in the event of one converter failing, because in general only one band would be lost. The converters are 8-bit 'successive approximation' converters, running at approximately 40 mbit/s. They are extremely sensitive, enabling pre-amplification to be kept to a minimum. This very sensitivity caused considerable problems during testing however, because it was difficult to avoid excessive pick-up when the additional ground equipment was connected. This particular design of digitizer also requires very great care in setting up, if a strictly linear response is to be obtained. Unfortunately after the setting up was completed, using various fixed voltages as test inputs, it became clear that the digitizers behaved slightly differently when presented with changing voltages. As a result, there are slight differences in 'bin size' (step size) over the range of grey levels. Although the effect is small, it is important for the most demanding applications and is discussed qualitatively in Hughes (1984).

The entire sampling sequence is controlled by a clock built into the multiplexer. This clock controls the scan mirror motion, while the start-of-scan signals from the scan mirror control the operation of the rest of the multiplexer and all other scan-related functions. These include the operation of the scan-line corrector, the calibrator and the data-processing electronics to the scan mirror. Dynamic inter-actions between these three components caused some problems during development.

As its final act, the multiplexer encodes the signals, as well as various items of housekeeping data, into a single 85 M/bit/s data stream. From the point of view of the TM instrument, it would have been more economical to have had two parallel data streams, each running at half the total data rate. However, the savings in electronic developments costs would have been more than outweighed by increases in overall spacecraft and ground segment costs, because of the need to handle two data streams. Other trade-offs are described in Hughes (1984).

The main power supplies provide relatively unregulated power to the individual circuits. Fine regulation is done individually on each board. This enables each to provide the level of smoothing that it needs, and gives an added level of isolation for the prevention of cross-talk. The only exception to this is the multiplexer, which uses power straight from the power supply. The multiplexer is the major single heat source in the TM instrument. It and the rest of the electronics dissipate most of the power used by TM.

14.8   THE DOWNLINK

There are no recording facilities of any kind aboard the Landsat spacecraft. This decision was presumably made because of their poor reliability track-record. The decision, combined with high data rates of TM, forces Landsats 4 and later to employ a uniquely complex downlink system. The spacecraft itself transmits data directly, at the real-time rate of 85 Mbit/s, to one of three satellites of the Tracking and Data Relay Satellite System (TDRSS), which occupy complementary positions on the geostationary orbit. TDRSS relays the data, also at 85 Mbit/s, to the White Sands receiving station, where they are recorded on high-density magnetic tape. The data are then relayed to the Goddard Space Flight Center, at 50 Mbit/s via the Domsat link. Processed TM data are copied to computer-compatible tape (CCT) and film, and delivered a few miles down the road to the offices of Eosat Inc. (Copies are also flown to the EROS Distribution centre for archiving.)

A consequence of this communication system is, of course, that the global coverage of TM can only be guaranteed while a complete TDRSS network exists. For example, until TDRSS-3 was launched in the spring of 1989, a small canoe-shaped region remained out of reach of the link. It took in the whole of India, and extended North and South to regions of Central Asia and of the Indian Ocean. This gap, being foreseen, was covered by two local ground stations, at Hyderabad and Bangkok, which were able to receive data from within their line of sight. This included the bulk of the populated parts of the exclusion zone. During this period, data from these stations were transmitted to Goddard on magnetic tape. Other suitably equipped ground stations can receive data from Landsat whenever it passes within line of sight range. Many such stations exist around the world. However, the failure of any part of the TDRSS link will inevitably also put part of TM's coverage in jeopardy.

14.9 CALIBRATION AND IN-FLIGHT CHECKS

The turn-round time at the end of each scan is used for the DC restoration of the detector amplifiers, as discussed in Section 14.7, and for on-line calibration. Radiance-controlled tungsten filament lamps are used for the calibration of the reflective bands, and a controllable black body is used for the thermal band. A black surface of known temperature is used to ensure that each active scan starts from a fixed level.

The calibration assembly is shown in Fig. 14.6. It is mounted in front of the prime focal plane, and therefore excludes the optics (except for the relay optics) from onboard calibrations. The original intention had been to supplement the once-per-scan onboard calibrations with a once-per-orbit calibration against the Sun. However the latter was deleted, at NASA's direction, because of a conflict with other aspects of the spacecraft design (Hughes, 1984). The lamps are mounted away from the light path, illumination being fed to the detectors through fibre optics. The fibre optic terminations are mounted on the downstream face of a shutter, or 'flag', which swings in front of the prime focal plane during the turnround periods of the scanning mirror. The lamps may therefore be left on permanently, and switched only when setting changes are required. The tendency of tungsten filament lamps to age means that an additional mechanism had to be introduced to maintain their outputs constant. This takes the form of an individual silicon photodiode for each lamp, which drives a feedback circuit controlling the lamp current. The radiance of each lamp is controlled to a level equivalent to 1900 K. (Compare this with the system employed on the more recent IRS-1 instruments (Chapter .18). Developments in light-emitting diode technology permitted their use as calibration sources, and enabled the IRS-1 calibration assembly to be much simplified.) In front of each lamp is a different neutral density filter, attenuating the radiance by 0, 25 or 50%. In automatic operation, the lamps are switched approximately every 40 sweeps (20 complete scans), all possible permutations (including all off) giving eight possible calibration levels. The three filters are mounted on a wheel, so that lamp/filter combinations can be changed. This covers the possibility that one lamp may fail, and also facilitates any checks that may be indicated on the relative stability of the different lamps.

A system of folding prisms merges the outputs from the three lamps, before feeding the illumination to six fibre optic bundles, one for each of the six reflecting bands. Each bundle is terminated with a field lens designed to illuminate an area of the focal plane covering one array of 16 detectors. Thus, while the shutter is at rest in front of the focal plane, all the detectors of the six reflecting bands should receive an equal illumination, although their individual spectral responses will affect their degree of excitation. How evenly the detectors of any one band are actually excited will depend on the quality of the field lens. How evenly the different bands are

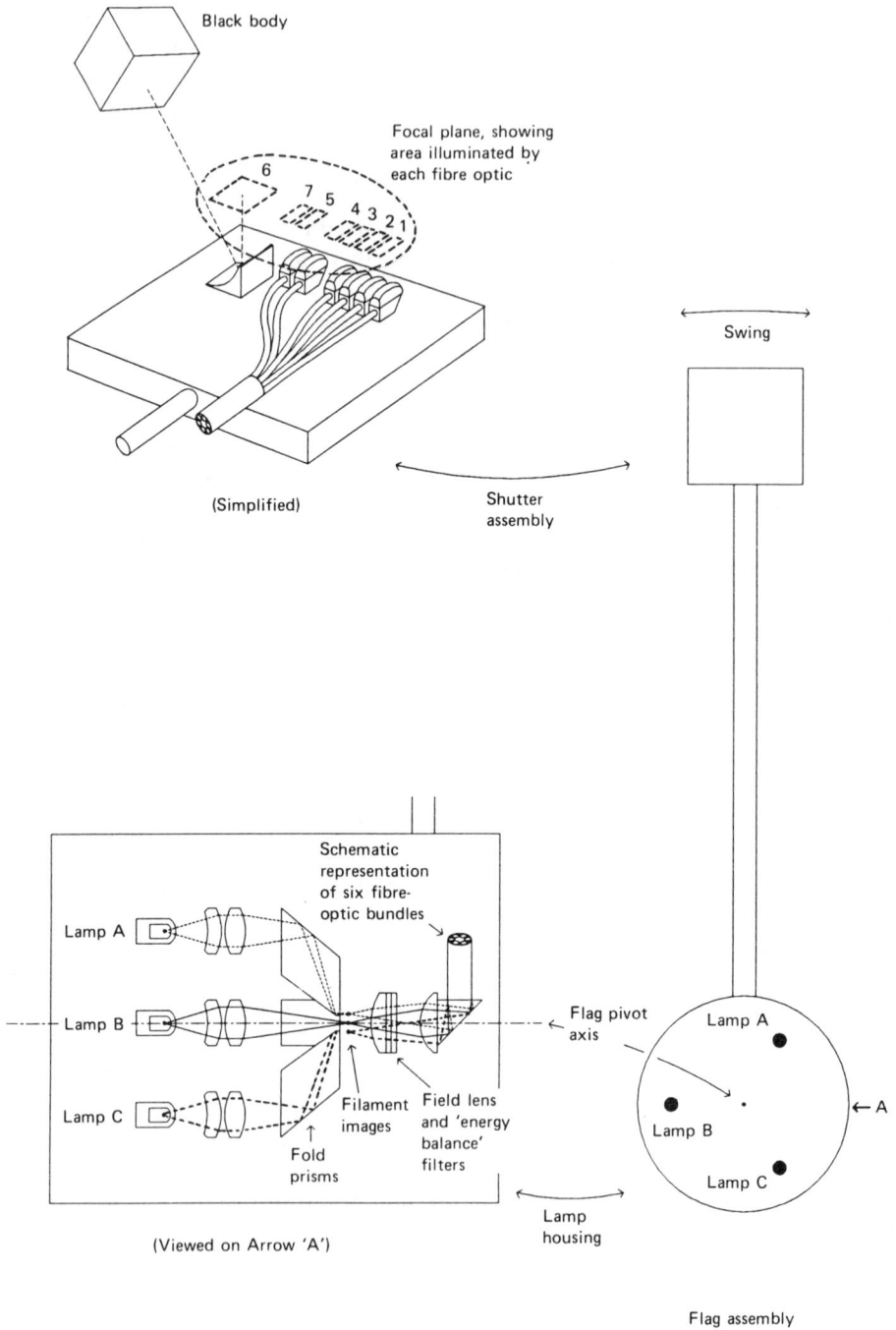

Fig. 14.6 — TM calibration unit.

excited will depend on whether all the bundles collect and transmit an identical proportion of the radiance from the lamps. The design of the field lens was not a problem, but equalizing the energy collected by the various bundles turned out to be a major challenge. Hughes (1984) describes the difficulties encountered in obtaining interband uniformity to within 10% at each intensity level.

The radiance from the black body is reflected on to the Band 6 detectors via a 'toroidal' mirror on the shutter, and hence (as with that for Bands 5 and 7) via the relay optics. (A toroidal mirror is convex on one axis, and concave on the other. It takes its name from the fact that its profile is similar to the inner surface of a toroid.) Three selectable temperature settings are provided (Engel & Weinstein 1983), plus the shutter's dark surface.

A dark area on the shutter precedes the fibre optic terminations, and provides each detector with a brief period of zero illumination prior to the next active sweep. It is during this period that the amplifier outputs are restored to zero, as discussed in Section 14.5.

The wand carrying the shutter is mounted on a spring; which is designed to resonate naturally at twice the scan rate under the guidance of an electronic control loop. Problems were encountered with the dynamics of the system, including the overshoot provisions.

Details of the post-processing use of these data in generating radiometrically corrected imagery are given in NASA (1983). Also described are the procedures for producing geometrically corrected data; for applications where such interference with the fidelity of individual pixels is justified.

The comprehensive pre- and post-launch radiometric calibration tests that were carried out are described in the next section.

14.10   DATA QUALITY (AND QUALITY CHECKING)

One detector (unidentified) on the Landsat 4 TM failed prior to launch. In routine image products, its output is replaced by the corresponding pixels from a neighbouring detector. However, two teams (Bernstein *et al.*, 1984; Fusco & Treverse, 1984) have both shown that it is generally better to use the same detector from a different band.

A major problem with satellite-borne instruments is that, in general, all comprehensive testing and calibration must be carried out in an environment quite different from that in which the instrument will actually be used. No matter how thoroughly it has been tested on the ground, the sceptical user will only be really interested in the results of in-flight checks. These, however, are difficult to do. The literature is replete with experiments which attempt to estimate, from the ground, the radiance that was probably reaching the instrument at the time that specific imagery was being acquired.

For example, Castle *et al.* (1984) describes a preliminary attempt to calibrate the entire TM system in a single operation. This of course is always the best way to calibrate any instrument — if it can be done with confidence. An attempt was made to predict the radiance reaching the 'entrance pupil' from ground measurements and from atmospheric predictions (see Chapter 2). On the day of the experiment, an

8-cm layer of two day-old snow was covering the flat gypsum surface of the White Sands Missile range. Ground measurements were taken around the time of an overpass, as were suitable atmospheric measurements. From these the total radiance reaching the TM entrance pupil was estimated. Only a few pixels were studied, and only Bands 2, 3 and 4 were evaluated. (Not surprisingly perhaps, Band 1 had saturated under the intense illumination.) However, as a means of gaining confidence in the stability of the on-board calibrator the exercise was obviously well worthwhile.

More recently, Slater (1987) has underflown the satellite with aircraft-borne sensors. By using similar detectors both above and below the atmosphere, he is able to measure its transparency at the time of the overpass. By taking exhaustive meteorological measurements at the same time, he is also able to relate the measured transparency to the standard data from which the user would normally have to work.

14.10.1 Radiometric performance

The radiometric performance of a remote sensing instrument is either relatively insignificant, or of the utmost importance, depending on your point of view and on what you are trying to measure. The present author must register a personal interest here. After a lifetime of bitter experience (in a non-Space environment) it is his view that 'the canny experimenter never relies on anything outside his personal control'. The calibration figures that come with satellite-generated imagery clearly come into this category. Therefore he would always include, in his experiment, some way of cross-checking them; or better still, of calibrating them out altogether. The latter approach is, in fact, the norm over large areas of remote sensing practice. Many classification techniques rely on interband comparisons for their discrimination, which of course immediately sidetracks many of the uncertainties in the original data. Their results are then validated by cross-checking against 'ground truth' data, which calibrates out the rest.

None of the above is intended in any way to minimize the importance of providing accurate calibrations for remote sensing instruments. For it could, with equal truth, be said that 'no instrument has really come of age until it can be used with confidence to derive quantitative results' — irrespective of whether the experimenter chooses to avail himself of the facility.

The radiometric performance of TM was specified in terms of hypothetical scenes, whose maximum and minimum radiances in each band were specified, and which are given in Table 14.2.

The columns showing watts per square metre per steradian per micrometre have been added to those given in Hughes (1984), because this is the figure that represents the actual intensity of the scene under study. (It is not known what allowance might have been made, in the orignal figures, for imperfections in the filter profile. Therefore the conversion was carried out simply by dividing by the nominal channel bandwidth as given in Table 14.1.) They illustrate graphically how the radiance levels fall off at the longer wavelengths. The figures in the previous pair of columns show the benefit to be gained by resorting to larger bandwidths. The figures also show why it was necessary to resort to cooling to obtain adequate performance from Bands 5 and 7. The radiometric performance of each reflective band was specified in terms of the maximum permissible signal to noise ratio at both the minimum and maximum scene radiance. Because users tend to think in terms of noise equivalent radiance

Table 14.2 — Radiometric performance of TM as flown on Landsats 4/5

Band	Hypothetical scene radiance				SNR (band average)						NEΔρ		
	(W/m²/sr)		(W/m²/sr/μm)		Scene minimum			Scene maximum					
					Measured		spec	Measured		spec	Measured		spec
	Min	Max	Min	Max	Landsat 4	Landsat 5		Landsat 4	Landsat 5		Landsat 4	Landsat 5	
1	2.8	10.0	40.0	143	52.0	60.2	32	143.9	143.2	85	0.16	0.16	0.8
2	2.4	23.3	30.0	291	60.0	59.7	35	279.0	234.9	170	0.18	0.21	0.5
3	1.3	13.5	21.7	225	48.0	46.25	26	248.0	215.1	143	0.20	0.23	0.5
4	1.9	30.0	13.6	214	35.0	46.2	32	342.0	298.7	240	0.19	0.22	0.5
5	0.80	6.0	4.0	30	40.0	35.8	13	194.0	175.4	75	0.23	0.25	1.0
7	0.46	4.3	1.7	16	21.0	28.3	5	164.0	180.6	45	0.41	0.37	2.4
6	300 K	320 K	300 K	320 K	0.12[†]	0.13[†]	0.5[†]	0.10[†]	0.11[†]	0.42[†]			

[†]These figures are NEΔT (°C).

level ($NE\Delta\rho$) this figure is also computed and included in the specification. For Band 6, performance was specified directly in terms of noise equivalent temperature difference ($NE\Delta T$). As can be seen from the table, the prelaunch radiometric performance of all bands well exceeded the specification. The performance margin for the thermal band led Hughes (1984) to suggest that a new instrument could have the 120-m IFOV reduced by half.

Markham (1984) reports that small leaks of white light were detected, probably associated with gaps between the filter mounts in the primary focal plane. These were investigated on the Landsat 4 instrument, and found to be never worse than 1% of the detector's response (although, when imaging a high-contrast zone, a dark pixel could be contaminated to a greater extent by light leaking from a neighbouring bright zone). Markham believed that the Landsat 5 TM instrument was suffering from a similar problem, and to roughly the same extent, although the details were different.

Both Landsats 4 and 5 experienced loss in detector performance in Space, on Bands 5–7 only, which was cured by short-term warming of the cooler. It was felt that the loss was probably a result of gradual icing up of the cold window, and that longer postlaunch 'outgassing' periods might usefully have been used (Hughes, 1984). Various other minor anomalies are also discussed in the same report.

In orbit, checks started immediately after launch, and have been continuing, with varying intensities, ever since. The early calibrations are discussed in detail in Engel *et al.* (1983), Barker *et al.* (1984a, b), Schueler (1983), Wukelic *et al.* (1985), Lansing *et al.* (1984), and Wrigley *et al.* (1984).

The subject of dynamic range is also discussed in some detail in Hughes (1984). The dynamic range of an instrument is the difference between the minimum signal to which it will respond and the maximum signal that it will follow faithfully. Many instruments go 'non-linear' outside a certain range, and, if a linear response is important, the quoted dynamic range may be considerably less than the range that the instrument will in fact handle. The range of a complete instrument is limited by that of the weakest element in it. In many cases, the detector element (whatever it might be) would handle a considerably greater range than is actually needed, and the electronics are artificially constrained to clip the output if the input signal exceeds stated limits.

In the case of TM, the practical lower limit was imposed by the levels of self-generated noise introduced by the detectors. The upper limit is dictated by the capability of the later stages of the electronics. However, this is adjustable. The actual measured dynamic range of both TM instruments (4 and 5) exceeded the specification. There is of course no virtue in having a wider dynamic range than required, particularly if the signal is to be digitized. In an ideal world the electronic gain would have been adjusted before digitization, to apply the span of 255 grey levels to just the range displayed by the incoming radiance (or perhaps that demanded by the specification whichever was the greater). In practice what is done is to 'stretch' the digital data, on the ground, so that the published imagery is in accordance with the specification.

If a channel is driven beyond its upper limit then the weakest link will 'saturate'. It will be driven hard against its metaphorical end stops, and will usually take some time to recover and to resume normal behaviour. It was specified that all TM channels should recover in less than four IFOVs. Since the dynamic range of TM was optimized for ground-based studies, all the bands can easily saturate when presented with clouds or snow. Band 1 channels suffer most from this effect, because their upper limits were deliberately set lower than the others in relation to the solar radiance. Hughes (1984) discusses a research programme designed to study the incidence of excessive recovery time, and to draw conclusions from it.

14.10.2 Spectral performance

When perusing the average sensor description, it is easy to imagine that the spectral response of the detectors is perfect: that all radiation within its pass band is transmitted and all that outside is rejected. In fact, it is seldom possible to achieve either of these desirable states. The design of bandpass filters for any kind of signal is difficult. It is usually possible to do moderately well in the case of electrical signals, because electrical components can be obtained with almost any required impedance (resistance, capacitance etc.), and they will be accurate and consistent in their performance. With optical components, it is much more a case of making the best possible use of available materials, and sometimes distinctly unsatisfactory compromises have to be accepted. Although the cut-off at the edge of the band is usually reasonably steep, there is often considerable variation in transmission between the two extremities of the pass-band.

The spectral response of a channel is also a function of the individual responses of the rest of the components in the light path. In the case of Bands 1–4, this comprises five mirror surfaces, and the response of the detector itself. For Bands 5–7, two more mirrors and two windows are interposed. The spectral performance of the system was computed from the measured responses of the individual components (NASA, 1983). The window transmittances were measured using the actual flight parts; those of the filters were measured on the sheet of material from which the actual filter was cut. The mirror reflectances were measured using 'witness' material, prepared at the same time as the flight component. The transmittance of the Band 6 filter material had to be measured at room temperature. However, a piece of witness material was used to derive a factor of correcting the transmittance figures to the operating temperature of 90 K.

Each individual detector has its own unique spectral response. However, the response of silicon and InSb are relatively flat over the region of interest, and it was found possible to use a single average figure to represent the response of all 16 channels. But the response of HgCdTe is not flat in the region of 10 μm, and so for Band 6, individual measurements have to be used. (For Bands 5 and 7, it was found necessary to base measurements on spare arrays made from the same wafer.) Apart from Band 6, the same figures are used for both Landsats 4 and 5. Fig. 14.7 shows typical spectral performances.

'Out of band response' is the effect of any radiation that might get through the filters (or round them!) at wavelengths outside the transmittance band. Calculations were carried out by the filter vendors.

The figures in Table 14.3 are Landsat 4 prelaunch measurements. Band edges are defined as being the wavelength at which transmission is reduced by one half. The 'difference' is that between the observed and specified values (a negative value means that the specified value is greater). The units are micrometres.

It is virtually impossible to measure the spectral response of an instrument once it

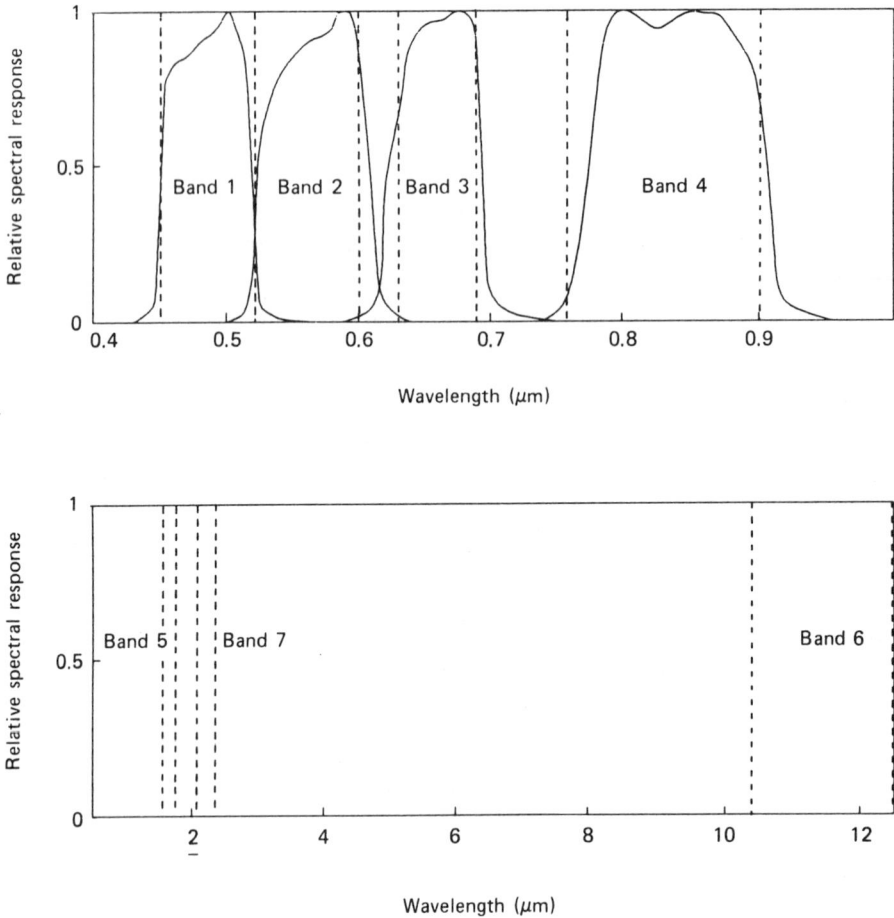

Dotted lines show nominal band limits.
Vertical scales differ for each band.

Fig. 14.7 — TM Spectral responses (Bands 1–4) (after NASA 1983).

is in orbit, although inferences can be drawn by statistical analysis of large numbers of scenes, and from multiple repeat visits to the same sites. At the time of writing of Hughes (1984), radiometric stability logs had been carried out covering the first 400 days. The logs covered regular visits to areas with very different spectral characteristics. The variation came to less than 1%, and it was inferred from this that the spectral characteristics were stable.

14.10.3 Spatial frequency response
The effective IFOV figures were calculated for TM, on the basis of pre-launch component figures (Markham, 1984). As Table 14.4 shows, the IFOV figures

Table 14.3

	Lower		Upper		Bandwidth	
Band	Observed	Diff.	Observed	Diff.	Observed	Diff.
1	0.452	0.002	0.518	−0.002	0.066	−0.004
2	0.529	0.009	0.610	0.010	0.081	0.001
3	0.624	0.006	0.693	0.003	0.069	0.009
4	0.776	0.016	0.905	0.005	0.129	−0.011
5	1.568	0.018	1.784	0.034	0.216	0.016
7	2.097	0.017	2.347	−0.003	0.250	−0.020
6	10.422	0.002	11.661	−0.839	1.239	−0.861

normally quoted (which are the geometric figures based on the aperture sizes) are slightly optimistic as compared with the actual effective figures. The along-track figures are predictably better than those for the along-scan direction because, in accordance with theory, the effect of the low-pass anti-aliasing filters were included in the latter Fig. 14.8 shows the component design MTFs for selected wavebands, together with the pre-launch measured values. In accordance with theory, the amplitude first falls to zero when one spatial cycle exactly matches the detector aperture, i.e. is equal to one IFOV. It can be seen that, at the shorter wavelengths, the resolution of the telescope is almost too good, and contributes virtually nothing to the spread. The effective field of view is controlled almost entirely by the size of the detector apertures, with an additional secondary effect, in the along-scan direction only, generated by the anti-aliasing filters. At the longer wavelengths, the telescope's margin of performance is less.

The MTF figures in Table 14.5 are quoted in NOAA (1984). According to Markham, however, they were not measured directly, but were derived from closely sampled step response measurements. The standard deviations indicate the response differences between the different detectors within each band.

Note that these figures are not identical to the MTF figures given in Table 14.1, which were obtained from Markham.

The interesting point is made (Markham, p. 30) that the best estimates of the line spread functions are less well defined for TM than for MSS 'primarily because more information is available on the TM spatial responses', and they do not all agree.

14.10.4 Interband registration

Interband registration is a function of focal plane layout and scanning and sampling accuracy. The original specification had been for a maximum mis-registration of 0.2 pixels between bands on the same focal plane, and of 0.3 pixels between bands on different focal planes. The actual mis-registration measured was only 0.1 pixel.

Interband registration was measured at 29 locations on the image. The figures in Table 14.6 are given in terms of fractions of an IFOV, and are the worst observed.

A major preoccupation, which emerges clearly from Hughes (1984), is with

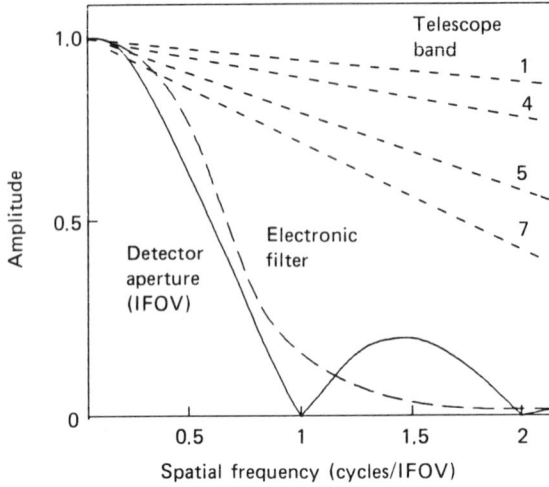

Component design MTFs — reflective bands

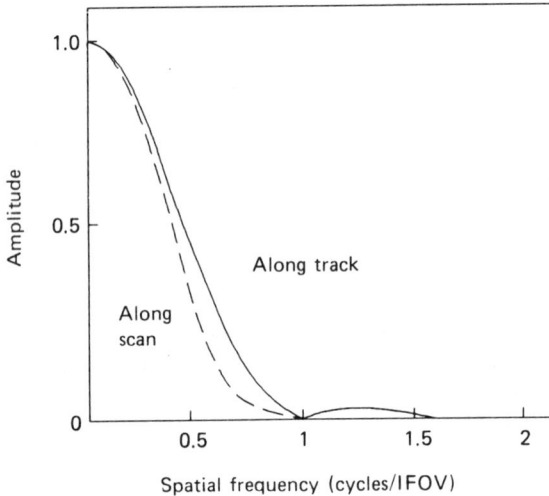

Net MTFs from measurements — Bands 1–4

Fig. 14.8 — TM spatial responses (after Markham 1984).

temporal registration. It is regarded as important that the same pixel should represent exactly the same patch of ground, not only on all the bands on one image, but also on imagery taken at different times. It is clearly inconceivable that the spacecraft should be flown with sufficient accuracy to achieve this with raw data. The question is, can it be achieved by monitoring (and correcting for) the various known

Table 14.4 — Pre-launch figures for EIFOV (Markham, 1984).

Bands	Along scan	Along track	IFOV
except 6	36	32–33	30
6	141	124	120

Table 14.5 — Square-wave responses (SWR) (NOAA, 1984)

	Bar width (m)							
	30		45		60		600	
Band	SWR	SD	SWR	SD	SWER	SD	SWR	SD
1	0.46	0.01	0.76	0.03	0.94	0.02	1.0	0
2	0.44	0.02	0.72	0.04	0.96	0.03	1.0	0
3	0.41	0.01	0.72	0.02	0.91	0.02	1.0	0
4	0.43	0.01	0.76	0.03	0.95	0.03	1.0	0
5	0.42	0.02	0.76	0.02	0.92	0.03	1.0	0
7	0.44	0.02	0.76	0.02	0.92	0.02	1.0	0

	Bar width (m)						
	120		180		240		2000
6	0.44	0.04	0.78	0.01	0.94	0.02	1.0	0

Table 14.6 — Interband registration (NOAA, 1984)

	Along scan	Across scan
Within primary focal plane	<0.1	<0.13
Within cooled focal plane	<0.08	<0.1
Between focal planes	<0.19	<0.27

errors and deviations, and by using realistic geometric correction algorithms? It was specified that such temporal registration errors should not exceed 0.3 pixel (after geometric correction) and this has been achieved.

Finally, the question arises, how well can TM imagery be corrected to fit a map? The specification was to 0.3 pixel, and this appears to have been achieved also. Hughes (1984) considers, in some detail, the strategies involved in achieving good geometric correction.

14.10.5 Relative information content

Considerable study is described in Hughes (1984), devoted to comparing the usefulness of TM imagery with that of MSS. A wide range of studies was carried out, more than one of which concluded that a TM scene contained twice as much information per pixel as MSS. Perhaps the most interesting study described was that of Williams *et al.* (1984). Williams and colleagues successively degraded TM data towards MSS quality in order to see which of the improvements contributed most of TM's improved performance. They found that the improved quantization (255 grey levels instead of 64) had the most powerful effect, followed by the improvements in spectral performance. Surprisingly, they found that TM's improved spatial resolution actually degraded the results. (It must be assumed from this, however, that they were working with large fields. Other workers have found, inevitably, that TM is better at handling small fields.) The explanation given for the effect is that a fine resolution picks up variability within a field that a coarse resolution smoothes out, and that this variability was confusing the classifiers used. No doubt more recent classifiers are able to cope better with this effect.

14.11 THE HOST SPACECRAFT

While the first three Landsat spacecraft were essentially the same vehicles as those used for the Nimbus series of Weather satellites, a new and larger vehicle (the multimodular spacecraft — MMS) was designed for Landsats 4 and 5 (Freden & Gordon, 1983). The weight has gone up from 953 kg to 1996 kg and the power capacity from 0.75 to 1.5 kW. The TM instrument itself consumes 3.33 kW, while MSS uses 0.05 kW.

14.11.1 Attitude control

Details of the attitude control systems were not made available; however, the basic strategy is described in Chapter 12.

Pointing accuracy was specified to be within $0.01°$, with a stability of $10^{-6}°/s$ (both at one standard deviation probability). This compares with $0.7°$ and $0.04°/s$ for the earlier, MSS only Landsats.

14.11.2 Temperature control

The main body of the instrument is maintained roughly at ambient temperature by controlled radiation to deep space. A bank of adjustable louvres is built into the radiative cooler housing. Behind these louvres, emitting surfaces radiate away the heat from the electronics and other sources. A control system adjusts the position of the louvres to maintain the temperature constant within reasonable limits. This design worked well in Space, but it caused considerable problems on the ground because of the proximity of the two coolers. Some 110 W were being radiated away by the ambient cooler, and if even a fraction of one per cent of this reached the 'cold' cooler (described below) then it would be completely swamped. To combat this problem, a Space-background simulator with exceptionally low reflectivity was devised, involving the use of open-faced honeycomb material 'cut on the bias'

The detectors for Bands 5–7 are mounted on the cold patch of a two-stage radiative cooler (discussed in more detail in Chapter 12). Hughes (1984), when reporting on the development of this item, continually emphasizes the importance of adopting a *low-risk* approach. The strategy in this case was to borrow heavily from the design tried and tested on the VISSR instrument flown on the early GOES missions (Chapter 9). Trade off studies were carried out for each of the main design features, but in most cases the VISSR solution was adopted. The cooler's Earth shield was built of a honeycomb aluminium material, for stiffness and light weight, but had to have a highly polished low-emittance surface to minimize scatter onto the second-stage radiator. This proved problematic for the material suppliers. The solution adopted was to bond commercially available kitchen foil onto the Earth shield's reflecting surfaces.

One significant environmental difference between the two coolers is the fact that the TM cooler is oriented at right angles to the Earth, rather than directly away from it. This means that the intermediate stage has to cope with considerable thermal loading from earthshine; and also that the radiating surfaces see considerably less than the full hemispherical view of deep space. Not without significance also is the fact that the TM cooler carries far more detectors than does that of VISSR. Nevertheless, the TM cooler was designed to produce minimum temperatures of less than 87 K under full load.

The trade-off studies, reinforced by the VISSR experience, led to the following choices. The mirror-finish radiation shields were of polished nickel, finished with vacuum-deposited aluminium. The surfaces separating the cold cavity from the intermediate cavity were coated with electroplated gold on electroplated gold, while those separating the intermediate cavity from the ambient zone were coated with multilayer insulation coated with electroplated gold. The structural members which support the intermediate and cold cavities have conflicting requirements placed on them. They have to be stiff, stable, and able to cope with large thermal gradients, but at the same time they must offer very low thermal conduction and radiation. After the trade-off study, the VISSR solution was again adopted, namely the use of filament-wound fibreglass.

A degree of control is provided by the use of a thin film heater, maintaining the temperature within ±0.2° of three selectable temperatures, namely 90, 95 or 105 K. Thin, low-conductance cables are used to provide electrical connection between the cooled zone and the warm components. Some of the design parameters of the coder are given in Table 14.7.

Table 14.7

Field of view (°)	Horizontal 160	Vertical 114
Radiator areas (cm^2)	intermediate 660	Cold 430
Min temp @ max load (K)	intermediate 147	Cold 84.4
Heat load (watts)	intermediate 2.2	Cold 0.117
Radiation surface	Black-painted honeycomb	

14.12   SOME TYPICAL APPLICATIONS

Table 14.8

Band	Range	Principal applications
1	0.45–0.52	Coastal water mapping; soil/vegetation differentiation; deciduous/coniferous differentiation
2	0.52–0.60	Green reflectance by healthy vegetation
3	0.63–0.69	Chlorophyll absorption for plant species differentiation
4	0.76–0.90	Biomass surveys; water body delineation
5	1.55–1.75	Vegetation moisture treatment; snow/cloud differentiation
7	2.08–2.35	Hydrothermal mapping
6	10.4–12.5	Plant heat stress measurement; other thermal mapping

14.13   PRODUCTS

This topic has been covered too thoroughly elsewhere to warrant inclusion here.

REFERENCES

Barker, J. L., Abrams, R. B., Ball, D. L. & Leung, K. C. (1984a) 'Characterisation of Radiometric Calibration of Landsat-4 TM Reflective Bands' Landsat Early Results Symposium, NASA GSFC May 1984.

Barker, J. L., Ball, D. L., Leung, K. C. & Walker, J. A.(1984b) 'Prelaunch Absolute Radiometric Calibration of the Reflective Bands of the Landsat-4 Protoflight Thematic Mapper' Landsat Early Results Symposium, NASA GSFC May 1984.

Bernstein, R. *et al.* (1984) 'Analysis and Processing of Landsat-4 Sensor data using Advanced Image Processing Techniques and Technologies'. *IEEE Transactions: Geoscience & Remote Sensing*, **GE-22** (3), 192–221.

Castle, K. R., Holm, R. G., Kastner, C. J., Palmer, J. M., Slater, P. N., Dinguirard, M., Ezra, C. E., Jackson, R. D. & Savage, R. K. (1984) 'In-flight absolute radiometric calibration of the Thematic Mapper' *IEEE Trans on Geoscience & RS.* **GE-22** 3 May 1984.

Chipaux, C. (1990) Matra Space Branch, Paris; enhancement, in draft, to Chapter 3.

Engel, J. L., Lansing, J. C. Jr., Brandshatt, D. G. & Marks, B. J. (1983) 'Radiometric Performance of the Thematic Mapper', 17th Int. Symp. on R. S. of Env. Ann Arbor, Michigan, May 1983.

Engel, J. L. & Weinsteino (1983) 'The Thematic Mapper — An Overview' IEEE trans. on Geoscience & RS **Ge-21** 3 July 1983.

Freden, S. C. & Gordon, F., Jr. (1983) 'Manual of Remote Sensing' Chapter 12, American Society of Photogrammetry.

Fusco, L. & Treverse, P. 'TM Failed Detector Replacement' Landsat Early Results symposium NASA (GSFC) 1984.

Hughes (1984) 'Thematic Mapper: Final Report'. Santa Barbara Research Center, Hughes Aircraft Company, October 1984.

Lansing, J. C. & Barker, J. L. (1984) 'Thermal Band Characteristics of the Landsat-4 Thematic Mapper' Landsat Early Results proceedings NASA GSFC May 1984.

Markham, B. L. (1984) 'Characterisation of the Landsat Sensors' Spatial Responses'. NASA Tech. Mem. 86130.

Murphy, R. (1989) Chief, land processes branch, NASA, correction in draft.

NASA (1983) 'Landsat-4 Science Characterisation Early Results' Goddard SFC 2/83. NASA Conf. Pub. 2355.

NASA (1984) 'A Prospectus for Thematic Mapper Research in the Earth Sciences' NASA Tech. Memo 86419.

NOAA (1984) 'Landsat 4 (& 5) Data Users Handbook' U.S. Geological Survey, 1984.

Reulet, J. F. (1988) CNES, private communication.

Schueler, C. F. (1983) 'Thematic Mapper Protoflight Model Line Spread function' 17th Int. Conf. on RS, Ann Arbor, May 1983.

Slater, P. N., Biggar, S. F., Holm, R. G., Jackson, R. D., Mao, Y., Moran, M. S. Palmer, J. M. & Yuan, B. (1987) 'Reflectance and radiance based methods for the in-flight absolute calibration of multispectral sensors'. *Remote Sensing of Environment* **22**.

Williams, D. *et al.* (1984) 'A Statistical Evaluation of the Advantages of Landsat TM Data in Comparison to MSS Data' *IEEE trans: Geoscience & Remote Sensing* **GE-2** (3).

Wrigley, R. C., Card, D. H., Hlavka Christine, A., Hall, J. R., Mertz, F. L., Archwamety, C. & Schowengerot, R. A. (1984) 'Thematic mapper image quality, registration noise and resolution' *IEEE Trans. on Geoscience & RS.* **GE-22** (3), May 1984.

Wukelic, G. E., Foote, H. P., Petrie, G. M., Barnard, J. C. & Eliason, J. R. (1985) 'Final Report, Landsat TM image Data quality analysis for Energy-related Applications'.

15

IRAS: the Infra-red Astronomical Satellite

Consultant: **Dr Helen Walker**†
Source: IRAS (1988)

15.1 INTRODUCTION

IRAS is the odd one out among the instruments studied in this book because its telescope was pointed *upwards*. IRAS was a short-life mission designed to survey the entire sky at four far infra-red wavelengths. The wavelengths ranged from 8 to 120 µm, representing black body peak temperatures from 240 to 30 K. To obtain the extremely high sensitivities required, it was necessary to cool the entire instrument with liquid helium. The mission came to an abrupt end when the helium was exhausted. Although the IRAS instrument was technically very similar to the other instruments in the book, it was actually used mainly in a *non*-imaging capacity. That is to say, the positions and brightnesses of (mainly) point sources were listed in a catalogue, rather than being presented visually in a two-dimensional image (in fact, sky images were also produced as a byproduct). The only part of the instrument obviously affected by this difference in use is the focal plane (Section 15.4), although the detector specification will also have been different, as will the scanning and analysis strategies.

The two originators of the IRAS project, the USA and the Netherlands, started work independently in 1974. They came together in 1975, with Britain joining as a minor partner in 1976, although official agreements were not signed until 1977. The consortium was built around the internationally recognized expertise of each collaborator. Thus the telescope and detector system was the responsibility of NASA. The design of the satellite, and the integration of the telescope into it, was the task of the Netherlands Agency for Aerospace Programmes (NIVR); while the British Rutherford Appleton Laboratory (RAL) organized the tracking and data acquisition facility, provided the non-real-time control software and carried out the

† Rutherford Appleton Laboratory, UK.

preliminary data analysis. Final data processing was carried out by NASA. The satellite was launched on 26 January 1983, becoming operational two weeks later after extensive check-outs. The last observations were taken on 22 November of the same year.

Even before IRAS flew, it was already regarded as the first of a series. IRAS was the survey satellite, and its successor(s) would have the task of following up the interesting phenomena that would certainly be discovered. ISO (Infrared Space Observatory) is an ESA project, initiated in March 1983. Its objective is to provide long-exposure (stare) capability with finer spatial resolution over a wider spectral range (2.5 to 200 µm). Spectroscopic capability will also be incorporated. The main instrumentation will be non-imaging, but a 'camera' will be included, featuring an array of infra-red detectors (Walker, 1990).

Many of the features of IRAS will be familiar to readers who have persevered this far. The orbit was conventional, Sun-synchronous at 900 km. However, there was no requirement for accurate time-table keeping, and therefore no active orbit control. A Ritchey Chretien telescope was used, similar to those of Thematic Mapper and MSS, although a modified form of pushbroom scanning was employed. Flexibility was given to the surveying programme by swinging the entire satellite to point to the required area of sky.

The detector apertures were elongated in the across-track direction, and there were relatively few of them. The detectors of the four bands were staggered, so that the position of a broad-band source may be accurately computed by comparing the outputs from the different bands.

The feature that readers may find the most unfamiliar is the cooling system. To obtain the extremely high sensitivities required, the detectors needed to be cooled to cryogenic temperatures. To obtain the high stabilities required, the structure had to be maintained at a constant temperature. Both these problems were solved at one stroke, by enclosing the entire instrument in an open-necked dewar which was cooled with liquid helium.

15.2 THE ORBIT

The orbit of IRAS was conventional, Sun-synchronous, with a published altitude of 900 km, an inclination of 99° and a period of 102.9 min.

Because the instrument was surveying the fixed star background, rather than the rotating Earth, an analysis of the orbital cycle and the ground-track pattern is relatively uninformative. Nevertheless, for consistency, Fig. 15.2 is offered as usual. IRAS did in fact execute a nominal 14.00 orbits per day, which meant that it overflew exactly the same tracks every day. This was convenient, because it enabled the ground station to follow a consistent routine.

For the same reason, the scanning strategy was quite different from others presented in this book. Whereas successive orbits follow ground tracks that are about 25° apart, the same orbits follow 'sky tracks' that are no more than 0.07° apart. (Note that we use the term 'sky track' to denote the path of the satellite's 'zenith' point; not the aiming point of the telescope.) Whereas a Sun-synchronous orbit enables the Earth to be scanned completely in one day; it takes six months for such an orbit to get round the entire sky. Scanning of the sky is inevitably basically sequential, although,

Fig. 15.1 — (a) The IRAS Spacecraft. Courtesy: RAL.

as is discussed in the next section, certain liberties may be taken if the instrument is well designed.

Unlike most Earth-resources missions, the IRAS mission did not require adherence to any kind of published timetable. Therefore the satellite was placed in the

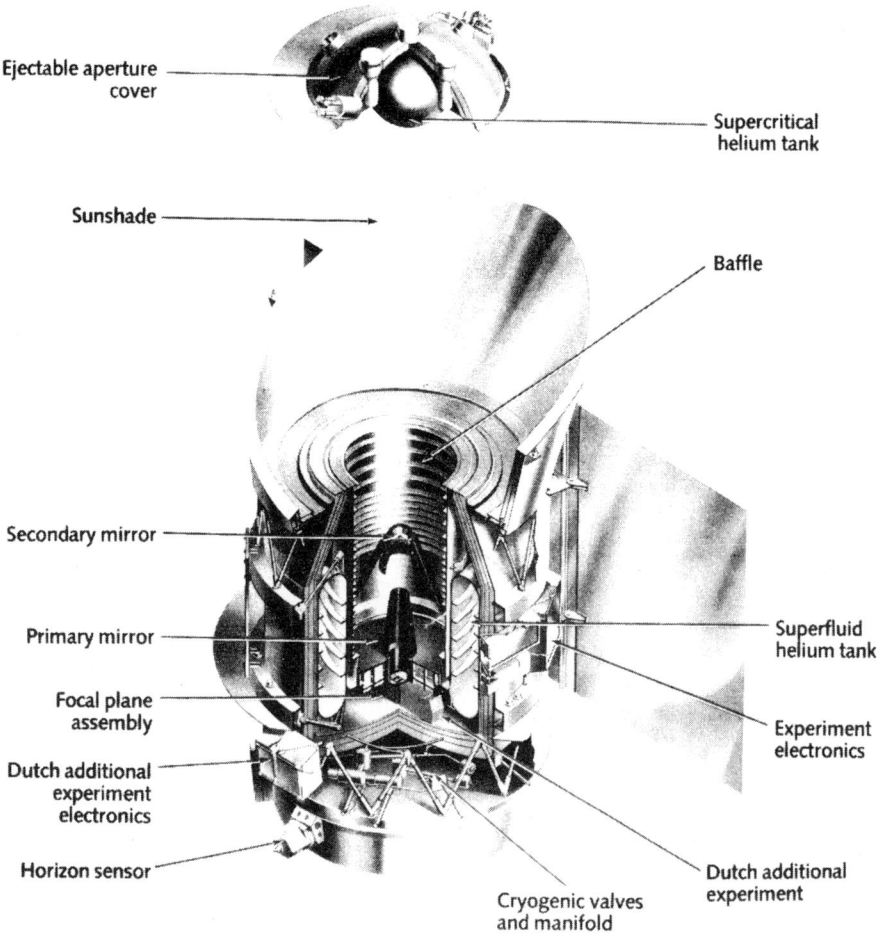

Fig. 15.1 — (b) the IRAS Instrument. Source: IRAS (1988).

most accurate possible orbit, and then left to fly free, with the position monitored at all times, but with no corrections applied, except of course to the attitude. The reason for using a Sun-synchronous orbit was purely to facilitate the screening of the telescope aperture from sunlight and earthlight. Entry into the telescope of either of these would have caused the liquid helium to boil off rapidly enough to cause the telescope to explode.

15.3 SCANNING

The cross-track scanning method employed by IRAS was basically pushbroom, although with considerable variations, as discussed in Section 15.5. Along-track scanning was obtained by rotating the satellite so that it remained looking more or less directly away from the Earth, just as Earth-resources instruments are rotated to

Table 15.1

Spacecraft				
Mass — satellite (kg)	809	Launched	1.83	
— TM (kg)	173			
		Target	Sky	

Orbit			
Altitude @ Equ.(km)	915	Orbits/day	14.00
Semi-major axis (km)	7272	Orbits/cycle	14 (referred to Earth)
Eccentricity (%)	—	Days/cycle	1 (referred to Earth)
Period (min)	102.9	Shift/orb (°)	25.71 (referred to Earth)
Inclination(°)	99.0	km @ Equ.	2863 (referred to Earth)
Descending node	—	*Sky* track spacing (°)	0.0714 (satellite path)
Ground speed (km/s)	6.47	km @ Equ.(km)	—

Telescope			
Type	Ritchey Chretien		
Focal length (m)	5.45	80% blur c. d. (μrad)	121–484
Aperture (m)	0.57	F.P. dia (mm)	102
f-number	9.56	Field-of-view (°)	±0.537
1y mirror dia (m)	0.60	IFOV (μrad)	1469 by (881 to 221)[†]
2y mirror dia (m)	0.24	(μm @ F.P.)	8004 by (4802 to 1204)[†]
Total obscuration (%)	18	(m on ground)	n/a
Clear area (m^2)	0.210	EIFOV(m on ground)	n/a
Exit pupil		Interband reg. (pix)	n/a

Scanning			
Method	Complex pushbroom		
Rate (double scans/s)	n/a	Mirror swing (°)	±3.85
(rad/s)	0.0011		
Period (ms)	n/a		
Active scan time (ms)	n/a		
Efficiency (%)	60		
Sampling interval (ms)	—		
Integrating time (ms)	190/190/390/780		

Channel details

Band	Wavebands (μm)	Detectors	NEFD‡ (W/m^2/ $\sqrt{\text{Hz}}$)×10^{26}	Noise equiv. power (W/$\sqrt{\text{Hz}}$)×10^{16}	MTF @ pix rate trk/scan
12	8–15	SiAs	0.03	3	—
25	17–30	SiSb	0.025	0.6	—
60	40–80	GeGa	0.046	1.0	—
100	83–119	GeGa	0.21	0.6	—

Cooling
Superfluid helium

Electronics	
Consumption (kW)	—
Tape recorder?	Yes: housekeeping + data
Compression	Yes

Downlink	
Data rate (Mbit/s)	0.004/1
(GHz)	—

Data	
Word length (bits)	16
Grey levels	65 536
Image width (pix)	6000
Image width (km)	n/a

†See text.
‡Noise equivalent flux density.

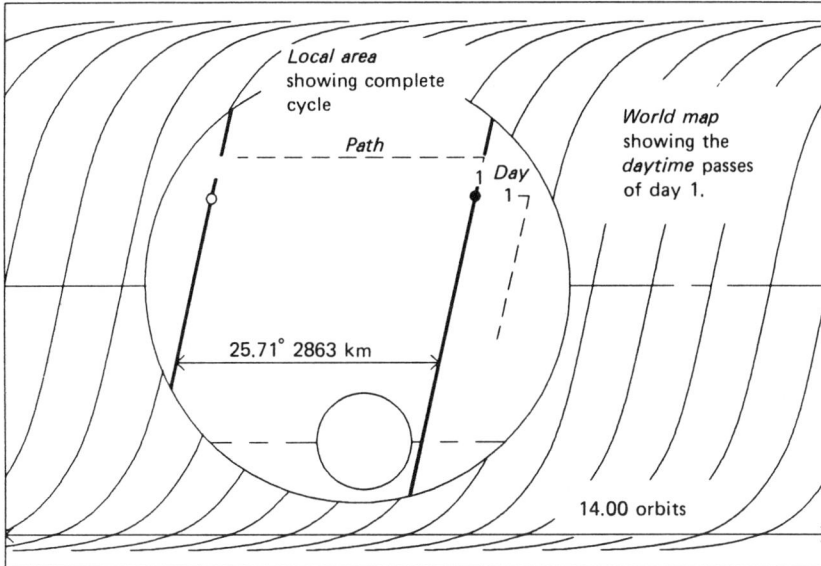

Fig. 15.2 — Ground-track pattern for IRAS.

look in the opposite direction. This was an important feature of the cooling arrangements, because it protected the aperture of the dewar from earthlight.

In practice considerable liberties were taken with telescope alignment in order to provide a degree of flexibility for the surveying strategy. The strategy for each orbit was planned to keep the aiming point a safe angular distance away from bright sources such as the Sun, the Earth or the Moon. Smaller sources such as Jupiter were often simply sidestepped.

As a general practice, the satellite was rotated in pitch some 10% faster than the orbital rate of $0.058°/s$, giving a scanning rate of $0.064°$ (1.12 mrad)/s. This meant that small amounts of lost time could be made up without leaving gaps in the coverage. (Compare this method of 'off-zenith' viewing with the 'off-nadir' viewing *mirror* employed by the SPOT HRV instruments (Chapter 16); and the method of rotating *half* the telescope employed by Meteosat (Chapter 11).) A good example of the use to which this freedom could be put would be the sidestepping of Jupiter. By adjusting the roll angle at the start of each sweep, the spacing of the 'aiming paths' (traced out by the aiming point) was made considerably coarser than the sky track spacing of $0.0706°$ (1.23 mrad) per orbit. As is discussed in Section 15.5, the width of the detector array is 0.5° (8.7 mrad). Successive scans were normally separated by just under half this angle, namely 0.237° (4.14 mrad), to enable each source to be confirmed by an 'hours confirming' second pass. Sources which had moved within this time were rejected as being certainly local in origin. (Other, longer-period, re-observations were used to distinguish between slower-moving sources such as planets and asteroids, and fixed stars.)

An interesting constraint that would have left a significant area uncovered had it not been for this flexibility, was created by the 'South Atlantic anomaly'. This is a depression in the Van Allen belts that occurs over the South Atlantic, and increases substantially the radiation levels at the altitude of the mission. The radiation temporarily increases the sensitivity of the detectors, and renders invalid any data acquired during passages through it. The detectors were therefore switched off during such passages. The gaps in coverage were filled during later scans, when the sky areas concerned were clear of the anomaly. It is interesting that none of the other missions chronicled in this book have reported any problems with this anomaly. This could be either because their altitudes are normally some 100 km or more lower, or because their sensitivites are very much less.

15.4 THE TELESCOPE

The layout of the IRAS telescope is shown in Fig. 15.3. It is a straightforward Ritchey Chretien design, as discussed in Chapter 3. However, because of the high contrast between the brighter sources and the background — and the wide range of source-intensities explored — stray radiation was an unusually serious problem. Other systems may, or may not, include baffles to minimize the possibility of reflections allowing stray light to reach the focal plane. However, if so, they are certainly not accorded the importance given to them in the IRAS design. For this reason, it has been thought desirable, for IRAS only, to indicate their presence in the figure.

The telescope was made entirely of beryllium (Devereaux, 1983). This is a material that features a good deal in telescope design reports. Its basic properties are light weight, coupled with more than adequate strength, and corrosion resistance. It possesses good thermal conductivity, which minimizes thermal gradients, and high thermal capacity, which reduces the temperature changes generated by a given thermal input. Since the telescope had to be built and tested at ambient temperatures, and then cooled down to 2° K for operational use, it was important that a single material should be used throughout. The secondary mirror was coated with aluminium to enhance its reflectivity at visual wavelengths. Sufficient signal strength was obtained without coating the primary mirror also. All non-reflecting surfaces were coated with optical black.

All space-borne telescopes have to be designed to retain their dimensions in two different gravity environments, as is discussed in Chapter 3. However, with the IRAS telescope there was the additional problem of coping with two widely differing temperature regimes. No attempt was made to render the telescope usable under ambient conditions; instead all testing was done at cryogenic temperatures. However, this meant that, not only did the structure have to withstand frequent temperature excursions, as the instrument had constantly to be returned to ambient

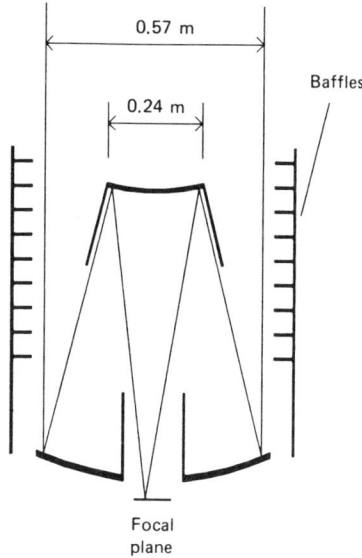

(a) Telescope

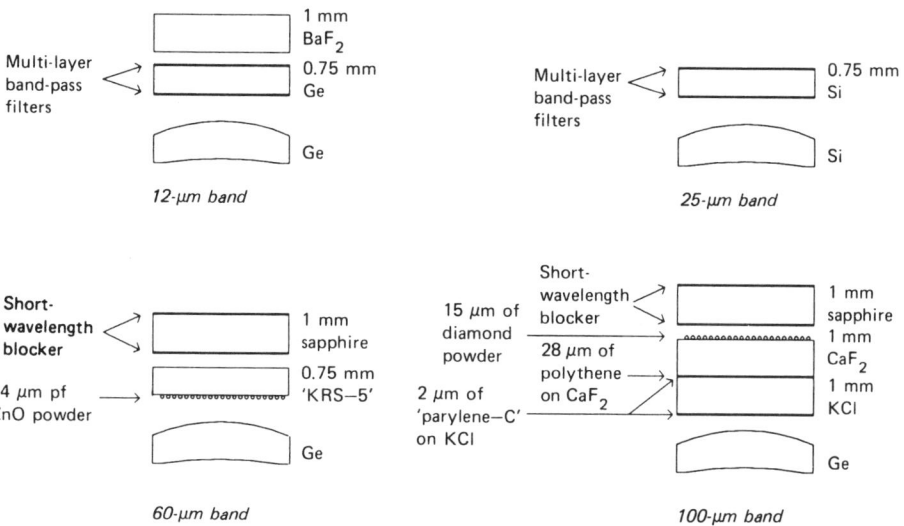

(b) Filter construction

Fig. 15.3 — The IRAS telescope.

for adjustment, but that the profiles to which the mirrors were cut figured and polished had to be modified slightly so as to be correct at 2°K.

15.5 THE FOCAL PLANE

The layout of the IRAS focal plane is shown in Fig. 15.4. Its design is discussed in some detail in Bamberg (1984) and Darnell (1984). At first sight, it is somewhat

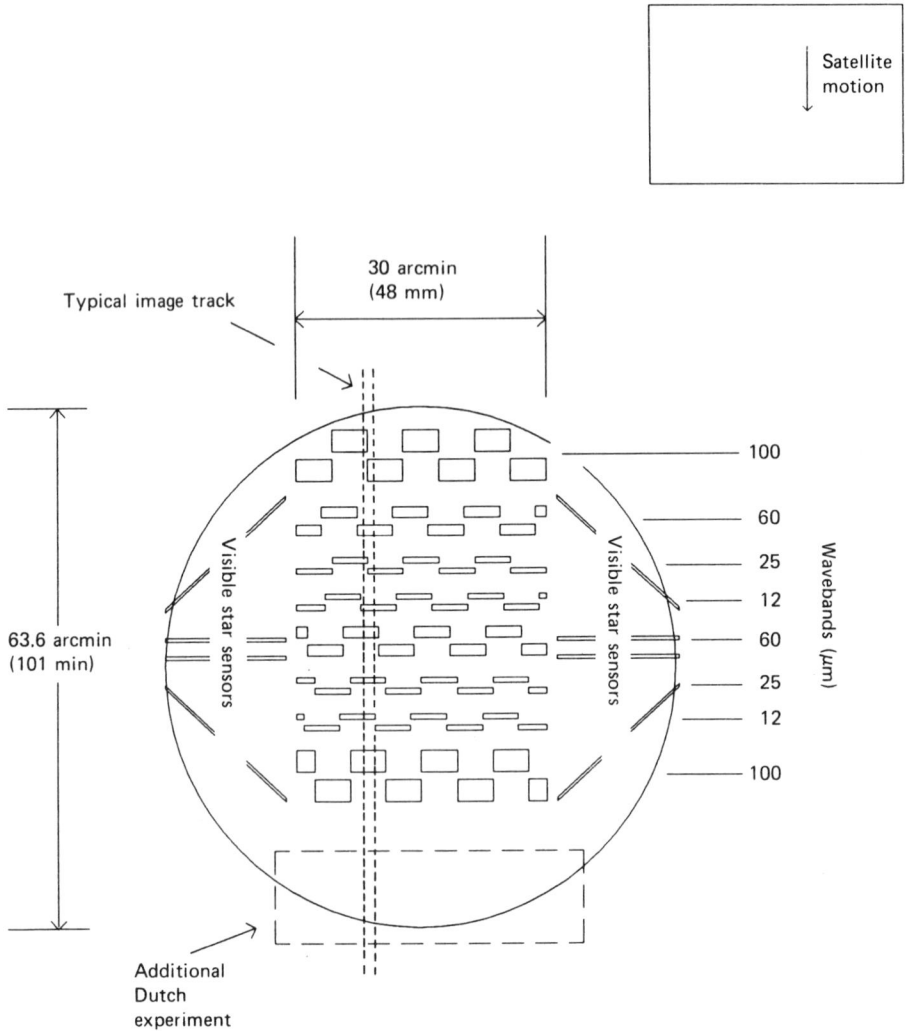

Fig. 15.4 — The IRAS focal plane.

reminiscent of those of Thematic Mapper (Fig. 14.4) and MSS (Fig. 8.4), which may appear surprising since the scanning system has been claimed to be basically 'pushbroom'. However, it must be remembered that IRAS was not designed strictly as an imaging instrument, in spite of its technical similarity to instruments that are. Its main function was to identify the positions and flux densities of point sources, most of the more important of which could be expected to radiate (albeit to different extents) in more than one band.

Again at first sight, we have eight arrays of seven or eight elements each; with each waveband being duplicated. The detectors are elongated, to enable the full 30′

(8.73 mrad) swath to be swept with, effectively, seven detectors. (The reason for there actually being eight, in most cases will become plain very shortly.) Thus economy of instrumentation might appear to have been bought at the expense of cross-track resolution. Closer inspection will reveal, however, that the detector apertures for the various bands are staggered in an apparently haphazard way with, as it were, the seventh detector often split between top and bottom. The focal plane is in fact a hybrid design. From the point of view of target *evaluation*, it is a moderately conventional, if rather coarse, multi-band array of four by seven elements. For broad-band target *detection* purposes, however, it is one large 62-element array. The pair of dotted lines on the figure represent the image track of a typical source.

Thus from a geometrical viewpoint it would appear that a source, visible in all four bands, could be located to within 30'/62 minutes (29" s) or 141 μrad. For a source visible in fewer bands, the resolution will, of course, be less. In practice, however, the image of a point source was relatively large, and depended on the wavelength. For the two longer wavelengths, the image was too large to contribute significantly to the location accuracy. However, those at the two shorter wavelengths overlapped onto adjacent detectors in such a way as to enable the higher temperature sources to be positioned to within 1 or 2" s (50 to 100 μrad). This was very much better than had been expected (Walker, 1990).

The design is an interesting example of a trade-off analysis arriving at a unique solution to meet a unique requirement. In essence, a relatively simple in-flight hardware design was bought at the expense of a much increased post-flight analysis load.

Duplicating each waveband enabled fast-moving sources to be eliminated from the subsequent processing as being local in origin, even on a single-band analysis. A source that passed this test was regarded as having been 'single-band seconds-confirmed'. The aiming paths observed were arranged to be just under half the swath width apart (i.e. 4.14 mrad), giving a guaranteed 100% overlap. This enabled more slowly moving objects to be rejected, and enabled candidate sources to be 'hours confirmed'. Note that the arrays for the three shorter wavebands are duplicated in the same order, but that the two 100-μm arrays are placed as far apart as possible to counter the reduced resolution at this wavelength.

The specification had demanded that the instrument be 'diffraction-limited' at all wavebands. In the terminology used in this book, this means that the detector apertures should be smaller than the Airy disc (which approximates to the 80% blur circle). While the lengths of the apertures was dictated by the considerations already discussed, their width had to be as large as possible, within this constraint, to maximize radiance capture. Effectively, this meant that widths should approximately halve between one waveband and the next. In practice, it did not prove possible to meet this requirement for the 12-μm band, whose width is the same as that for the 25-μm band. Thus the resolution is the same for both these bands. Note that the term 'resolution' is used here in very nearly its proper context, namely the ability of the instrument to resolve two closely spaced point sources.

The much elongated apertures either side of the main array are visible star detectors. Each array is one track-separation wide, so that known fixed stars are picked up by one visible array, then twice by the main infra-red arrays and then again by the other visible array. Each A-shaped pair is in fact a single unit. The track followed by the image of a source can be computed from the transit time between the two. The visible arrays are also attitude detectors. The yaw angle is computed from the difference in transit time between the two A-pairs on one side of the focal plane. The yaw angle must be known accurately if errors in the position computation of infra-red sources are to be avoided. Rotations about the other two axes could be inferred — if the precise position of the instrument was known — by comparing the apparent (measured) positions of fixed stars with their known actual positions. A pitch error would be detected by a source being picked up in the correct position across-track but at the wrong time. A roll error would cause the source to be picked up at the correct time but in the wrong position across-track. Once again, in-flight simplicity was bought at the expense of post-flight analysis complexity.

It is characteristic of most reflecting telescope designs, including the Ritchey Chretien, that the zone of perfect focus is in fact saucer-shaped. However, even at 12 μm, the depth of focus was ±1 mm, which was sufficient to enable a plane detector array to meet the specification.

15.6 THE DETECTORS

Fig. 15.3(b) shows the design of the different filters, while Fig. 15.5 shows the construction of the focal plane assembly. Accurately machined apertures control the raw IFOV of the instrument. (The term 'raw' is used to highlight the much greater difference, between the (cross-track) IFOV of a single detector and the useful resolution of the instrument, than is the case with most Earthward-looking instruments (Section 15.5).) Behind the apertures come the spectral filters and, behind these, an array of cylindrical 'field' lenses. The field lenses are designed to concentrate the radiance captured by the aperture on to as small an area of photocell material as possible. As is discussed in Chapter 5, increasing the radiance flux on a given area of material increases the signal to noise ratio and improves the signal quality. In particular, it helps to swamp the effects of ionizing radiations.

Because of the low radiance levels of stellar infra-red sources, it was important to have the transmittances of the optical elements as high as possible. This was made difficult by the fact that the optical properties of most materials is affected by temperature. It proved possible to find materials for the field lenses, whose optical and physical properties at cryogenic temperatures were well-known. Both germanium and silicon were suitable, with germanium having the edge, for all bands except Band 2. Both materials have absorption bands in this region (15–30 μm); but on balance silicon was felt to be the least compromised and was chosen for this band.

No similar information was available on the transmission properties of candidate filter materials at these temperatures, and so a considerable amount of information had to be amassed before any design decisions could be made. For the two short-wavelength bands, multi-layer interference filters were built up which gave good sharp cut-offs, as Fig. 15.7 indicates. But at the longer wavelengths it was necessary to rely on the bulk properties of several different materials, with results which were less satisfactory. The problem of out-of-band exclusion exists for all systems, but was far more acute than usual for IRAS because of the extremely low in-band radiance

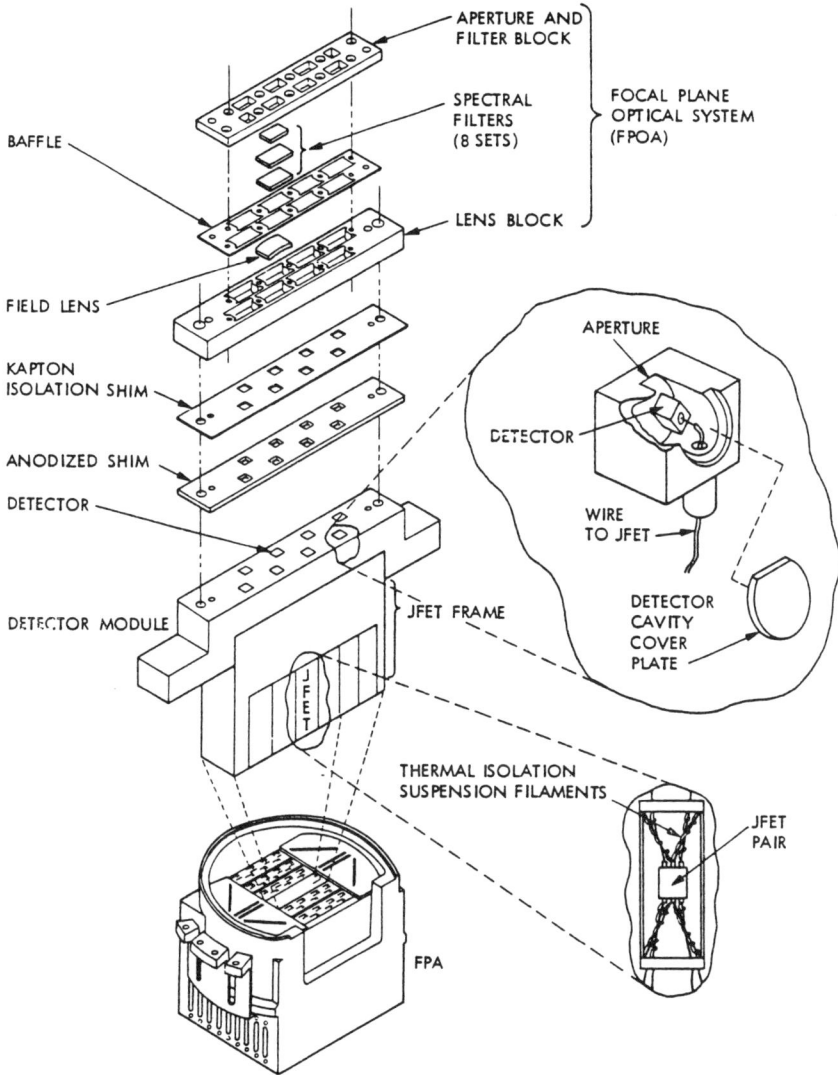

Fig. 15.5 — Exploded view of FPA assembly.

levels. Not only did great care have to be taken with the filter design, but the housings had to be totally light-tight around the field lens edges (a minor failure in this area, on Thematic Mapper, is described in Chapter 14).

The detector cavities themselves were also the subject of considerable design effort. The cavities were made from gold/platinum alloy to shield the photocells from gamma rays; and those for Bands 3 and 4 were designed as 'reflecting integrating' cavities. These cavities are the optical equivalent of a reverberation chamber. Quanta which miss the target are reflected and re-reflected from the cavity walls until they strike it.

Very few performance measures of the completed assembly were carried out before launch. This was largely because of the extreme difficulty of so doing with a cryogenic telescope. (The problems encountered in testing other instruments, with considerably less extreme requirements, have been chronicled elsewhere.) It was left largely to the in-flight checks (Section 15.9) to evaluate both the spectral and the spatial responses.

15.7 THE DETECTOR ELECTRONICS

The detectors were photoresistors of various materials. The job of the electronics is to translate the change in resistance with photon flux into a voltage, and eventually into a digital signal.

The radiation levels are very small, so the resistance of the diodes remains very high. This renders the circuit highly susceptible to noise, dark current, and pick-up problems. Pick-up was avoided by good screening. The other problems were countered by placing the first stage electronics in the cold area. The individual detectors were each backed by a matched pair of junction field effect transistors (JFETs), housed within the cold block. The JFETs were connected to the warm electronics by special low thermal-conductivity wires. The electrical dissipation of the JFETs (about 200 μW) was sufficient to maintain them at an adequate working temperature of 60–70 K. An integral heater was provided for cold-start purposes.

The detector electronics are illustrated in Fig. 15.6. A small constant voltage was applied to one side of the diode (the bias supply). This input voltage could be set to any of three levels; low, nominal and high. This was the means whereby the 'gain' of each detector could be adjusted. The JFETs effectively formed the differential input stage of a unity-gain source follower amplifier, converting the high-impedance output of the photodetectors to a low-impedance signal of the same voltage. The low-impedance signal could then be handled safely by the warm electronics.

The rest was basically standard operational amplifier practice. The output of the operational amplifier was fed back to the other end of the detector, through a 2000-MΩ feedback resistor. The feedback current maintained a 'virtual earth' at this point by exactly matching the current through the detector. The feedback current was proportional the amplifier output voltage, which was thus maintained proportional to the current through the detector. A 1-pF capacitor across the feedback resistor reduced the gain of the system above some 80 Hz.

To minimize noise generation, the feedback resistor was mounted in the cold part of the electronics. This caused problems because the resistance of even the best available resistors varied somewhat with temperature. At 2 K the resistance of the

$0.1\mu F$

2×10^{10}

$70\ ^{\circ}K$

$806k$ $806k$

$-12\ V$

Bessel filter
(150 μm)

Anti-aliasing
filter network

To A/D
converter

D.C.
compensation

InAs
photo-
resistor

$3M$ $15\mu F$ $0.01\mu F$

$+11\ V$

$133k$

Spike
detection
circuit

Multiplexer

Liquid He cooled

Bias supply

Warm

Note

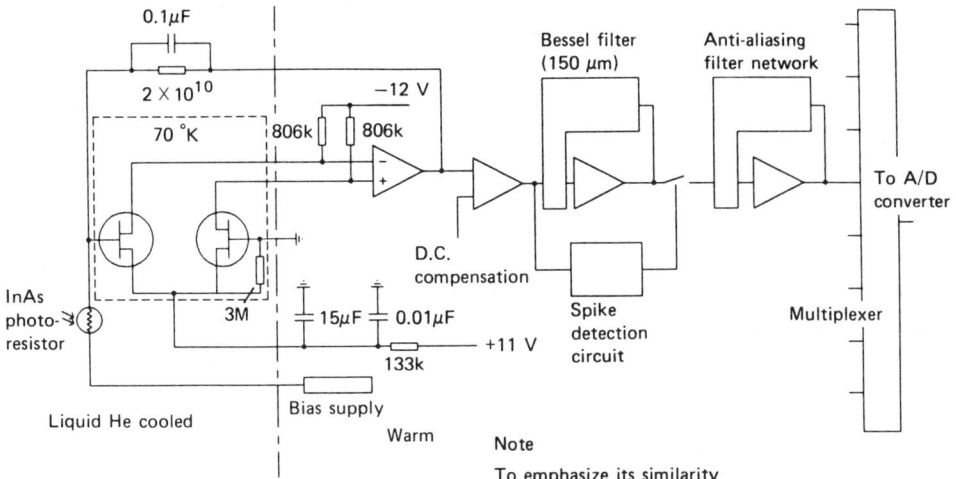

To emphasize its similarity
with other circuits, this
pre-ampliier stage has been
drawn with the positive and
negative rails inverted.

A typical infra-red channel

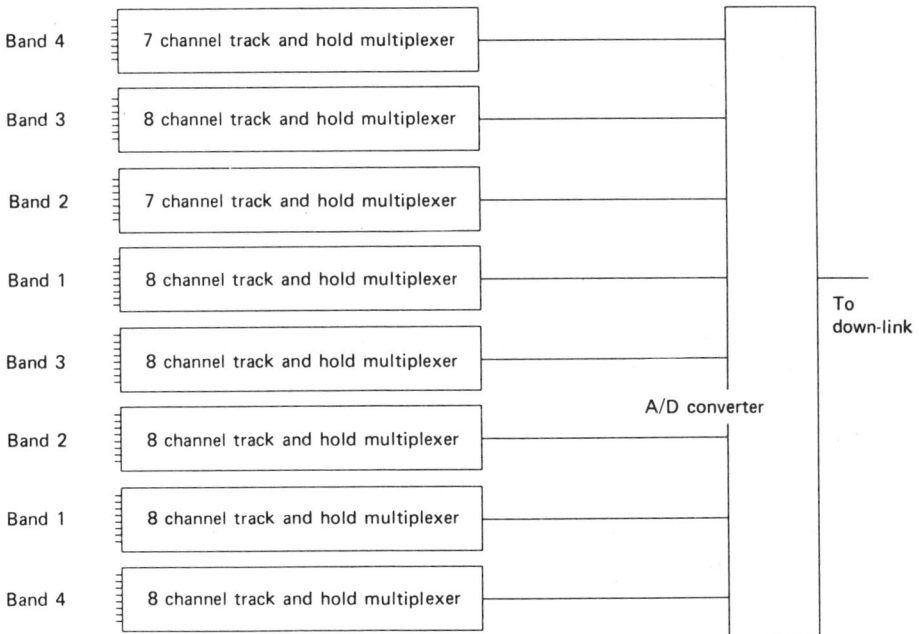

Band 4 — 7 channel track and hold multiplexer

Band 3 — 8 channel track and hold multiplexer

Band 2 — 7 channel track and hold multiplexer

Band 1 — 8 channel track and hold multiplexer

Band 3 — 8 channel track and hold multiplexer

Band 2 — 8 channel track and hold multiplexer

Band 1 — 8 channel track and hold multiplexer

Band 4 — 8 channel track and hold multiplexer

A/D converter

To
down-link

Fig. 15.6 — IRAS electronics.

nominal 10^{10}-Ω resistor selected doubled. At this temperature, its resistance also varied with voltage — presumably because of its effect on the resistor's internal temperature. Since this resistor controls the overall gain of the amplifier, the effect had to be modelled, and corrected for, during the data analysis process.

On one of the detectors, the frame became shorted to earth. A modified circuit was designed, in which the faulty detector was energized in the opposite direction from the normal. By applying the bias voltage to what would normally be the earthed side of the JFET pair, it was possible to convert the virtual earth to a virtual fixed voltage, and thus reverse the operation of the amplifier.

In addition to three different gain settings, it was possible to introduce seven different offset voltages in parallel with the pre-amplifier output. Because the detectors were so cold, there was no problem with dark current, but voltage offsets could still be produced by the DC-coupled electronics. The offset was adjusted, along with the gain, to ensure that the output of the analogue electronics remained within the range of the A/D converter.

Because of the extreme sensitivity of the instrument, it was necessary to take steps to block the sharp spikes produced by cosmic ray strikes. This was done by passing the signal through a Bessel filter designed to generate a time delay of 150 µs. This gave time for a spike detection circuit, driven by the undelayed signal, to flip a fast-acting switch, and to prevent the spike from reaching the A/D converter. A track/hold amplifier provided a constant signal to the following circuits while the switch was open.

A further amplifier provided an additional facility for gain adjustment. The anti-aliasing filters rolled off at 12 dB/octave, with a cut-off frequency of 6 Hz for the 12- and 25-µm bands; and 3 and 1.5 Hz respectively for the other two. The signals then passed to multiplexers (one per band), and then to a single A/D converter, operating at 125 µV per 'grey level'. The wavebands were sampled at rates proportional to their spatial resolution, namely 16, 16, 8 and 4 Hz.

The visible detector electronics were similar, except that a MOSFET pre-amplifier was added after the source follower amplifier, and the system was AC-coupled at this point.

The infra-red detectors were sampled at different rates for the different wavebands.

After the ADC, the digital electronics took over. These processed the digitized data, collected telemetry information from various additional sensors, received and executed commands from the onboard computer, and transmitted the data to the onboard computer. From there, the data were compressed (see next section) and read out to one of two onboard tape recorders.

Two identical computers were installed, each being able to access the other's memory. Each had 32K of 16-bit RAM and 3K of ROM. The ROM was used to store routines and data essential to the safety of the mission.

15.8   DATA COMPRESSION

The quantity of data to be stored and transmitted was reduced by one half, by reducing each 16-bit digit to 8-bits. The technique was basically to transmit the full 16 bits at intervals (in fact, once every 4 s). The remaining data are recorded as differences from the preceding value. The technique is discussed in detail in IRAS (1988). Briefly however, it operated as follows:

— The 16-bit value of the previous datum point was reconstructed by reversing the compression step.
— The difference was evaluated between the current 16-bit value and the previous 16-bit value.
— An 8-bit word was constructed from this difference in a modified floating point form. The most significant bit gave the sign of the difference. The next 3 bits were the 3 most significant bits of the difference, and the remaining 4 bits were a code representing the exponent.

The accuracy of this compression technique is clearly critically dependent on the rate of change of the incoming signal. However, in practice it appears that the errors introduced were small. This system may be compared with that adopted by SPOT (Chapter 16). The SPOT strategy produces a smaller degree of compression, with no loss in certainty, but with a loss in radiometric resolution which varies with the interpixel contrast.

15.9 CALIBRATION AND IN-FLIGHT CHECKS

Unlike most other instruments described in this book, all the serious evaluation and calibration of the complete system was left to be carried out in orbit. Readers of earlier chapters will already be aware of the problems encountered in ground testing other systems. In particular it is recorded, in Chapter 14, that Hughes found it as difficult to create test equipment capable of checking out Thematic Mapper as to produce the instrument itself. And this was for an instrument dealing with relatively high light levels; the bulk of which is operated at ambient temperatures. To have carried out convincing pre-launch tests on IRAS would have been very much more difficult still.

Once in orbit, however, both relative and absolute calibrations were attempted. The objective of the relative calibration was to reconstruct the flux-induced current in the detector at each significant point in time. An internal reference source was used for this purpose, mounted behind a small hole in the centre of the secondary mirror. It was able to input flashes of constant brightness in each of the four wavebands. This permitted confidence in the repeatability of the system to be maintained. The planetary nebula NGC 6543 was used as an external check. Being near the north ecliptic pole, NGC 6543 is observable on every orbit, and it is visible in every waveband. It is also small enough to count as a point source, at least for the relatively large detectors of IRAS. The stability of the internal source was shown to be stable to within 2% throughout the mission. For diffuse sources, an area of sky was used. An area was chosen which had smoothly varying brightness, was free from point sources and which was also close to the north ecliptic pole.

The absolute calibration attempted to relate the detector currents to absolute source brightnesses, as obtained by other techniques. This was more difficult, because IRAS was flown to study wavelengths that do not penetrate the Earth's atmosphere. Also there were no reliable alternative measurements made directly at these wavelengths. It was necessary therefore to extrapolate from brightness measurements made at shorter wavelengths. For the two shorter wavelengths, stars were used. However the ratio of measured flux to extrapolated flux was very large, and the precision of the calibration remained a little uncertain. For the longer wavelengths, asteroids were used. Being much cooler, the extrapolation required was much less. Unfortunately asteroids are irregular objects, whose albedo is constantly changing. Fortunately the principle objective of IRAS was not affected by minor uncertainties in the absolute flux measurements.

The spacecraft attitude and position were checked at the beginning and end of each scan by reference to visible stars. Where possible, a fix was also obtained near the middle of a scan.

The basic performance of a detector is, in effect, its ability to pick out a weak genuine signal from its own background noise. Its sensitivity in terms of volts per unit flux is just one of the factors contributing to its performance. The performance of the detectors were defined in terms of their noise equivalent flux density (NEFD), which is the amount of light that would give the same signal as that channel's own internal noise. The NEFDs of all the channels were evaluated by analysing 5-min samples of data, taken while the telescope was observing a dark area of sky. While the background sky was dark, point sources were always present. So the algorithm that evaluated the noise signal had to discriminate against sharp peaks. The NEFDs from six representative samples were averaged to give a single final estimate. The total scatter, for a given detector, was in fact less than 25%. Some detectors were, however, considerably worse than the average. A mean noise level for a band was therefore evaluated, leaving out the few detectors that were particularly bad.

Note that each point source contributes to three data samples, two of which have a weight of 0.5.

Throughout the mission, the performance of the detectors was very stable. Most continued to exhibit their pre-launch behaviour. Although some were affected by the launch, they maintained their new calibrations thereafter.

Many of the detectors suffered from a small amount of 1-Hz cross-talk from the temperature sensors mounted in the focal plane. This was revealed by evaluating the power spectra of the fluctuations in the data streams. However for only two detectors did the cross-talk exceed the r.m.s. noise. In addition, however, one detector picked up some 0.25 Hz from an engineering data multiplexer.

Having found this cross-talk, it was possible to correct it to a considerable degree with the ground software. The noise level of one detector increased by a factor of 3 for two periods (one of a few days and one longer) during the flight. No explanation for this behaviour was found.

The detectors were not entirely linear. Their responsivity could be affected by the total amount of light falling on them, and also on the rate of change of the incoming light signal (frequency response). The effect of varying the total flux was checked by observing asteroids as they approached the Moon. As they did so, stray light reached the detectors by extraneous paths, and increased the background illumination at a smooth and progressive rate. The effect of frequency response was measured by slewing the telescope at various rates, from the nominal (3.85'/s) down to 1/16 of nominal. It was possible also to stare at selected sources, with designated detectors, for up to 120 s. The internal reference source was also used to give a constant illumination for up to 120 s. The 12- and 25-μm detectors turned out to be linear, with a uniform frequency response, independent of background levels. The 60- and 100-μm detectors give much less clear-cut results, being particularly affected by high illumination levels.

The stability of the electronics was checked by monitoring a region of the sky near the north ecliptic pole. Stability proved quite good on a short-term (less than a day) basis.

After the mission was completed, it was discovered that the sensitivity of the 100-μm detectors increased after receiving a large signal, such as passage through the galactic plane. This was discovered by comparing upward and downward scans. Analysis of special calibrations over Saturn confirmed the effect (more details are given in IRAS (1988)).

When a bright source passed close to the field of view, stray light tended to reach the detectors and cause 'optical cross-talk'. When a bright source actually crossed the focal plane a pattern of residual reflections from the secondary mirror struts produced a diffraction pattern which could confuse the point-source detector algorithm. A number of tests were carried out to verify the predicted rejection of such out-of-field sources. These tests confirmed the size of the no-go zones round Jupiter, the Moon, the Sun and the Earth.

Out-of-field performance was monitored throughout the flight, because of concern that gases escaping from within the materials used could re-deposit on the structure and increase the light scatter. No degradation in out-of-field rejection was found. This is interesting, because many instruments had great problems with the re-deposition of escaping gases.

NGC 6543 was used to check the stability of the internal reference source. Overall stability was within 2% throughout the mission.

15.10 DATA QUALITY AND QUALITY CHECKING

In the case of IRAS, the term 'data quality' means mainly the probability of a genuine candidate reaching the catalogue, and of a spurious peak being rejected. Of slightly less importance is the accuracy of the data recorded for each entry. Sources with but 'moderate' quality flux estimations were included.

The completeness of the catalogue was improved significantly, for the lower flux level sources, in areas where a third survey was completed. The third survey covered 78% of the sky before the helium ran out (Walker, 1990).

15.10.1 Radiometric performance
As we have already mentioned, this is not a subject that is of great interest to users of IRAS data.

The noise equivalent flux density was measured by scanning areas of high galactic latitude (which should be reasonably free from stars), and away from known extended sources. To evaluate the NEFD from the electronic signals, a 'Gaussian noise estimator' was used, which was designed to discriminate against sharp peaks. Thus a complete absence of stars in the sky region used was not essential. The results from six separate scans were evaluated. There was less than 25% difference between the various evaluations for each detector. An average NEFD was evaluated for each detector, and then a mean noise for each band was evaluated, omitting detectors which were particularly noisy. The results are given in Table 15.2.

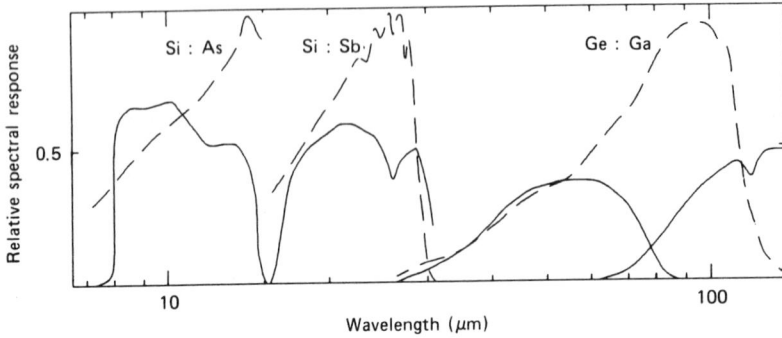

Component reponses

Solid lines show transmission filter/lens combination

Dashed lines show response of detector materials

Note logarithmic wavelength scale

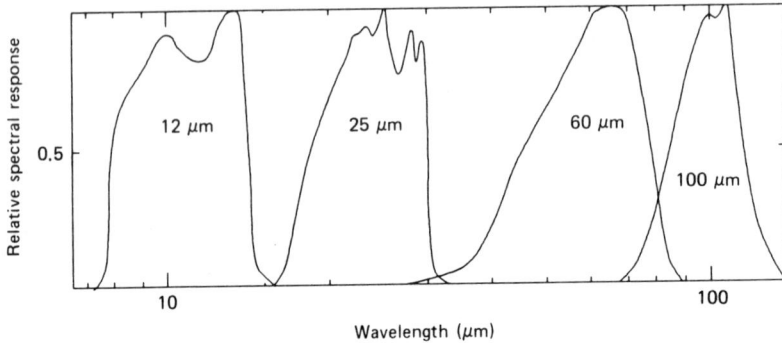

Channel responses (summation of above curves)

Fig. 15.7 — IRAS spectral responses.

Table 15.3 was obtained from Low (1983).

15.10.2 Spectral performance

Fig. 15.7 shows the pre-launch spectral response curves, as well as the component curves from which they were predicted.

Tests were carried out to check the consistency of the spectral response of all the

Table 15.2

Channel (μm)	NEFD (10^{-16}W/m^2)	Uncertainty (±)
12	132	51
25	62	8
60	43	4
100	59	12

Table 15.3

Band	Optical efficiency	Responsivities		NEP	NEFD
		System	Detector		
(μm)				$(W/\sqrt{Hz}\times10^{-16})$	$(W/m^2/\sqrt{Hz}\times10^{-26})$
12	0.47	0.115	0.35	3.0	0.03
25	0.44	0.53	1.7	0.6	0.025
60	0.55	0.34	1.0	1.0	0.046
100	0.16	0.24	1.8	0.6	0.21

detectors. First, all the detectors were carefully calibrated against NGC 6543 which, for the purposes of this test, may be considered as a point-source 150 K black body with spectral lines. Then each detector was used to measure a number of sources both much hotter and much colder than NGC 6543. Any abnormal spectral response will generate an abnormal reading from either the hot or the cold sources. All results agreed to within 16%, and most agreed within 10%. This implied that no faults had developed with individual detectors, such as delaminations of the filter elements, leaks around the filter mounts etc.

But, because it had not been possible to check the spectral response out thoroughly before launch, there was still the possibility that the actual response differed from that predicted from the component tests. A semi-quantitative test was arranged to probe this problem. Sources with a wide range of temperatures were observed, and the ability of the detector to *reject* radiation outside their passbands was studied. In general it appears that the results were satisfactory.

15.10.3 Spatial response
The spatial performance or 'resolution' of most instruments covered in this book was evaluated in terms of their modulation transfer function, or spatial frequency

response. For the SPOT instrument (Chapter 16), it was evaluated by assessing the degree of blurring on imagery of densely populated areas. For Thematic Mapper, greater emphasis was placed on the evaluation of component performance (Chapter 14). Spatial frequency response is effectively a measure of the degree of blurring, and of consequent loss of detail in an image. While it would be perfectly legitimate to evaluate a frequency response from an image derived from a series of points, it is but moderately appropriate to the problem of resolving and positioning such points. Any information about positional errors is in fact lost from a frequency domain presentation, unless the spatial phase lag is also presented, which it normally never is. (It follows from this that high-contrast features on terrain imagery are likely to be slightly displaced in the down-scan direction, but that is another story.)

It is hardly surprising, therefore, that the IRAS team chose not to go down that path. Markham(1984) gives results for Thematic Mapper, both in terms of spatial frequency response, and the time-domain-oriented 'impulse response'. The latter are effectively the same as the strategy adopted for IRAS. What is important to IRAS was the ability of the instrument to resolve closely spaced point sources, and also the precision with which such point sources can be located. The spatial performance of each IRAS detector in the critical across-track direction was evaluated by arranging for the image of a particular point source (our friend NGC 6543) to be scanned across it several times at different positions along its length. By reducing the aiming path spacing, the spatial performance of each detector could be evaluated at four or five positions along its length. The responses of individual detectors was found to vary quite widely, as Fig. 15.8 shows. The responses at the two longer wavelengths are relatively smooth, as is to be expected if the instrument is truly 'diffraction-limited' at these wavelengths. The response at 12 μm is, however, for most detectors, quite sharp and steep-sided, again as one might expect if the instrument is IFOV-limited at this wavelength (i.e. if the Airy disc/blur circle is significantly smaller than the size of the aperture). Interestingly, however, the response at 25 μm is very similar, whereas, according to the specification, the response should be diffraction-limited at this wavelength. The effect has been attributed to the cylindrical convex field lenses, which have significant transmission loss at these wavelengths. This leads to improved performance at each end of the detectors, where the lens thickness is less.

In the along-scan direction the spatial response of the detectors proved to be much more uniform, and it did not prove necessary to use individual profiles in the subsequent analyses.

15.11   THE HOST SPACECRAFT

The 'host satellite' in fact comprised various units mounted integrally with the telescope but thermally isolated from its cold parts by heat-shields and by low-conductivity mountings (nine fibreglass straps). A detailed description is given in Pouw (1983). Briefly however, it comprised the two on-board computers, the fine Sun and horizon sensors, the S-band antenna, the power units, the battery (for use during eclipses), reaction wheels, gyros etc. The satellite major dimensions (with solar panels deployed) were 3.60 m (height) by 3.24 m (width) by 2.04 m (depth).

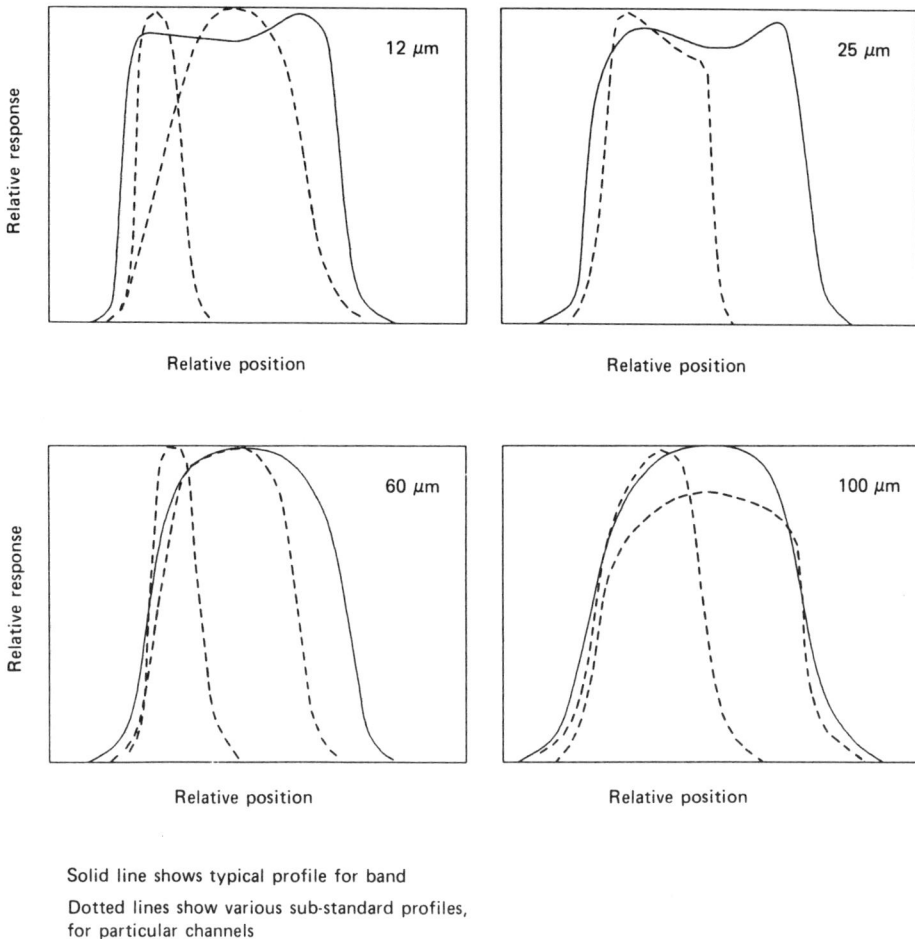

Solid line shows typical profile for band

Dotted lines show various sub-standard profiles,
for particular channels

Fig. 15.8 — IRAS spatial responses (across track).

15.11.1 Attitude control

Attitude was sensed, in the conventional manner, by a combination of a two-axis horizon detector and three orthogonal gyros. During normal operation, the Y-axis is kept pointing towards the Sun, using the Sun detector. This ensures that the sunshade and the solar panels are both at their most effective. Rotation about the remaining axis (the Z-axis) was detected by the Z gyro, to provide the required scanning rate across the heavens. For reasons discussed in Section 15.10, the telescope was in fact swung forwards slightly faster than the orbital speed would indicate. Because of its importance, the Z gyro was duplicated. The other two gyros were basically for back-up purposes only, and gave considerably less accuracy than the Sun detector. They were mainly used during eclipses, when no scientific data were taken.

The spacecraft attitude was controlled, again conventionally, by three orthogonal reaction wheels, with excess angular momentums being dumped to the Earth's magnetic field. Details of the technique were not given, but the general procedure is discussed in some detail in Chapter 12.

15.11.2 Temperature control

One of the unique features of the IRAS mission was the extremely low illumination levels with which the detectors had to contend. As is discussed in Chapter 5, solid-state devices of all kinds produce their own internal noise, which is a direct function of their absolute temperature. To meet the specification, the IRAS detectors had to be maintained at a temperature of some 2 K.

The rest of the telescope had to be maintained at an even temperature, if its dimensional stability was not to be compromised. Constant temperature enclosures of all kinds are easiest to control if they are either hotter or colder than their (variable) environment. A temperature-controlled oven requires the constant input of heat, which is obviously undesirable if a part of the system has to be kept cold, and therefore a temperature-controlled refrigerator was chosen instead.

The entire telescope was enclosed in a toroidal tank of superfluid liquid helium†. This in turn was enclosed in an open-necked dewar flask, surrounded by many layers of insulating blanket. The telescope aperture was protected from solar radiation by means of a sunshield. The temperature of the shield was maintained at about 95 K by a three-stage radiator. This is the logical extension of the two-stage radiator discussed in previous chapters. The inner surface of the sunshield was mirrored, to avoid scattering stray radiation into the telescope. The solar panels provided a degree of extra shielding to the body of the spacecraft. These precautions enabled the telescope assembly to be maintained at a temperature of 2–5 K, while the outer skin of the spacecraft varied about a mean of some 195 K.

As the liquid helium boiled off, it was vented to space through two nozzles mounted symmetrically on the dewar's exterior. The alignment of the nozzles was however less than perfect. During the mission, the satellite lost altitude at a rate of 10 m per day. This downwards drift stopped, however, as soon as the helium was exhausted.

The amount of helium nominally carried was 73 kg. Flowmeters were fitted to monitor routine usage. However, there was no charge indication in the tank, and no way of knowing how much helium had boiled off during launch. With hindsight, this is recognized as something of a mistake. Attempts were made to estimate the amount of helium left by carrying out 'sloshing' experiments involving moment of inertia evaluations. However, they took place too late to predict the end of the mission before it occurred.

To protect the telescope during launch, and during the first week in space, a self-contained cryogenic cover was fitted. The cover used supercritical helium, and operated between 6 and 15 K.

---

† **Superfluid helium** is the form of helium normally used for cryogenic applications, particularly in space. Its anomalous surface tension properties ensure that it adheres to the walls of tanks and thus remains accessible.

Additional temperature control was provided by keeping the line of sight away from hot sources such as the Sun, the Earth or the Moon at all times.

15.12 SATELLITE OPERATION

The ground station was able to make contact with the satellite on nearly every orbit. Because of the high data rates, only a few minutes' visibility was required to receive a day's collection of data. However, in practice the acquisition program was split up into nominally half-day satellite observation periods (SOPs). One high-zenith pass per SOP was selected as the prime pass, and it was during this that data were collected and instructions for the next SOP were transmitted. The passes chosen were nominally the closest to 0800 and 1800. The duration of a SOP varied between 10 and 14 hours.

If the link failed, then the next orbit was used as a back-up for the transmission of SOP instructions. Each tape recorder could hold the results of a complete SOP, and the two were used alternately. This meant that any overpass during the next SOP could be used for data reception.

During the mission, an apparent tendency of things to wrong at weekends, when key staff were absent, was statistically confirmed.†

After the mission, experiments were carried out on the fail-safe devices, a good many of which failed to work†.

An unexpected increase in the heat load, during the northern summer, was eventually attributed to reflections from the polar ice cap†.

Close to the end of the mission, the orbit entered an eclipse phase. It was necessary to programme the instrument to switch to battery stand-by mode during the period of each eclipse. If this was not done, a fail-safe procedure operated, which could only be reversed from the ground. Should this occur, some half an orbit would be lost. Unfortunately the ground-based eclipse prediction program was in error and the eclipses started a week before they were expected. As the situation was recovered, the helium supply failed†.

15.13 IRAS PRODUCTS

The satellite was controlled from the ground station at Rutherford Appleton Laboratory in England. And it was here that the data from the satellite were received. However the analysis was done at the Jet Propulsion Laboratory and Caltech. The data were obtained and processed one SOP-worth at a time.

Four different types of source were defined. The first were point sources up to about 2′ (0.6 mrad) in apparent extent, depending on wavelength. The second were small extended sources up to about 8′ (2.3 mrad) in the along-track direction, and the third were anything larger. Note that the selection criterion was based solely on the dimensions of the source in the in-scan direction. A source that is extended in one direction only would receive different treatment according to the direction of its

† (Walker, 1990).

extension. The fourth category were moving objects, such as asteroids and comets, which received a different type of processing. Basically the rejects, from the hours, weeks and seconds confirmation filtration, were scanned for objects with appropriate spectral characteristics. Those which passed were written to a special file for further analysis at a later date.

The additional Dutch experiment, indicated in Fig. 15.4, was a small slitless spectrograph. When a bright source was detected by the principle instrument, its spectrum was extracted from the spectrograph data and put through the same kind of confirmation testing (Walker, 1990).

The original data, as digitized, were reconstructed as closely as possible by reversing the data compression algorithm described in Section 15.6. The success of this reconstruction depended very much on the accuracy of the preceding data. Therefore very stringent tests for transmission accuracy were applied. If any one-second frame contained a single parity error then not only was that frame rejected completely, but so was the following frame and preceding frames. In general fewer than 5 s of data were rejected from each pass.

The path of each scan was reconstructed by using data from the Sun sensor and gyros, and by the observation of bright stars that lay in the scanning track. Infra-red observation of visible stars was used to check the accuracy of the reconstruction. Most survey observations contained two or more visual star observations. These enabled in-scan pointing errors to be kept to typically less than 5″, and cross-scan errors to less than 10″. However, for the few scans where one or fewer visual star observations were possible, the in-scan errors could rise to 1–2′.

Candidate point sources were identified by running a simple low-pass filter over the data stream from each detector. Only peaks greater than about 2.5 times the local r.m.s. noise level were picked out. In view of the enormous amount of data to be processed, the method of estimating the noise level had to be somewhat crude, and it tended to reflect the recent noise history of the detector, rather than its current experience. A real-time filter of any kind will inevitably suffer from the same problem, because the future is, as yet, hidden from it, whereas the past is known. It is possible to design a symmetrical filter for use on recorded data which takes equal account of future and past excursions. However, the computational demands are very large. The strategy adopted introduced large errors when the noise level was changing rapidly. In particular, it resulted in the loss of a number of genuine sources in the 'shadow' region down-orbit of the galactic plane.

A proportion of the point sources detected in this way could be expected to be spurious. They were either local (fast moving) objects or solar-system (slower moving) objects such as asteroids or comets. One asteroid passed so rapidly that it failed the 'seconds-confirmed' test, to be described.

Fast-moving objects were eliminated by comparing the detections from the two arrays at each wavelength. If both arrays did not score a hit at the right time interval to represent a stationary object, and at roughly the same intensity, then the candidate was immediately rejected. Those that remained became 'single-band seconds-confirmed' detections.

Attempts were then made to combine the single-band detections into 'band-merged' sources, by comparing the hits from the different wavebands. As has already been discussed, the arrays at the different wavebands were staggered, in order to improve point source location. So both in-scan and cross-scan tests had to be applied before a 'seconds-confirmed band-merged' source could be declared.

At this point each detection was checked against a file containing at least 32 000 known sources. Identification numbers were assigned to successful matches. The next step was to run an 'hours-confirmed' test. This was run on data from three successive passes in at least one band. Sources that passed this test received more accurate estimates of position and brightness. The final step was 'weeks confirmation'. Here, however, only agreement in position was sought, to avoid rejecting variable sources. (Note, however, that there is some residual bias against widely varying celestial sources, with periods of less than a few days.)

Only objects that passed the weeks-confirmation test obtained an entry into the catalogue database.

Candidate small extended sources were identified using a simpler algorithm. Square-wave filters of various sizes were applied to the data stream from each detector. If one of the filters indicated a larger flux than had the point source detector at that point, then the object was considered to be a possible extended source. A seconds-confirmation test was applied, similar to that for point sources. The single-band hours-confirmation test was also similar, except for slight additional complications due to the finite size of the sources. The weeks-confirmation tests, however, were carried out on individual bands. Only at the final stage of processing was any attempt made to look for multi-band small extended celestial sources.

Known asteroids were used to test the rejection procedure. No known asteroids got through.

To provide data for the study of asteroids all seconds- and hours-confirmed detections with the right infra-red signature were written to a special file. A similar procedure was adopted for comets. Neither were included in the 1984 catalogue. Instead, a separate catalogue was complied at a later date.

In addition to the small extended-source catalogue, maps were prepared of the total infra-red emission over the whole sky. After calibration, the data were re-sampled to produce images at a pixel size of $2' \times 2'$ (with map resolutions of $6'$). There were 212 images prepared, each $16.5° \times 16.5°$.

Obviously flawed data were rejected, but no systematic rejection process was carried out. However, each coverage was processed into a separate image, so that moving or variable sources could easily be picked out, and background variations caused by the Sun illuminating solar-system dust did not confuse the maps.

The final catalogue was prepared from the database by carrying out a final search for unreliable sources and by applying final calibration corrections. In addition, other astronomical catalogues were compared to find associations to known objects.

REFERENCES

Bamberg, J. A. & Zaun, N. H. (1984) 'Design and performance of the cryogenic focal plane optics assembly for the Infrared Astronomical Satellite (IRAS)' SPIE 509 Cryogenic Optical Systems and Instruments.

Darnell, R. J. (1984) 'Cryogenic refractor design techniques' SPIE 509 Cryogenic Optical Systems and Instruments.

Devereaux, W. P. (1983) 'Cryogenic infrared imaging beryllium telescope for Infrared Astronomical Satellite (IRAS)' Ball Aerospace Systems, Boulder, Colorado.

IRAS (1988) 'IRAS Catalogs & Atlases — Explanatory Supplement' Joint IRAS Science Working Group.

Low, F. J., Beichman, C. A., Gillet, F. C., Houck, J. R., Neugebauer, G., Langford, D. E., Walker, R. G. & White, R. H. (1983) 'Cryogenic Telescope in the Infrared Astronomical satellite (IRAS)'.

Markham, B. L. (1984) 'Characterisation of the Landsat Sensors Spatial responses'. NASA Tech Memo 86130.

Neugebauer, G., Habing, H. J., Duinen, van R., Aumann, H. H., Baud, B., Beichman, C. A., Beintema, D. A., Boggess, N., Clegg, P. E., Jong, de T., Emerson, J. P., Gautier, T. N., Gillett, F. C., Harris, S., Hauser, M. G., Houck, J. R., Jennings, R. E., Low, F. J., Marsden, P. L., Miley, G., Olnon, F. M., Pottasch, S. R., Raimond, E., Rowan-Robinson, M., Soifer, B. T., Walker, R. G., Wesselius, P. R. & Young, E. (1984) 'The Infrared Astronomical Satellite (IRAS) Mission' *The Astrophysical Journal* **278**.

Pouw, A. (1983) The IRAS spacecraft *Journal British Interplanetary Soc.* **36** 17. Vol. 36, No. 1, pp. 17–20.

Walker, H. (1990) Rutherford Appleton Laboratory, U.K., Private communication.

16

SPOT: the French commercial Earth-resources satellite

Consultant: **A. Péraldi**†

16.1 INTRODUCTION

The purpose of the SPOT series is to provide a commercial Earth-resources imaging service, in competition with Landsat, at visible and near infra-red wavelengths (the 'reflective' bands). The spectral range is considerably more limited than that of Thematic Mapper, but the resolution is improved, and a unique variable angled-viewing facility is offered giving the prospect of a form of stereo-imagery.

A series of five spacecraft is planned, with further developments under consideration. SPOT 1 was launched on 22 February 1986 and, at the time of writing, is still working almost perfectly. SPOTs 2 and 3 are identical in design to SPOT 1 (Matra, 1989). SPOT 2 was launched in January 1990, and will be operated in parallel with SPOT 1 for as long as they both last. SPOT 3 is planned for launch in 1993. SPOT 4, due for launch in 1995, will contain an additional band and other improvements.

SPOT 1 carries two identical 'high-resolution visible' instruments — HRV1 and HRV2. Each instrument covers three narrow bands, almost identical in range to those of MSS — the 'XS' bands — and one wider 'panchromatic' or 'PA' band. Individual images are nominally 60 km square. The multispectral bands have a pixel size of 20 m, giving a nominal 3000 pixels along each side of an image. For the panchromatic bands, the equivalent figures are 10 m and 6000 pixels along a side. Each instrument carries an angled-viewing mirror, which enables it to view any track within ± 27° of the ground track. When both instruments are set to nominal nadir viewing, a double swath of scenes is acquired, either side of the ground track. However, off-nadir viewing is used quite regularly in normal operation to improve

† Mr Armand Péraldi was Scientific and Technical Director of Matra until his tragic death in late 1989.

Fig. 16.1(a) — SPOT. © CNES.

flexibility of coverage. It is also used to acquire imagery of particular areas from different viewing angles, to produce stereo-pairs for terrain modelling purposes. The two images are inevitably obtained several days apart, and sometimes much more than that. Lighting conditions may well be different, crops may have been harvested or snow may have fallen, and so this form of 'time-delayed' stereo is somewhat less than ideal. However, coupled with the 10-m pixel size of the PA band, the stereoviewing feature of SPOT has opened up new possibilities for the mapping of inaccessible areas.

   Another feature of SPOT which distinguishes it from many earlier instruments is that it is a pushbroom system. A line of 6000 individual charged coupled device

Angled viewing
mirror

Calibration
system

Telescope

Focal plane

Electronics:
detector and
services

to downlink

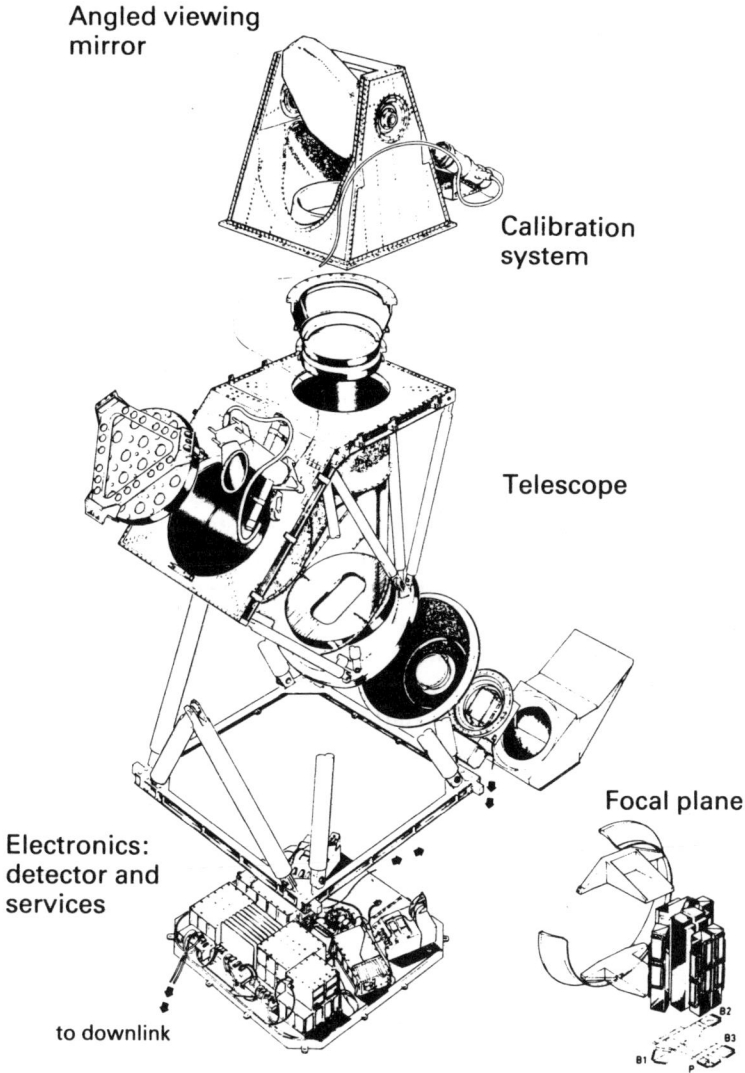

Fig. 16.1(b) — HRV. © CNES.

detectors (CCDs) is swept over the ground as the satellite moves forward on its orbit. While this system introduces considerable problems in the area of calibration and performance checking, it eliminates all moving parts from the scanning system and thus improves pointing accuracy considerably.

The SPOT satellite is designed as a modular system. The ancillary services are all housed in the host satellite, the SPOT 'bus', to which the observing instruments are bolted as self-contained packages. The SPOT bus is therefore available for hosting a wide range of projects including, for example, the ERS1 synthetic aperture radar project developed by the European Space Agency.

The SPOT project was initiated, with heavy government backing, in 1976. At least four satellites were scheduled from an early date, to provide promise of an assured service for at least 10 years. It was fully realized that this continuity was vital if commercial customers were to develop the new techniques, applications and markets which were essential if this imaginative venture was to succeed. Although basically a French project, both Belgium and Sweden joined early as junior partners. Sweden operates the only non-French 'official' ground station, at Kiruna in Northern Sweden. Being above the Arctic Circle, Kiruna complements admirably the main ground station near Toulouse, the home of SPOT. While many countries receive SPOT data, only Toulouse and Kiruna may transmit to it, or activate the on-board tape recorder.

Even before SPOT 1 was launched, it was clear that design modifications were both practical and necessary. While SPOTs 1 to 3 are nominally identical, an improved design for SPOTs 4 should improve its usefulness significantly. The PA band, which, on SPOTs 1 and 2 looks several kilometres ahead of the three XS bands, is being replaced by a short-wave infra-red (SWIR) band at $1.6 \mu m$ with a pixel size of 20 m. The focal plane is reconfigured to ensure that all four bands are fully registered. The red band, XS2, (0.61 to $0.69 \mu m$) is being modified to offer either 20-m or 10-m pixel size fully registered. (As will emerge in due course, this is not quite such a drastic modification as might appear.) The addition of the SWIR band is a highly welcome development, because its absence is certainly limiting the usefulness of the data from SPOT 1.

The SPOT satellites are built by Matra Espace, at Toulouse in France, under the auspices of the French Space Agency, CNES. However a separate marketing and sales organization, SPOT Image, has been created to handle the commercial aspects of the operation. CNES claim copyright on all SPOT data, and on all derived products. Where however a customer has added value by *introducing additional information* to a product then the royalty claimed by CNES is reduced. Where the form of the data is changed so much that its pixel nature is destroyed then no royalty is claimed. The operation of SPOT's copyright policy is constantly developing as SPOT Image strive to maximize their returns without unduly constraining the freedom of the market to develop new initiatives. Therefore no attempt is made in this book to cover it in detail.

16.2   THE ORBIT

The SPOT orbital period is 369 orbits in 26 days. This leads to a value of k (see Chapter 1) of $14\frac{5}{26}$ (14.192) orbits per day. The altitude emerges as 822 km at the Equator (832 km at 45°N). The launch time was chosen to give an orientation between the orbital plane and the Sun of 22.5°. This in turn gives a nominal descending node (southward passage over the Equator) of 10:30 h.

Residual atmospheric drag leads, at this altitude, to a daily loss in altitude of one or more metres, depending on the level of solar activity. Other, less predictable, perturbations are caused by solar radiation pressure and by the pull of the Moon and the Sun. The altitude is boosted, every one to two months, by between 50 and 200 m. The inclination is corrected once a year. These adjustments, together, ensure that

Table 16.1

Spacecraft			
Mass — satellite (kg)	1750	Launched	2.86, 1.90
— single HRV (kg)	250		
		Target	Land

Orbit			
Altitude @ Equ. (km)	822	Orbits/day	$14.1923(14\frac{5}{26})$
Semi-major axis (km)	7206	Orbits/cycle	369
Eccentricity (%)	0.1	Days/cycle	26
Period (min)	101.4	Shift/orb. (°)	25.37
Inclination (°)	98.7	(km) @ Equ.	2824
Descending node	10:30	Gnd track spacing (°)	0.976
Ground speed (km/s)	6.56	(km) @ Equ.	108.6

Telescope			
Type	Modified Schmidt		
Focal length (m)	1.082	80% blur c.d. (μrad)	12
Aperture (m)	0.33	F.P. dia. (mm)	78.8
f-number	3.3	Field-of-view (°)	± 2.08
1y mirror dia. (m)	0.3	IFOV (μrad)	12.2[†]
2y mirror dia. (m)	—	(μm @ F.P.)	13.1[†]
Total obscuration (%)	15	(m on ground)	10[†]
Clear area (m²)	0.073	EIFOV (m on ground)	—
		Inter-band reg. (pix.)	0.3 (PA band unregistered)

Scanning	
Method	Pushbroom
Rate	n/a
(rad/s)	—
Period (ms)	n/a
Active scan time (ms)	n/a
Efficiency (%)	—
Sampling interval (ms)	1.504[†]
Dwell time (μs)	1.504

Channel details (nominal)

Band	Wavebands (μm)	Detectors	Dynamic range[‡] (W/m²/sr/μm)	Noise equiv. reflectance (% F.S.)	MTF @ pix. rate trk/scan
XS1	0.50–0.59	Si CCD	8.56–350	—	—
XS2	0.61–0.68	Si CCD	4.00–355	—	—
XS3	0.79–0.89	Si CCD	2.36–262	—	—
PA	0.55–0.75	Si CCD	5.40–346	—	—

Electronics	
Consumption (kW)	0.615 (0.750 with recorder)
Tape recorder?	Two
Compression	PA band only

Downlink	
Data rate (Mbit/s)	25×2
(GHz)	8

Data	
Word length (bits)	8
Grey levels	256
Image width (pix)	3000/6000
Image width (km)	60/60

† For XS bands, effectively doubled, by pairing of sensors.

‡ At 'nominal' gain, giving grey level 238 for the maximum radiance quoted.

the descending node time is maintained to within ± 15 min, and that ground track accuracy is maintained to within ± 0.7 km.

The large number of ground tracks is a direct result of the small swath width of the HRV instruments, which in turn is dictated by their high resolution. It was regarded as essential to keep the data rates within the capabilities of the large number of Thematic Mapper-oriented ground stations around the world, which limits the number of pixels per second that can be acquired. In addition, however, as is discussed in the next section, it would have been impossible to have designed a high-quality telescope with a significantly wider field of view. However, on the plus side, having a large number of ground tracks provides an opportunity for an unusually imaginative ground-track pattern. The angled viewing capability of SPOT gives rise to a need for a progressive fill-in of the coverage pattern, so that all parts of the globe can be visited as frequently as possible (Fig. 16.2). As discussed in Chapter 1,

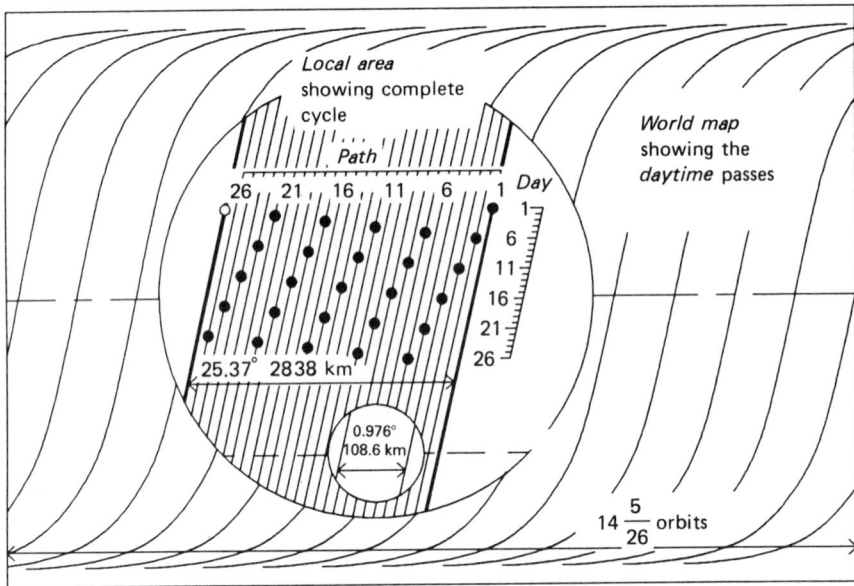

Note that Day 2's orbits are five tracks to the east of those of Day 1.

Each track represents two scenes on the SPOT world reference system.

Fig. 16.2 — Ground-track pattern for SPOT.

tomorrow's orbits are five ground tracks to the east of today's. Since the angled-viewing facility enables the system to see across a total of nine ground tracks, even at the Equator, any part of the globe can be visited at least every 3–5 days.

The ground tracks are not to be confused with the columns of the SPOT World Reference System (see Section 16.12). In fact, each ground track spawns two columns on the WRS. The columns follow the centre-lines of the two image swathes acquired by SPOT when operating in 'twin vertical' mode.

At a given latitude, the altitude may vary by up to about 1 km. During a single orbit, the altitude may vary by up to 11 km, as a result of the combined effects of the oblate shape of the Earth and the residual orbital eccentricity.

SPOT 2 occupies the same orbit as SPOT 1, at 180° from it.

16.3   ANGLED VIEWING AND STEREOSCOPY

Off-nadir viewing is a concept that SPOT pioneered, originally as a means of obtaining stereo-imagery (Matra, 1989), and as a strategy for permitting more frequent repeat imaging of particular areas. However, in practice, it has been used even more to obtain the most flexible possible coverage, for the fulfilment of specific orders. It is quite common for the viewing angle to be changed more than 10 times per orbit.

It is possible to obtain a degree of stereoscopic modelling from Landsat imagery, though only from the overlap areas at the edge of a swath. The difference in viewing angle across the breadth of even a Landsat swath is relatively small, and so the stereoscopic effect obtainable is also small. By contrast the SPOT system is able to view at up to 27° either side of vertical, and the effect in this case is dramatic. The SPOT facility does not provide perfect stereoscopy, because the two images may well be acquired some considerable time apart. (Note that these time differences should be improved during the period while both SPOTs 1 and 2 are operational.) Problems can arise if the atmospheric conditions are different, or if snow has fallen, or if large areas of terrain have been harvested in the interim. The most important application areas, however, are mapping and geological studies (Matra, 1989), which have not proved too sensitive to these problems. Future plans include the rotation of the rotation axis to be along-track. This will provide almost perfect stereo-pairs, but at the expense of coverage flexibility.

Fig. 16.3 shows clearly the action of the viewing mirror. It is operated by a stepping motor, thus providing accuracy and stability while stationary, and is tiltable in 0.3° steps, giving a change in line of sight angle of 0.6°, which, close to the vertical, is equivalent to 8.7 km on the ground. In 'twin vertical' viewing mode, the scan mirrors are offset slightly from the true vertical, so that the two adjacent swaths overlap by a nominal 5 km. The system enables SPOT to image a swath up to 475 km either side of the nadir, which is equivalent to about 4.5 ground tracks. This means that, even at the Equator, there are eight or nine opportunities for observing any particular area in every orbital cycle. At 60° latitude, the number of opportunities is doubled.

In oblique viewing, the pixels become elongated in the across-track direction, and the width of a scene increases from 60 km to up to 80 km. Note, however, that the scene does not become trapezoidal-shaped, as would a photograph. This is because the instrument is moving forwards as it scans, whereas a camera images the entire scene from one point.

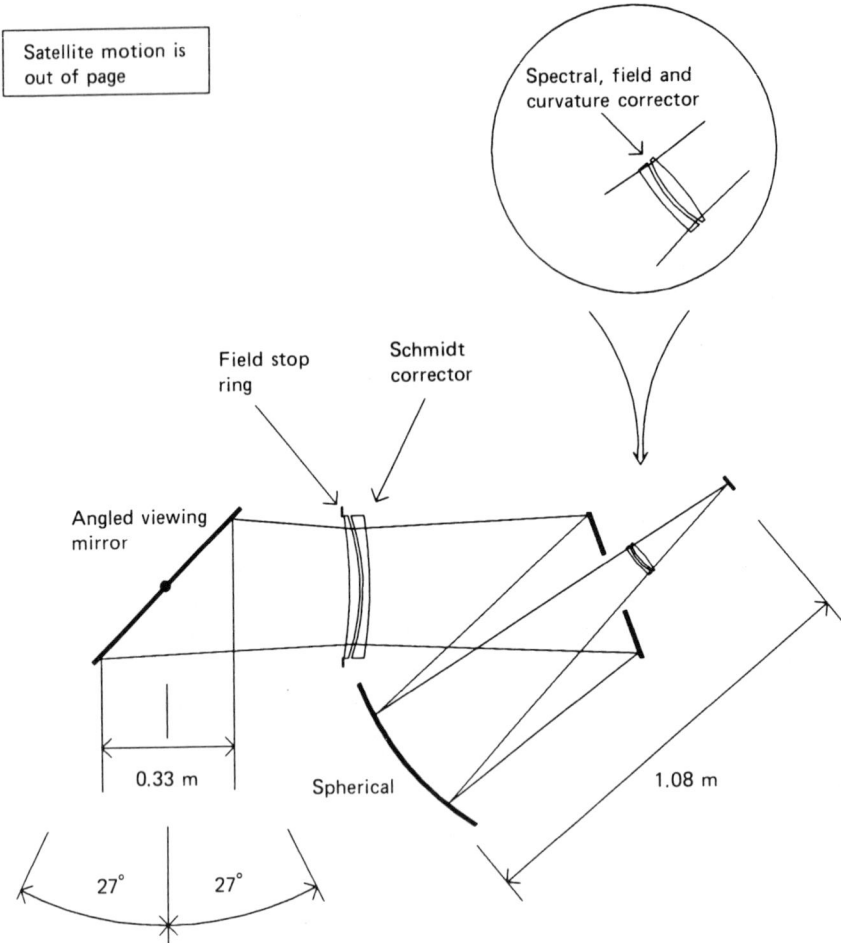

Fig. 16.3 — The SPOT telescope.

16.4 THE TELESCOPE

The layout of the SPOT Telescope is shown in Fig. 16.3. It is a modified Schmidt design. In discussing its development, and the reasons for making that particular choice, CNES have drawn attention to an important limitation in the options open to them. In a long-term project such as a new satellite, most design decisions have to be based on the technology and Space-proven components available at the start of the project. Those of us who come to wonder, some ten years later, why they did not use something more up-to-date must remember how rapidly technology advances, and what a long time ten years is.

The two basic requirements that the telescope must meet were (Reulet, 1988) (a) to provide precise radiometric measurements, in three spectral bands and (b) to avoid geometric distortion of the resulting imagery. Thus it was necessary to design for excellent registration between bands, good resolution and a distortion-free image at the focal plane.

The decision to go for pushbroom scanning, using CCDs, had two important effects for the design of the telescope. The first was that it introduced the requirement for a wider field of view than previous systems had employed. In particular, the 60-km swath width specified led to the need for a field of view of ±2.1°. This immediately eliminated the classic Ritchey Chretien design whose maximum field of view is ±1°. However, pushbroom scanning substantially increases the time that an individual detector may spend gathering light from a particular pixel. The 'integration' time is increased, which means that the total flux required to achieve a given signal/noise ratio can be obtained from a smaller and lighter telescope. One other fully reflective system, the three mirror (Baker & Paul) design, was rejected because the large blockage at the centre of the image unduly compromises the MTF, and also because the focal plane is buried deep within the optics, making physical access difficult.

There remained the Schmidt hybrid system. The principles of the Schmidt telescope are described in Chapter 3. Its main features are its simplicity of construction and its wide field of view. The SPOT telescope is basically to the Schmidt design. However the thin single-element corrector plate, that is normally used to compensate for the second-order deficiencies of a simple spherical mirror, is replaced by a more robust convex/concave doublet. A further smaller doublet is mounted in front of the focal plane. Its purpose is to correct for the curvature of the focal plane that is generated by the preceding optics. All telescopes suffer from this problem, although its severity in practice will depend partly on how far the field-of-view limits are pushed. In the report on the IRAS mission (see Chapter 15), the curvature of the Ritchey Chretien focal plane is discussed as a problem although, in the event, no correction was required.

The side-effects of these lenses all interact. For example, the high-resolution and generally high-performance specification, demand the virtual elimination of chromatic aberration over the frequency range covered. This added a new dimension of complexity to the problem; and the achievement of a satisfactory overall design is regarded as one of the main achievements of the SPOT design (Reulet, 1988). The addition, on SPOT 4, of a mid-infra-red band will render the problem more difficult still. It is being tackled by the use of 'exotic' glasses (Chipaux, 1990).

Every aspect of this design highlights the need to conceive the entire instrument as a single entity. Had the original specification been more relaxed, had the choice of detectors been different, had the decision been made a few years earlier or later, then the whole design might have been quite different.

The focal length (focus) of the telescope can be adjusted from the ground by axial movement of the field corrector doublet. This particular lens was chosen because its effect on the focal length is relatively small, making it unnecessary to design for microscopic movements (Begni, 1984).

An ideal telescope with a circular entrance pupil, no central obstruction, and a f-number of 3.3, would give an 84% blur circle diameter at the focal plane of about 6.8 μm in the region of SPOT band XS3. For a focal length of 1.082 m, this corresponds to an angular diameter of 6.28 μrad, or 5.2 m on the ground. A central obstruction has little effect on the size of the central spot, but reduces slightly the percentage of the energy which falls within it. The SPOT specification demands that the 80% blur circle diameter shall be no greater than 13 μm at the field edge. This, as we shall shortly discuss, is equal to the IFOV. In design terms, this demanded an astigmatism distance of less than 30 μm at the field edge. This specification placed heavy demands on the design and manufacture of the telescope elements. The angled-viewing mirror is a particularly critical component, because of the wide range of incidence angles over which it must operate.

The telescope structure is basically a tubular framework, to which the assemblies of optical and other components are bolted. The framework is designed to be particularly stiff, with a first natural resonance above 45 Hz. This is comfortably above the frequency of any serious excitations that might occur during launch, and well above anything that the satellite components might generate in flight. No movements of any critical components were detected after launch, and the structure has proved particularly stable thereafter. The framework is built of graphite epoxy trusses with titanium attachments, which combine a high modulus of elasticity with a low coefficient of thermal expansion.

The HRV design is also modular, with the telescope, the angled-viewing mirror and the electronics assembly all being separable. This enables each to be built, tested and calibrated independently of the others.

16.5   THE FOCAL PLANE AND BEAM SPLITTER

The SPOT Focal plane is shown conceptually in Fig. 16.4. At a first inspection, the main diagram appears to depict two arrays of 6000 detectors each, one for the panchromatic (PA) band and the other for the multispectral (XS) bands. However, each light ray headed towards the XS array is in fact split into three separate rays by means of dichroic beam splitters (optical filters), and directed to one of three arrays of 6000 detectors. This is shown schematically in the two insets. The physical arrangement is somewhat more complex than that depicted, although the effect is the same.

The light path is further complicated by the fact that each array is made up from four separate chips, which cannot be butted directly together. Therefore each array (including this time the PA array) has an additional 45° 'optical line divider' to enable adjacent chips to be offset. The function of this device is to transmit radiation destined for chips 1 and 3, and to reflect that destined for chips 2 and 4. Some authors have suggested the use of a 'mirror' which is only silvered over the areas where reflection is required. This solution was rejected for SPOT however, because of difficulties that occur at the transitions. Instead, the mirrors are half-silvered all over. For SPOTs 1 to 3, the transmission and reflection are both 27%, which represents a very considerable light loss. For SPOT 4/5, however, improvements in coating techniques are expected to improve both figures to 45%. For the future, 'buttable' chips are now available which leave a gap of just a few pixels at the join. These are being seriously considered for the next generation of instruments.

The detector arrays are oriented across-track, so that the orbital motion completes the scanning of the two-dimensional imagery. Those of the XS bands are scanned in pairs, so that the nominal pixel width of 10 m is doubled. The sampling rate is also adjusted, to produce 20 m-square pixels.

The individual detector apertures are nominally 13 μm square which, with a telescope focal length of 1.08 m, leads to an IFOV of 12.2 μrad, or 10 m at 822 km

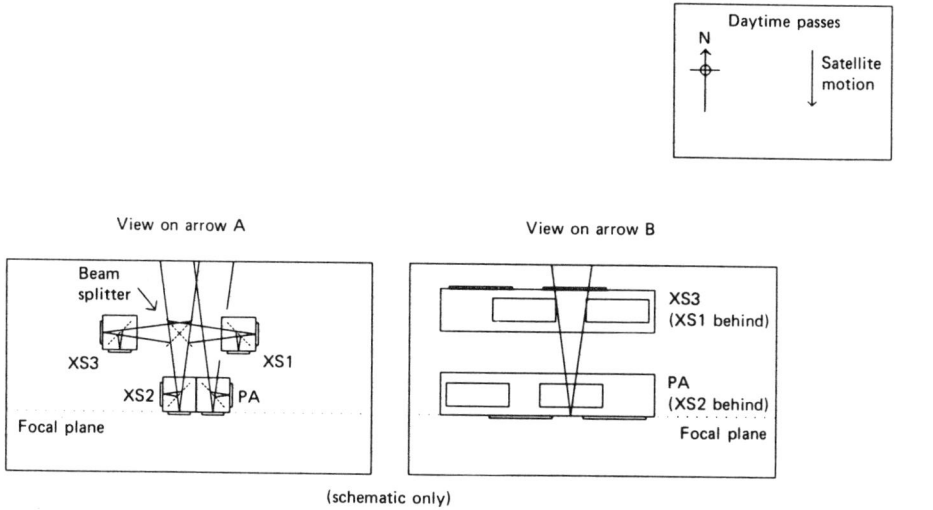

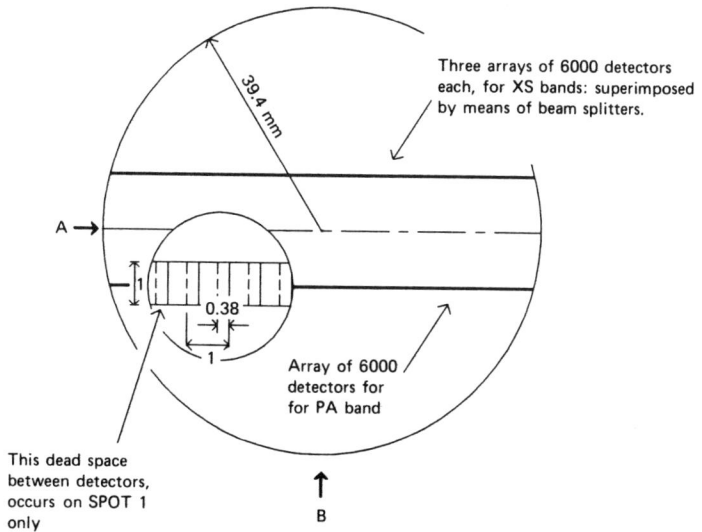

Daytime passes

N

Satellite
motion

View on arrow A

Beam
splitter

XS3   XS1

XS2   PA

Focal plane

View on arrow B

XS3
(XS1 behind)

PA
(XS2 behind)

Focal plane

(schematic only)

39.4 mm

Three arrays of 6000 detectors
each, for XS bands: superimposed
by means of beam splitters.

A →

1

0.38

1

Array of 6000
detectors for
for PA band

This dead space
between detectors,
occurs on SPOT 1
only

B

Units in IFOVs (= 13.1 μm or 12.2 μrad) d)

(XS band detectors are summed in pairs,
to generate 20-m pixels)

Fig. 16.4 — The SPOT focal plane.

altitude. For the XS bands, this IFOV is effectively doubled by the pairing process already mentioned. (In practice, there is a 5-μm blanked zone between the detectors, so strictly the IFOV should be quoted as 10 m by 6.15 m.) The detector arrays are positioned on the focal plane (either actually or apparently) symmetrically either side of the centreline. At any moment, the XS arrays are looking approximately 7.5 km behind the nadir point, while the PA array is looking roughly an equal distance ahead. Thus, where the two are scanning together, the PA-band image is positioned some 25% ahead of the XS image.

16.6   THE DETECTORS

The problem faced by all purveyors of multispectral imagery is to provide perfectly registered images of the same scene in different spectral ranges. To obtain the maximum value from an expensive image, a given pixel must represent the same area of the ground — to an accuracy of, say, 0.25 of a pixel — on each of the single-band images that make up one scene. (Thematic Mapper claims to do much better than this (Chapter 14).) Only then can useful computations be carried out, comparing the brightness of an area of terrain at the various wavelengths.

The SPOT solution to this problem as we have seen, is to split a single incoming beam using dichroic mirrors, which are fabricated to transmit one frequency band and to reflect another. The importance of this can be illustrated by considering the SPOT panchromatic band. The multispectral images are registered to well within the required limits, while the Panchromatic image is not. Sophisticated ground processing is required to recover the situation, but at a cost both in terms of time, money and effort and in terms of loss of detail. The price that is paid for using this method to obtain good registration is, of course, much reduced light levels at the detectors.

The SPOT satellites use a pushbroom scanning system, which means that a total of 6000 individual detectors are required for each band. (For the XS bands, pairs of detectors are combined electronically, to provide 3000 pixels per row (see Section 16.7).)

Silicon charged-coupled devices (CCDs) are used, the operation of which is described in more detail in Chapter 5. Strictly, a CCD is not a photodetector, but a device for transferring charge. In the current context, however, it comprises an array of individual detectors fabricated onto a single strip of silicon, or other suitable semiconductor material. Free electrons, released by the arrival of incident photons (or by any other effect), are captured by the nearest of an array of electrodes. A charge thus builds up on each of these electrodes which is proportional to the flux incident on the local area of the silicon. At the end of the scanning period the charges are shunted, in sequence, to one end of the array, where they are amplified and emerge as an analogue signal. The size of the 'potential well' generated at each electrode is limited by the voltage applied to it. This limits the number of electrons that can be captured, and hence the maximum flux energy that can be detected. Local excedance of this limit will result in a flaring effect akin to that observed in photography. Thus the dynamic range of a CCD array is more severely limited than that of an individual detector, where the charge is immediately conducted away to more robust parts of the circuit.

A CCD is thus an *integrating* detector, in which the charge passed on represents the total flux seen by the detector during the period of a scan. This contrasts with most of the discrete-detector systems that we have covered. These feature track/hold amplifiers, which detect the nominally instantaneous readings which they see at the time that they are switched to 'hold'. The difference may not be that large in practice, however, because such amplifiers must be positioned downstream of anti-aliasing filters. By their nature, low-pass filters have an effect akin to integration.

The CCDs for SPOTs 1 to 3 are basically commercial devices, designed for office copiers, and carefully selected to produce matched spectral responses; whereas those for SPOTS 4/5 are being specially developed. These come in array chips, comprising 1728 individual detectors, of which a number at either end of the chip are not used. The arrangement used to obtain a single 6000 element array from four separate chips is shown in Fig. 16.4.

The final design limits for the wavebands were set largely on the basis of minimizing atmospheric attenuation. The two main bands to be avoided were the $0.93\,\mu$m, water vapour band, centred on and the ozone band, which occurs below $0.33\,\mu$m. In particular, the upper limit of Band XS3 was set below $0.9\,\mu$m to avoid the water vapour band. The molecular oxygen band, around $0.76\,\mu$m, does not affect any of the SPOT bands.

The spectral range of a CCD detector is virtually the same as that of undoped silicon, namely from 0.4 to $1.1\,\mu$m, with a maximum at $0.8\,\mu$m. The upper frequency limit is likely to be at least several megahertz, so that the practical limit on scanning rate is likely to be set by the mode of readout. When a silicon detector, CCD or otherwise, is operated well below its frequency limit, its dynamic range is better than 1000. Its response is about $0.7\,\mu$V/photon at $0.8\,\mu$m, or $5\,$V/μJ/cm^2.

Not shown in the diagram are the band-pass filters, which of course reduce the light levels at the detectors still further. These are fairly conventional units. Bulk absorption techniques are used for the short-wavelength cut-off ends, but it was necessary to resort to 'dichroism' (interference techniques) for the long-wavelength cut-off. With the relatively large variation in incidence angle generated by the SPOT optical path, the interference filters were less successful than might have been hoped. Fig. 16.6 shows the frequency responses obtained.

Problems were experienced with XS Band 2, which seemed to suffer the most from all the problem areas, while being the most critical from the application point of view. This may be the cause of the very high correlation that is found in practice between Bands 1 and 2. Much of the problem arises from the fact that the dichroic coatings are slightly porous, and trap oxygen molecules during fabrication. These molecules affect the optical performance of the coatings. The layering of the coatings is optimized with these molecules in place, so that, as they evaporate away into Space, the characteristics of the coating gradually change — to the detriment, particularly, of Band 2. New coatings are under development which will limit, or even eliminate, this problem. However, at the time of writing, they have yet to be Space-qualified (Matra, 1989).

The 'Line period' is 1.504 ms for the PA band and 3.008 ms for the XS bands.

16.7 THE DETECTOR ELECTRONICS

The design of the SPOT electronics was severely constrained by the requirements of a Space application, namely the need to minimize mass, volume and heat dissipation.

In addition, however, it was necessary to rely exclusively on components which were space-qualified, or could be made so, at around 1976/78. The main problem area quickly emerged as being the A/D conversion, at a rate of more than a million points/ second to 8bits quantization. No single converter that met the environmental requirements could match that speed and so a considerable degree of parallelization was introduced. For the panchromatic band, each of the four CCD chips is processed by an individual chain of electronics. For the multispectral bands, each digitizer serves two chips.

The electronics for a single HRV instrument are shown in Fig. 16.5. It will be

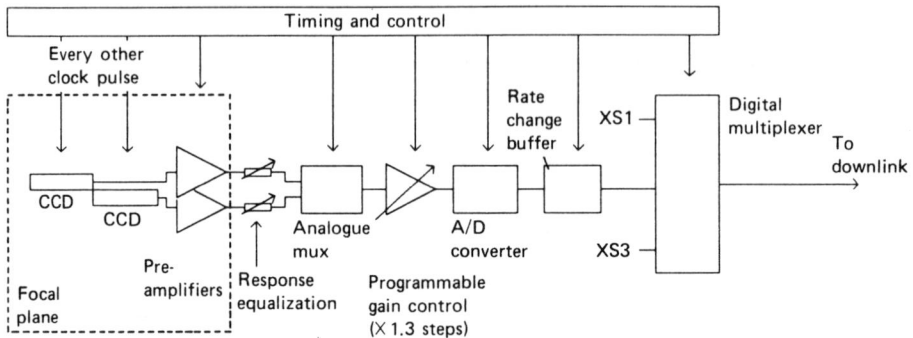

Multispectral bands

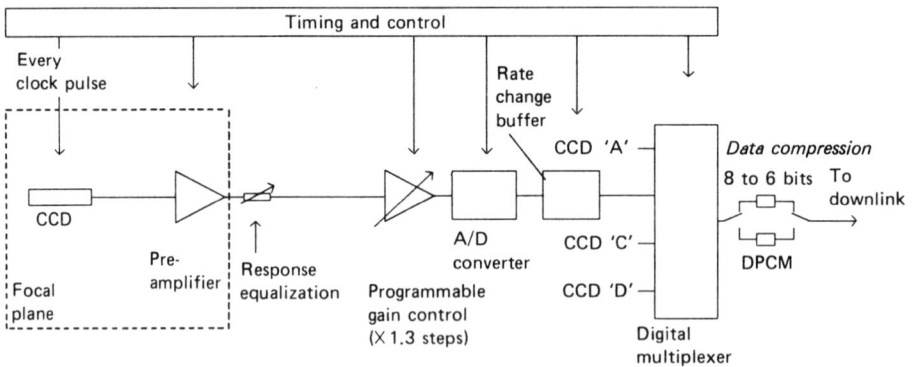

Panchromatic band

Fig. 16.5 — SPOT electronics (single HRV).

observed that no anti-aliasing filter is provided. The reason is this. The function of an anti-aliasing filter, as discussed in Chapter 6, is to prevent frequencies higher than half the sampling frequency reaching the sampling stage. In a discrete detector system, each detector generates a continuous stream of data, which is sampled by suitable electronic circuitry further downstream. In a CCD instrument, however, sampling effectively takes place at the focal plane, i.e. while we are still in the 'space domain'. Each CCD element detects the average flux over a cross-track dimension equal to the IFOV (we assume that the use of CCDs will normally imply pushbroom scanning). In principle, therefore, it will be fooled by regular features that are separated by less than two IFOVs. From then on however, the risk of further degradation from this cause vanishes. The CCD readings are shipped out into the electronics as a continuous analogue signal. This signal is sampled, by the A/D converter, at intervals timed to catch the true reading of each element. There is no further loss of data and no further risk of aliasing. In the along-track direction similar considerations apply, with the exception that contiguous samples are now well spaced out along the data stream, so that there is no realistic prospect of further space-borne filtering even if such were desirable.

Anti-aliasing filtration, because of its very function, has to take place while the data are still in unsampled form. By now it is clear that, in this case, that means at or before the focal plane. The finite resolution of the optics is of course the ideal mechanism for this. Matching the Airy disc size to the IFOV will ensure that ground features whose linear separation is significantly less than two IFOVs will be smeared out by this low-pass (spatial) filtering effect.

Thus the CCD chip effectively incorporates the functions of bias voltage generator, and (in this case) 1728 detectors, pre-amplifiers and anti-aliasing filters, together with a single analogue multiplexer. Under the control of the timing clock, each chip samples each of the elements, and outputs them in sequence as a burst of analogue signal. The basic timing-pulse rate allows the spacecraft to advance by 10 m per sample. The panchromatic-band CCDs receive every pulse, while those for the multispectral bands receive only every other pulse, thus synchronizing the two different sampling rates.

The CCD chips were carefully selected to offer matched spectral responses throughout any one band, but their calibrations remain different. The signal bursts therefore next pass through simple attenuators to balance out the differences. (There remain the differences between individual detector elements on each chip, but these are much smaller, and are addressed by on-line calibration). The XS-band signal bursts next pass through analogue multiplexers, which append the signal bursts from chips 2 and 4 to those from chips 1 and 3. The analogue signal bursts are further amplified using amplifiers whose gains can be adjusted, under ground control, in steps of times 1.3. By this means the expected signal level can be adjusted to fill the dynamic range of each digitizer. The digitizers for the XS bands also sum the outputs from each pair of detectors to generate the 20-m pixel width.

The SPOT detectors are 1728-element CCD chips, of which only the middle 1500 elements are used. The next step therefore is to throw away the outputs from the unused detectors at the ends of each chip. This is done by the use of rate-changing buffers, which also smooth out the flow of data to the downlink. The memories store the 1728 readings from each chip (or, in the case of the XS bands, each pair of chips) at high speed. The readings from the 1500 active elements (remembering of course that, for each XS band, there is a sequence of false readings in the middle of the signal burst) are then read by the digital multiplexer in the correct sequence, and at a speed which just fills the scan time.

In its raw state, the PA band signal contains a third as much data again as all three XS band signals. The combined XS band signals from two HRV instruments generate a data rate which just matches the capacity of Thematic Mapper-equipped receiving stations. The PA band signal is therefore compressed to bring it within the same limits. Two alternative algorithms are implemented, as described in the next section. This compression takes place between the multiplexer and the transmitter.

Two on-board tape recorders are fitted, so as to permit observations of parts of the world out of reach of the two SPOT-controlled ground stations. (Customer ground stations may receive local SPOT data but may not activate the recorder.) Each recorder provides 22 m of recording time, which represents about half the daytime path of a single orbit. The capacities of the recorders are matched to that of the downlink. To save the energy and stability costs of high-speed rewinds, replay is backwards and at recording speed. The tape must be recorded to the end before replay is possible, and replay must be complete before any further recording can be carried out. These design features impose significant limitations on the flexibility of the tape recorder system.

One recorder failed four months after launch. At the time of writing, the second still works, although part of the tape is damaged and unusable. On SPOT 4/5 the capacity of each recorder is being doubled.

16.8   THE DOWNLINK AND DATA COMPRESSION

The downlink characteristics were designed to be compatible with existing Landsat systems. This was intended to minimize the costs of an existing Landsat ground station wishing to adapt to the receipt of SPOT data. The data rate was therefore limited to about 50 Mbits/s, or 8 GHz, which is about half the rate at which the SPOT twin-HRV system is capable of acquiring data.

The data rate was effectively halved by designing the system only to handle the output from two of the possible four instruments. For this purpose, the panchromatic band is considered to be a separate instrument, since (in raw form) it generates more data than the three multispectral bands together. Thus it is possible, on a single pass, to obtain XS and PA data from one HRV, or XS or PA data from both HRVs (or of course XS from one and PA from the other). It is not possible to store the remaining data on the tape recorder.

However, at 32 Mbit/s, the PA data rate is still excessively high. A degree of compression is necessary to reduce the rate to the 25 Mbit/s of the combined XS bands. Two alternative methods are built into the electronics. The first is to lose the 2 least significant bits of information, and to quantize the rest into 6 bits (64 grey levels). This is in line with the quantization of MSS data, but is unlikely to commend itself to modern customers. The alternative strategy is more complex. The basic technique is to encode the output from two detectors out of three into just 5 bits, the remainder remaining unchanged at 8 bits. Selection is made according to the following pattern,

$$\ldots\ 8\ 5\ 5\ 8\ 5\ 5\ 8\ 5\ 5\ 8\ 5\ 5\ 8\ \ldots,$$

so that each compressed datum is immediately adjacent to one that is uncompressed. However, the simple solution, of transmitting just the 5 least significant bits and of copying the 3 most significant bits from the adjacent datum was rejected. The reason is that it would introduce seriously misleading errors whenever the difference between adjacent pixels exceeded ± 32 of the 256 grey levels. In built-up areas, the difference between adjacent pixels could be substantially more than that, and the correct interpretation of a single pixel could be important. However, if the difference between adjacent pixels is large then they clearly represent a boundary of some kind, and one or both are likely to be 'mixels' (a pixel spanning two different ground-cover types). Such a pixel would normally be excluded from classification evaluations, but would be of vital importance in picking out roads, buildings and other detail. For these reasons, an accurate representation is only essential where the difference is small, but the representation must be in the right 'ballpark' at all times. The compressed pixels are encoded from a look-up table which ensures that, where the compressed pixel is within $+6$ to -8 pixels of its neighbour, the 5-bit number transmitted represents the exact difference. If the true difference is outside this range, but within about ± 20, then reconstitution is within about 3–4 pixels. Within ± 50, reconstitution is within some 17 levels. Differences above ± 50 are all reconstituted as ± 60.

There is one additional refinement to describe. Instead of encoding each compressed datum by reference to its immediate uncompressed neighbour, the mean of the two uncompressed data on either side of it is used. This refinement has a low-pass filtering effect. It reduces the noise, or the likelihood of error, over homogeneous surfaces such as fields, but it also reduces the clarity with which boundaries are picked out. However, the effect is clearly not serious because the excellent definition of the SPOT panchromatic band is well-known.

Because of the number of options available, the instruments do not simply record everything that passes under them, as do Thematic Mapper and MSS. Instead, each pass is actively programmed in advance, the necessary instructions being passed to the onboard computer during night-time passes over Toulouse. The programme covers the two instruments to be used (twin XS, twin PA or one of each), the viewing angle(s) required and the video chain configuration (gain and PA compression mode). Changing the viewing angle takes some time and, if it requires to be changed during a pass, then significant amounts of data can be lost. Nevertheless, as has already been discussed, the viewing angle is normally changed many times per orbit.

The onboard tape recorders are capable of storing about half of one complete orbital pass. But they have to be replayed at record speed. However, the satellite is only in direct range of any one station for about 2500 km, or one-eighth of a pass. Replay would normally be during a night-time pass over either Kiruna or Toulouse.

Any suitably equipped country may receive SPOT data, but only after the completion of suitable contractual arrangements. At the time of writing, about a dozen have done so; namely Canada, Spain, India, Brazil, Thailand, Japan, Saudi Arabia, Pakistan and South Africa. China and Australia are in the pipeline.

16.9 CALIBRATION AND IN-FLIGHT CHECKS

The HRV instruments are designed for economical and efficient operation which, in the current context, demands relatively infrequent calibration. Calibration is carried out nominally weekly, which makes it possible to take the instrument out of service while the task is performed. The calibration unit is mounted outside the angled-viewing mirror housing, and the process is initiated by rotating the mirror until it closes off the aperture and blocks all incident radiation. As an additional precaution, calibration is only carried out during night-time passes. In this position, a tungsten-filament halogen lamp illuminates all the detectors via the entire optical path. The level of irradiance varies over the focal plane, but the profile is known and can be corrected for. The lamp was originally intended for relative calibrations only, which meant that an accurate knowledge of its radiance was not required. However, experience has shown its stability to be very good (Matra, 1989), and it is now used for absolute calibration purposes, cross-checked against solar calibrations as described below.

Solar calibrations of selected detectors can be carried out using a fibre-optic sunlight collection system. This enables some of the CCDs to be irradiated by a known amount of solar flux. Three separate fibre-optic bundles are taken to the exterior of the satellite. A 'Selfoc' lens is bonded to the end of each fibre. The Selfoc lens has a constant collection efficiency, up to an incidence angle of 6°, at which angle it experiences a sharp cut-off. The bundle ends are so positioned that as the satellite emerges from eclipse (night-time), and whatever the season, just one of them will have the Sun within that incident angle. The other end of the bundles is taken to the calibration unit, where the radiance is fed into the light path via a beam splitter. The optical arrangement is such that only a small number of detectors in each spectral band receive sunlight.

Additional checks of relative responsivity are made by imaging, uniform landscapes such as flat, snow-covered, terrain. Statistical analyses are then carried out.

No specific operational procedures are required to get the calibration data to Earth.

Perhaps the most important of the subsequent procedures is the **relative calibration**, in which are measured the zero offsets (mainly dark current) of individual detectors and also the inter-detector variability. These variabilities are quantified as 'normalization coefficients' and relate the response of each detector to the nominal for the array chip. The normalization coefficients are defined such that their mean for any one array chip is unity. Interchip normalization coefficients are also evaluated, again defined such that they average out to unity. A relative calibration provides all the information required to enable a clean, internally consistent, single-band image to be constructed.

All imagery is 'normalized' by the application of the relative calibration before further processing is carried out. This degree of radiometric correction is all that is required for cartographic and related purposes. Many terrain classification techniques are insensitive to small differences in spectral radiance, and are also perfectly adequately served by the results of this process.

An **interband** calibration is derived by evaluating the responses from the lamp (whose spectral radiance is known) or — more easily — from the absolute calibration (see below) if it is available. Most terrain classification techniques rely on the relative difference in reflectivity of the surface in two or more wavebands. Theoretically, therefore, all imagery used for this purpose should have an inter-band calibration applied.

An **absolute** calibration builds on the relative calibration already described. A single coefficient is obtained for each band to enable the individual normalization coefficients to be converted to watts per square metre per steradian per micrometre bandwidth (W/m²/sr/μm). Note that the calibration contains no constant term. This is because the offset has already been taken out during normalization. The absolute calibration enables the radiance reaching the instrument to be derived. No allowance is made for the amount lost, or gained, during passage through the atmosphere.

A **multidate** calibration is derived from absolute calibrations taken on different dates. It can be used for a long-term check on the stability of the CCDs and the electronics, or to lend confidence to multidate terrain studies.

Begni *et al.* (1986) describe an absolute calibration of the system by the use of the White Sands test site in New Mexico. A similar calibration of Thematic Mapper was carried out at the same time. Agreement with pre-flight calibrations was good.

During the post-launch assessment, it was discovered (Begni, undated) that the calibration unit exhibited short-term (few days period) fluctuations in its output. These fluctuations took the form of small patches of increased output affecting just a few pixels. However, the positions of the patches varied slightly from calibration to calibration and reduced the validity of any stability checks on the pixels affected. The problem was attributed to the likely existence of small scratches or dust particles on the first lens of the calibration unit. This lens is at the focus of the second lens, and any such blemishes would be imaged straight on to the focal plane. The reproducibility of the angled-viewing mirror is of the order of 100 μrad, or 80–100 IFOVs and so the detectors affected would vary from calibration to calibration. It is reported that the use of the calibration unit was abandoned on this account in favour of the use of selected ground test sites. However, there is no mention of this problem in the more recent handbook, and so it must be assumed that it has been overcome. If the stability of the detectors is good then it should not have been too difficult to devise a statistical strategy for rejecting faulty calibrations.

16.10 DATA QUALITY (AND QUALITY CHECKING)

The radiometric characteristics of the HRV instruments were designed on the assumption of a standard 'Rayleigh' atmosphere, with the following characteristics:

— horizontal visibility 23 km (limited by aerosols),
— precipitable water vapour 2 cm,
— ozone 0.35 cm of atmosphere (if compressed to STP),
— air mass 3, corresponding to a pressure at the geoid of 1013 mbars.

Under these conditions, atmospheric effects may normally be considered negligible. However, as discussed in Chapter 2, conditions may well differ widely from this ideal.

16.10.1   Radiometric performance

Begni *et al.* (undated) quotes the top-end SNRs for HRV1 and HRV2, as evaluated soon after launch, as given in Table 16.2.

Table 16.2 — Full-scale signal/noise ratios

Band	HRV1		HRV2	
	Specified	Measured	Specified	Measured
PA	200	190	190	110
XS1	210	370	190	330
XS2	200	250	200	240
XS	270	410	260	380

It is interesting that HRV2 was predicted to be somewhat less good than HRV1.

16.10.2   Spectral performance

The measured responses are given in Fig. 16.6.

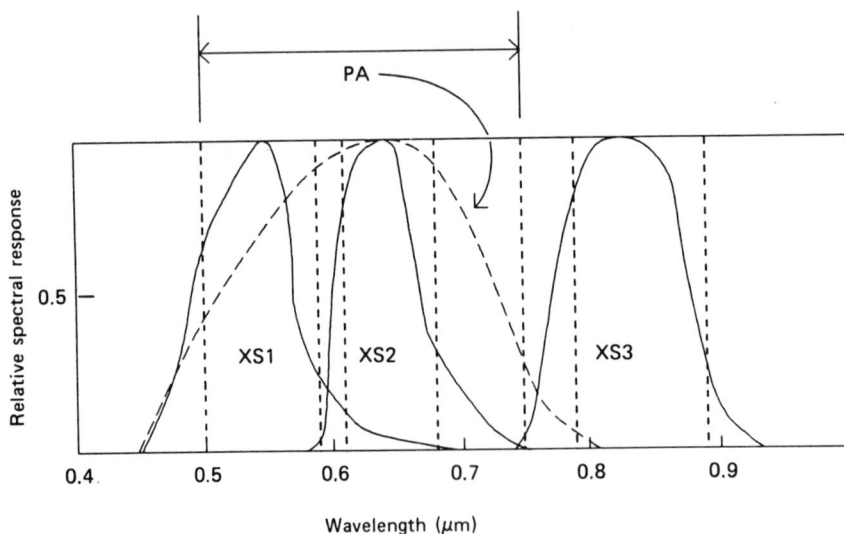

Dotted lines show the nominal band limits

Fig. 16.6 — SPOT spectral responses.

16.10.3 Spatial frequency response

A photo-interpretive technique was used (Leger, undated), based on aerial panchromatic photographs of selected cities. The strategy has already been discussed in Chapter 5. Briefly however it involved degrading the aerial photographs, to represent known MTFs; and then comparing them visually with PA imagery of the same areas. HRV1 turned out to be above specification, whereas HRB2 was judged to be on specification.

16.11 THE HOST SPACECRAFT

SPOT is designed on a modular basis, with communications and housekeeping facilities being concentrated in a host module, the 'SPOT Bus'. This module is equally capable of hosting alternative instrument systems such as ESA's synthetic aperture radar satellite, ERS-1.

The principle limitation on the useful life of a Sun-synchronous satellite such as SPOT is not, as might be thought, the availability of fuel for station-keeping. It is in fact the ability of the batteries to tolerate the rapid succession of charge/discharge cycles as the satellite passes into eclipse once on each orbit (CNES, 1988).

16.11.1 Attitude control

Attitude control is conventional; using three orthogonal reaction wheels, two magnetic torquers and one hydrazine thruster. The principles of the technique have already been discussed in Chapter 12.

There are three modes of operation: fine, coarse and safe hold.

In the **fine-pointing** mode, the satellite's attitude is sensed by a combination of rate-integrating gyros and Earth and Sun sensors. The position sensors are used to correct for the inevitable drift of the gyros. The Earth sensors are positioned so that they are normally pointing to the Earth's limb, where there is a sudden transition from the warm Earth to the cold of Space. They are used to provide a check on attitude about the roll and pitch axes. The yaw attitude is checked against the Sun sensor. Shortly before the satellite passes into eclipse (night-time), its Earth-pointing face passes from shadow to sunlight (see Fig. 1.1) and shortly after it emerges, the Earth-pointing face makes the opposite transition. The Sun's direction is known, and so any yaw bias can be detected at this point. All information is passed to the onboard computer, which generates commands to the reaction wheels without reference to mission control.

Coarse-pointing mode is used during initial orbit insertion, during orbit manoeuvres, and during any (hopefully temporary) malfunction of the fine-pointing actuation system. The same sensors are used, but control is by the use of hydrazine thrusters. These thrusters are also used, in different combinations, to produce torque-free thrusts for orbital correction.

The **safe-hold** mode would be used in the event of major failure to protect the satellite against permanent damage while awaiting the intervention of ground control. In this mode, the solar arrays are held fixed, and a separate set of sensors are used to sense the approximate position of the Sun. A separate electronics package utilizes the standard hydrazine propulsion system to keep the Spaceward side of the satellite facing away from the Sun. Excess heat is radiated away from this face, and it is essential, therefore, that it is never subjected to strong radiation. The hydrazine propulsion system is fully duplicated, and should therefore have a high degree of reliability. At the time of writing, the safe hold mode has never been used in anger (Matra, 1989).

16.11.2   Temperature control

The SPOT HRV design is based on a highly temperature-stable structure, and on relatively insensitive components. The only items that require really close temperature control are the CCD arrays. An operational standard of 20°C was selected for convenience during laboratory tests.

Structural stability was ensured by two methods. First, adjustable heaters are mounted on all important parts of the structure to stabilize the temperature within 2–3°C. Second, the main elements of the space-frame structure are made from carbon fibre tubes, which are designed to have a slightly negative coefficient of thermal expansion. The trusses have titanium end fittings, whose linear expansion per unit temperature rise almost exactly cancels out the contraction of the trusses.

The absence of thermal band sensors means that there is no requirement for the sophisticated cooling techniques employed by TM, AVHRR and others. The internal temperature is maintained within acceptable limits by careful design of the vehicle casing. In principle, the casing is designed to radiate away the design heat load when heated to the design temperature. In practice, it was necessary to allow the vehicle to heat up during active daytime passes, and to cool again during the relatively inactive night-time passes. To allow for uncertainties, however, a slight undercooling was built in. It was known, even before launch, that it would be necessary to switch off the electronics during part of every third orbit to keep the maximum temperature within its limits of 45°C.

The CCD arrays are mounted in a much more stable area. However, they suffer from the problem of having the pre-amplifier mounted at one end of each chip. The pre-amplifiers generate a certain amount of heat, which introduces the possibility of a temperature gradient down the chip. To maintain the temperature of the CCD chips as stable as possible, heaters whose dissipation matches those of the CCD chips are switched on whenever the CCDs are switched off.

16.12   SPOT PRODUCTS

The SPOT system has generated a World Reference System (WRS) similar to that devised for Landsat. The mesh is, however, considerably finer, on account of the smaller size of a SPOT image. In addition, because SPOT is capable of acquiring two images (nominally adjacent) on each pass, a change in terminology has had to take place. The term **column** is the equivalent of a Landsat **path**, in that it represents the centre line of an image swath. But, unlike a Landsat path, a SPOT column does not follow the satellite ground track. Each ground track in fact generates two columns, approximately 30 km on either side of it. The columns converge rapidly at the higher latitudes. At latitudes above the limits of northern France, the degree of overlap between the coverage of adjacent ground tracks becomes greater than 50%. At this latitude, the SPOT WRS ceases to define the columns relating to every other ground track. Imagery is still acquired on such orbits, but its imagery is identified by the number of the adjacent column. Further thinning out takes place at still higher latitudes. In the polar regions any attempt to identify imagery in relation to the orbit on which it was acquired is abandoned. Instead a grid of equilateral triangles is drawn, over a map of a suitable projection, as the basis for identifying the location of polar imagery.

SPOT **rows** are defined at 55 km intervals, to provide for scenes 60 km long, with a 5 km overlap. At the time of writing imagery is only sold as scenes whose centre lies precisely on a SPOT row, although it is understood that this arbitrary limitation is due for relaxation.

No specific provision is made in the SPOT WRS for obliquely acquired scenes. Such a scene is identified by the point on the WRS grid closest to its centre. This gives an indication of the coverage zone. The angle of view is also specified, which enables the scene corners to be pin-pointed to within a few kilometres.

The French Space Agency, CNES, claims copyright on all SPOT data, as well as on all products derived from SPOT data. While this policy is the cause of very considerable problems both for SPOT Image and for its customers; it is regarded as an essential part of the project's overall strategy for ensuring its long-term commercial viability. SPOT Image are keen to encourage as much use as possible of SPOT data, and also to encourage the pioneering of new applications. However, they are at all stages equally keen on maximizing their own income.

Royalties are payable to SPOT Image on the first sale, and on all selling-on of raw data. Data that has been enhanced by the application of sophisticated processing techniques count as raw data for this purpose. However, where a customer has added value by incorporating additional information into his product, then this is a 'derived' product and attracts a lower rate of royalty. Where the form of the data has been so substantially changed as to destroy the pixel structure of the data then no royalties are payable. Statistical studies, provided that no illustrations accompany the graphs and tables, are therefore exempt.

The level of royalty is computed on the amount of SPOT data incorporated into a product. For this purpose, however, a single 6000×6000 panchromatic scene is regarded as equivalent to the three 3000×3000 bands of a multispectral scene. While the charge for either full scene is not insubstantial, the amount payable on a 512×512 extract becomes quite small.

SPOT supply data at several different levels of processing. **Level 0** represents a raw scene, with no processing of any kind applied. It is not normally offered to customers. **Level 1a** has had applied the normalization process described in Section 16.9. Any missing data are set to zero. **Level 1b** may incorporate high-pass filtering, if such has been found to be necessary to combat sampling problems. Missing data are recreated by interpolation from adjacent data. Simple geometric correction is applied, to correct for Earth rotation and curvature and for viewing angle. Level 1b imagery has therefore been resampled, and may not be suitable for certain classification operations. **Level 2** has been fully geometrically corrected to one of several standard map projections. Other higher levels are also available. Data can be supplied on computer tape, on film or on paper. Recognized SPOT distributors, such as the National Remote Sensing Centre at Farnborough England, will supply their own range of derived products at similar prices. In some cases the quality of reproduction may be much superior, partly because of the lower throughput that they must commit to.

REFERENCES

Begni, G. (1984) 'An in-flight refocussing method for the SPOT HRV cameras.' *Photo. Eng. & R.s.* **50**, 12 12.84

Begni, G., Leroy, M. & Dinguirard, M. (undated) 'SPOT radiometric resolution performance evaluation: preliminary results' CNES and CERT.

Begni, G., Dinguirard, M. C., Jackson, R. D. & Slater, P. N. (1986) 'Absolute Calibration of the SPOT-1 HRV cameras' SPIE Vol 660 Earth R. S. using the Landsat T. M. and SPOT Sensor systems (1986).

Bodin, P. (1988) CNES, private communication.

Chipaux, C. (1990) Correction, in draft, to Chapter 3.

CNES (1988) CNES/SPOT IMAGE 'SPOT User's Handbook', English edition, 1988.

Leger, D. Leroy, M. & Perbos, J. (undated) 'SPOT MTF performance evaluation' CERT, Toulouse.

Matra (1989) Corrections in draft by A. Péraldi *et al.*, Matra Espace, Toulouse 1989.

Midan, J. P., Reulet, J. F., Giraudbit, J. N. & Bodin, P. (1983) 'The SPOT HRV Instrument' CNES report IAF 83–109.

Midan, J. P. (1985) 'The SPOT HRV Instrument: an Overview of Design and Performance' *Earth-Orient. applic. Space Technol.* 00.0.

Reulet, J. F. (1988) CNES, private communication.

SPOT Promotional Material — 1986.

17

MOS-1: the Japanese marine satellite

17.1 INTRODUCTION

The first Japanese Marine Observation Satellite was launched on 19 February 1987, from the Tanegashima Space Center, Hokkaido Island. MOS-1 is Japan's first foray into Earth observation satellite technology (NASDA, 1987). It is a second-generation experimental mission, the first in a five-mission series lasting some 13 years (ESA, 1987). Successive vehicles in the series will be oriented towards various application areas, the principal role of MOS-1 being to study marine phenomena such as ocean colour and temperature. However, the bands chosen are equally applicable to the traditional uses of ground cover and meteorology. The spacecraft are being developed and operated by the Japanese National Space Development Agency (NASDA).

The MOS project could be said to have started in February 1985, when the Space Activities Commission of Japan established a working group to make long-term proposals for the development of a Japanese Space effort. The group recommended the creation of a series of land and marine observation satellites, to be launched at a rate of one every other year — the programme to continue until, effectively, Japan was 'up with the best'. The group included specialists from a wide range of disciplines, and 'due to the fact that many users were unfamiliar with the limitations of space engineering, there were a variety of requirements in the proposed specification' (Tsuchiya *et al.* 1987) which could not all be met.

Four different satellite designs were recommended, covering marine, land and weather observation as well as geodetic studies. It was decided, however, that initial priority should be given to marine observation. In the event, it was found that the parameters of the MOS-1 instruments could be modified to make it equally applicable to land studies. For example, Band 2, of one of the three instruments on board, was changed from 0.64–0.71 μm to 0.61–0.69 μm (which is similar to MSS,

Fig. 17.1 — (a) The spacecraft. © NASDA.

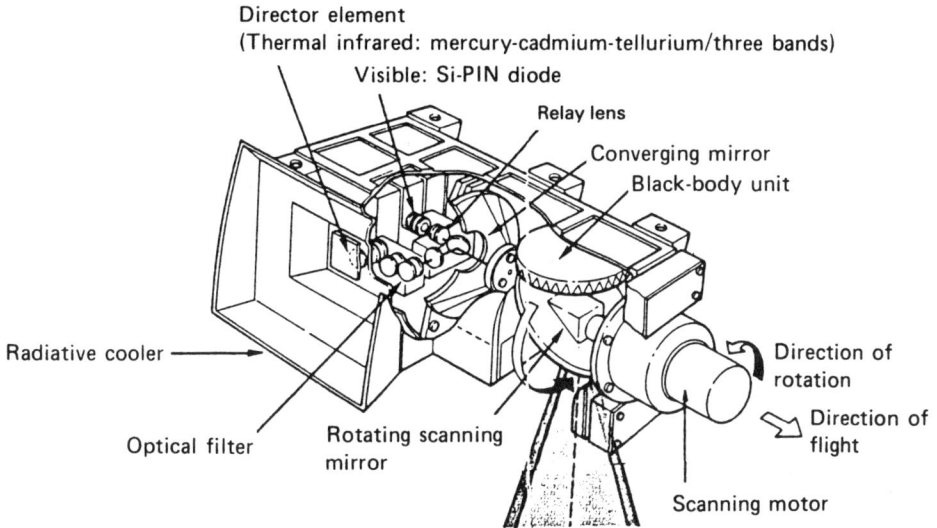

Director element
(Thermal infrared: mercury-cadmium-tellurium/three bands)
Visible: Si-PIN diode

Relay lens

Converging mirror
Black-body unit

Radiative cooler

Direction of
rotation

Direction of
flight

Optical filter

Rotating scanning
mirror

Scanning motor

MESSR

Detective section
[linear array CCD
2048 elements/band]

Spectral
section

Optics

No. 1 camera system
Band 1 and 2

No. 2 camera system
Band 1 and 2

Band 3 and 4

Band 3 and 4

Band 3 and 4

Hood

Field of view
(IFOV': 55μrad)

6.4°

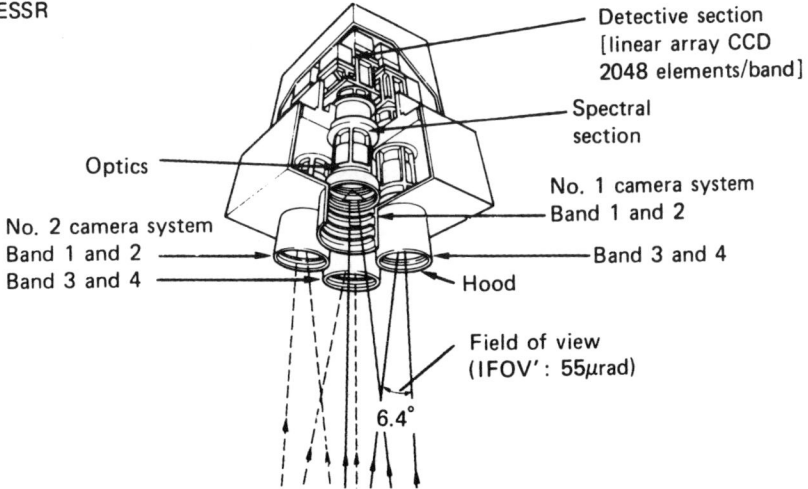

Fig. 17.1 — (b) The instruments. © NASDA.

but slightly narrower) in order to improve its response to the chlorophyll 'red edge'. The term **red edge** describes the phenomenon whereby active chlorophyll reflects much more strongly in the near infra-red than it does in the adjacent red band. Nothing else does this, and so the strength of the relative response from a well-chosen infra-red detector is a generally good measure of the amount or vigour of the vegetation. The report was completed in July 1977. The programme was approved in 1978, the basic design was completed in July 1981 and the original launch date was to be early 1984. However, even Japan is not immune from the problems that beset large technological projects, and the actual launch date (as already mentioned) was 19 February 1987.

The MOS-1 mission is intended as much as a vehicle for the development of techniques and of infrastructures, as for the provision of operational data. The initial specification covered the development of (a) a visible and near infra-red sensor (the MESSR) using CCDs and having a better resolution than MSS, (b) a visible and thermal infra-red sensor (the VTIR) with a ground resolution of 1 to 2 km and a temperature resolution of better than 1°C, (c) a microwave sensor (not covered in this book) and (d) a data collection system (also not covered). The planned life of the satellite was two years, although a longer life was expected.

In fact the VTIR, or 'visible and thermal infra-red radiometer' appears to have emerged from the same stable as AVHRR and the radiometer aboard INSAT, namely ITT Fort Wayne. It is very similar to AVHRR, and is thus based on a single beam being split and directed to just one detector per band (although, as is discussed later, the VTIR detectors are all duplicated). Cross-track scanning is by a continuously rotating mirror and is thus, as with most of the world's meteorological satellites, relatively inefficient. This 'waste' of available light is compensated for by the employment of a large IFOV.

The MESSR (multispectrum electronic self-scanning radiometer), however, bears no resemblance to its nearest American counterpart (the MSS). Instead it follows the increasingly fashionable multiple refractive concept also used on the more recent IRS-1, which we will be describing in considerably more detail in the next chapter. Pushbroom scanning is employed, with each telescope irradiating a single 2048-element CCD array. The observed swath is split between two fixed telescopes, so that (with overlap) the swath is divided into approximately 4000 pixels (the left and right hand sides of the swath are in fact normally observed on alternate orbital cycles, because of limitations in downlink capacity). The spectral range is also split between two telescopes to overcome the chromatic aberration problems introduced by the use of refractive optics. Unlike IRS-1, however, each telescope covers two adjacent bands. The total number of telescopes employed by the MESSR is therefore four.

Problems of language and distance have prevented any interaction between the author and authoritative Japanese sources. There is therefore a degree of conjecture involved in this short chapter.

17.2 THE ORBIT

The MOS-1 orbital period is 237 orbits in 17 days, leading to a value of k (see Chapter 1) of $13\frac{16}{17}$ orbits per day. These figures are achieved by means of an orbital altitude of

Table 1

Spacecraft			
Mass — satellite (kg)	750	Launched	2.87, 2.90
— MESSR + VTIR(kg)	100		
		Target	Sea, Land

Orbit			
Altitude @ Equ.(km)	909	Orbits/day	13.9412 (13⅙)
Semi-major axis (km)	7286.9	Orbits/cycle	237
Eccentricity (%)	0.4	Days/cycle	17
Period (min)	103.3	Shift/orb (°)	25.82
Inclination(°)	99.1	(km) @ Equ.	2874.6
Descending node	10:08	Gnd Track spacing (°)	1.52
Ground speed (km/s)	6.44	(km) @ Equ.	167.0

Telescopes			
Type	MESSR: twin multiple refractor		
	VTIR: Ritchey Chretien		
Focal Length (m)	—	80% blur c. d. (μrad)	—
Aperture (m)	—	F.P. dia. (mm)	—
f-number	—	Field-of-view (°)	$\pm (2 \times 3.2)/\pm 40$
1y mirror dia. (m)	VTIR: 0.015	IFOV (μrad)	55/1000/3000
		(μm @ F.P.)	—
Total obscuration (%)	—	(m on ground)	50/900†/2700†
Clear area (m^2)	—	EIFOV(m on ground)	—
		Inter-band reg. (pix)	—

Scanning			
Method	MESSR: pushbroom		
	VTIR: rotating mirror		
Rate (scans/s)	1/7.3		
(radians/s)	—		
Period (ms)	—		
Active scan time (ms)	—	efficiency (%)	—
Sampling interval (ms)	—		
Dwell time (μs)	—		

Channel details (nominal)

Band	Wavebands (μm)	Detectors	Dynamic range$^+$ (W/m^2/sr/μm)	Noise Equiv.MTF @ Reflectance (% F.S.)	pix rate trk/scan
MS1	0.51–0.59	Si CCD	— — —	—	—
MS2	0.61–0.69	Si CCD	— — —	—	—
MS3	0.72–0.80	Si CCD	— — —	—	—
MS4	0.80–1.10	Si CCD	— — —	—	—
VT1	0.50–0.70	Si diode	— — —	—	—
VT2	6.00–7.00	HgCdTe	— — —	—	—
VT3	10.5–11.5	HgCdTe	— — —	—	—
VT4	11.5–12.5	HgCdTe	— — —	—	—

Electronics	
Consumption (kW)	0.090/0.046
Tape Recorder?	No
Compression	No?

Downlink	
Data rate (Mbit/s)	8.78/1.6
(GHz)	8/2.2/1.7

Data	
Word length (bits)	8
Grey levels	256
Image width (pix)	2×2048/?
Image width (km)	185/1500

† at nadir.

909 km, which is towards the high end of the popular range. The orbital period is very close to the 233 orbits in 16 days of the later Landsats. However Landsat's altitude is nearly 200 km lower and so, at $14\frac{9}{16}$, Landsat achieves another half orbit per day. As discussed in Chapter 1, the ground track pattern is dictated by the fractional part of k, or k''. A k'' of $1/d$ would lead to tomorrow's orbits being one track to the east of today's, although none of the spacecraft covered in this book follow that pattern. What is more common is for k'' to emerge as $(1 - d)/d$, in which case tomorrow's orbits follow a path one track to the *west* of today's; and this is the pattern executed by MOS-1, as Fig. 17.2 shows. The early Landsats, with a k'' of $\frac{17}{18}$, followed a similar sequence (Fig. 8.2). Most, however, trace out a more complex pattern (see, for

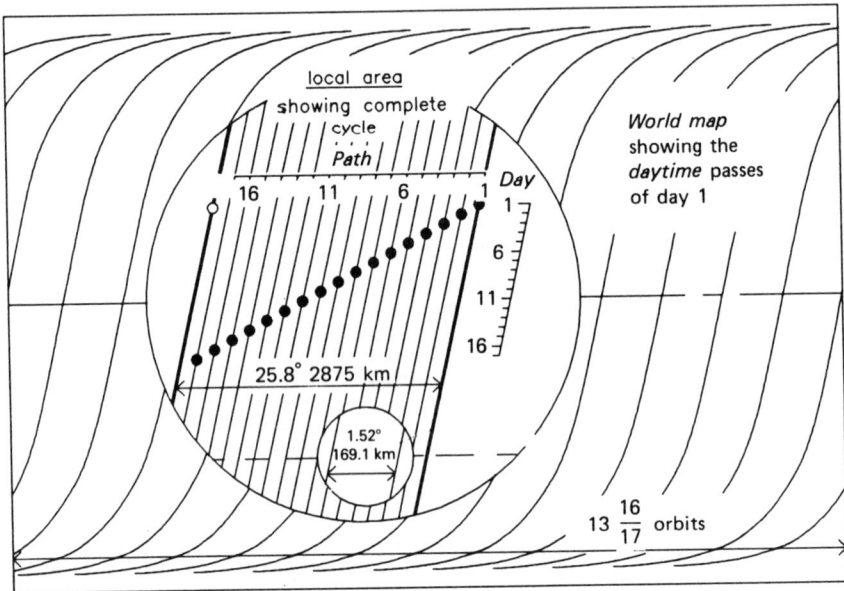

Fig. 17.2 — Ground-track pattern for MOS-1.

example, Fig. 14.2). The orbit is controlled to ensure that the executed ground tracks are within ± 20 km of the positions defined by the MOS-1 World Reference System (WRS).

Although MOS-1's MESSR instrument carries a side-by-side telescope arrangement, its WRS is defined in terms of the ground track positions. This is in direct contrast to SPOT, the originator of the side-by-side concept, whose WRS is defined in terms of the scene centres. With SPOT, each HRV is regarded as an entirely separate instrument, whose imagery is normally sold separately. It thus makes reasonable sense to introduce the considerable additional complication of divorcing the scene indexing from the orbital path. With MOS-1, however, even the MESSR data is normally disseminated with the outputs from the two halves of the swath combined into a single scene, and so there is no need for any such modification to accepted practice.

17.3 SCANNING

The MESSRs are pushbroom instruments, as pioneered by SPOT. Thus no cross-track scanning, as such, is required. Each half of the instrument employs a single 2048-CCD array, and scans a swath 100 km wide. The telescopes are aligned to provide an overlap of 15 km, so that the overall swath width is 185 km. Along-track scanning is arranged, in the conventional manner, by synchronizing the CCD sampling rate to the forward movement of the spacecraft. The IFOV of an individual detector is 55 μrad, or 50 m on the ground, and the sampling rate is matched to this to provide nominally square pixels with no overlap.

Downlink limitations mean that the output from both MESSR systems can only be acquired if data from the other two instruments (VTIR and MSR) are abandoned. In normal operation therefore, MESSR data from the two side of the overall swath are acquired on alternate orbital cycles.

The VTIR on the other hand, like AVHRR, uses a continuously rotating mirror. The mirror spins at 7.3 r.p.s., or one revolution in 0.137 ms. This is slightly faster than the 6 r.p.s., or 167 ms/rev, of AVHRR. During one complete revolution, the spacecraft advances on its orbit by exactly 1 km. As discussed in Section 17.5, this is approximately the same as the IFOV of the visible detector. Thus the nominal pixel size for this detector is 1×1 km. The IFOV of the three thermal band detectors is however 3 km on the ground. This means that the spacecraft only advances one third of an IFOV before each detector is sampled again. The nominal pixel size is therefore 1×3 km, with a considerable low-pass filtering effect in the along-track direction. Alternatively, two lines out of three can be rejected, to generate a traditional 3×3 km pixel, with no filtering effect.

During each mirror revolution, the VTIR detectors observe a patch of deep Space; and also a temperature-controlled black body incorporated into the instrument. Thus data are available to calibrate each line of infra-red data individually if required.

17.4 THE TELESCOPES

As with AVHRR, the VTIR telescope only requires a narrow angle of view. It is described as employing a Ritchey Chretien design, with an 'optical bench' arrangement for the filter and beam-splitter components. For further information on how it must be assumed to work, see Chapter 12.

The MESSR, being a pushbroom instrument, requires a relatively wide angle telescope system. In practical terms, this necessitates the use of refracting elements, as has already been discussed (Chapter 3). With the MESSR instruments, a fully refracting system was adopted. This is in contrast to the SPOT instrument where a Schmidt hybrid design is used, but is similar to the option chosen for IRS-1 (Chapter

18). As with these instruments, the field of view is split among more than one telescope, as Fig. 17.3 shows. The telescopes are described as being of 'Gauss' type, which means, as we saw in Chapter 3, that they bear some similarity to those of IRS-1.

MESSR scans each side of the swath separately. Unlike SPOT, (Chapter 16) however, the angle of view of each telescope is fixed, at 2.73° either side of vertical. As we will be seeing in the next chapter, this is in direct contrast to IRS-1, where the telescopes are aligned vertically to minimize distortion. The outputs from both telescopes are effectively disseminated as a single item.

The spectral range required is some 0.5 μm (0.51 to 1.10 μm). As discussed in Chapter 3, chromatic aberration problems make it impossible to achieve this spectral range, to the required level of performance, with a single refracting telescope, and so the spectral range is also split between two, making four telescopes in all. Bands 1 and 2 share one telescope, while Bands 3 and 4 share the other. Again this contrasts with IRS-1, where each band is provided with its own telescope.

17.5 THE FOCAL PLANES, AND BEAM SPLITTING

As with AVHRR, the concept of a 'focal plane' rests somewhat uneasily on the VTIR design, because each band effectively has its own focal *point*, which is physically widely separated from the other focal points. However, Fig. 17.4 is slightly less uninformative than its AVHRR equivalent (Fig. 12.4), because the VTIR detectors are all duplicated for reliability. As the figure shows, the duplicated detectors are placed as close as possible to each other. In the case of the three thermal band detectors, the gap is virtually insignificant.

Light destined for the two visible detectors (one plus its duplicate) follows a straight path, through a dichroic beam-splitting mirror, optical filters and a relay lens, to the silicon diodes positioned at the focal point of the telescope. Infra-red radiation is reflected by the beam-splitting mirror, and is directed to a system of further beam-splitters and filters, mounted in the vicinity of the radiative cooler (see Section 17.11). Six HgCdTe detectors (three and their duplicates) are mounted at the heart of the cooler, where their temperature can be controlled to about 100 K.

The MESSR, by contrast, is a pushbroom system, with each telescope incorporating two 2048-CCD arrays. Each array is fronted by a spectral filter to eliminate radiation outside its appropriate passband. Fig. 17.3 shows the arrangement. Each array comprises a single CCD chip, rather than four chips as employed by SPOT (Chapter 16). This simplifies substantially the design of the focal plane assemblies. However, unlike SPOT, there are four separate assemblies, one for each telescope.

The positioning of the CCD arrays on the focal plane, and/or the strategy for ensuring good registration between them, have not been described in the literature available to the author.

17.6 THE DETECTORS

Each band of VTIR is scanned by a single detector, each of which is conventional. The visible band employs a silicon diode, maintained at ground-ambient tempera-

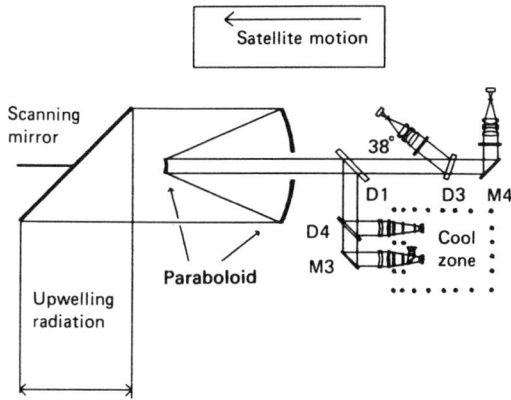

The VTIR telescope

The four MESSR telescopes

There is a degree of conjecture in these drawings

Fig. 17.3 — The MESSR and VTIR telescopes.

Band 1

Band 2

Daytime passes

N

Satellite motion

Direction of scan

0.75

3

0.3

3

1

1

1

3

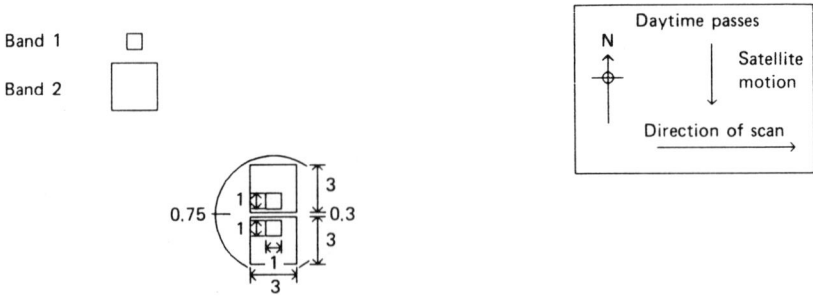

Units in IFOVs of Band 1

The diagram represents the 'logical' focal plane.

The beam-splitting arrangements are shown in Fig. 17.

VTIR — logical focal plane

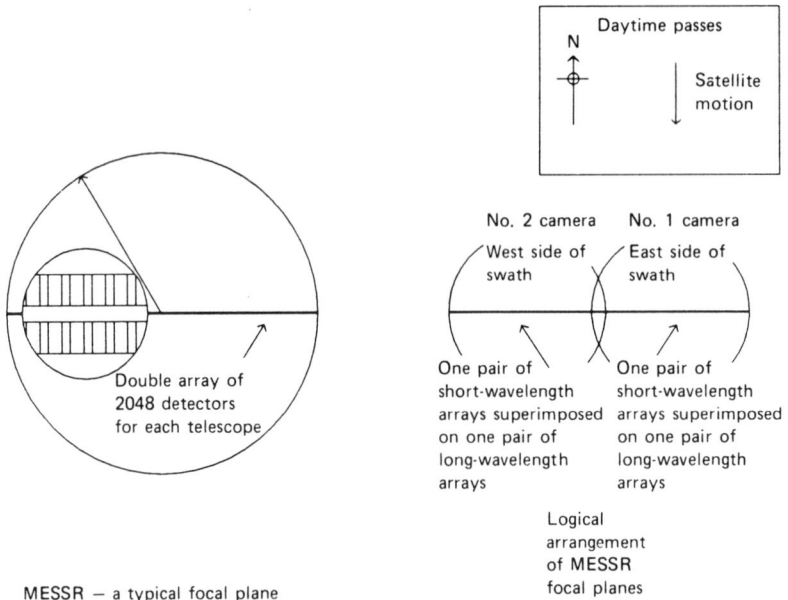

Daytime passes

N

Satellite motion

No. 2 camera

No. 1 camera

West side of swath

East side of swath

Double array of 2048 detectors for each telescope

One pair of short-wavelength arrays superimposed on one pair of long-wavelength arrays

One pair of short-wavelength arrays superimposed on one pair of long-wavelength arrays

Logical arrangement of MESSR focal planes

MESSR — a typical focal plane

There is a degree of conjecture in these two diagrams

Fig. 17.4 — The MOS-1 focal planes.

ture, whereas the three thermal bands are sensed by HgCdTe detectors, maintained at around 100 K.

The MESSR uses a single CCD chip for each (half) swath.

17.7 THE DETECTOR ELECTRONICS

There is no tape recorder aboard MOS-1, apart from a small machine for storing housekeeping data. Therefore data can only be acquired in the vicinity of a suitably equipped ground station.

The VTIR electronics must be assumed to be similar to that of AVHRR†, except for the fact that each VTIR detector is duplicated.

The VTIR electronics has 16 possible gain settings switchable by ground control, to enable the available dynamic range to be used to best effect over surfaces of differing albedos.

The MESSR detectors are not duplicated, because of the availability of a second entire telescope system. However the outputs from the two systems are integrated into a single electronics chain, and this is duplicated.

17.8 THE DOWNLINK

As mentioned in the previous section, there are no image data recording facilities aboard MOS-1. Therefore imagery can only be acquired of areas within the range of a suitably equipped ground station. However, at 8 Mbits/s, the data rate for MOS-1 is relatively low, and any station that can receive MSS should have no difficulty in adapting its receiving system to receive MOS-1 data. At the time of writing of Tsuchiya *et al.* (1987), the following organizations were preparing to receive MOS-1 data: ESA (Kirunna, Fucino, Trömsö, Maspalomas), NRCT: (Bangkok), Australian Landsat Station (Alice Springs), Japan (Showa Base).

17.9 CALIBRATION AND IN-FLIGHT CHECKS

The first six months of the life of MOS-1 were devoted to checks, calibrations and data-quality assessment.

17.10 DATA QUALITY

The calibration arrangements for VTIR are similar to those of AVHRR. During its 360° revolution, the scanning mirror looks for some time at deep Space, which provides a suitable fixed point for calibration purposes. In addition, a temperature-controlled black body is installed which provides another calibration point.

17.11 THE HOST SPACECRAFT

The spacecraft is basically a double-box structure, and embodies a modular concept of bus and mission modules. The instruments are installed in the mission module,

---

† Or not, according to the reader's preference.

while the power supplies, the attitude control and the downlink are housed in the bus module.

Unlike most of the other spacecraft described in this book, MOS-1 employs a single solar panel structure.

17.11.1   Attitude control

Attitude is controlled by two biassed momentum wheels and a magnetic torquer (see Chapter 12).

The specified pointing accuracies are $\pm 0.45°$ in roll and pitch, and $\pm 1.0°$ in yaw. Stability is required to be $\pm 0.016°/s$ in roll and pitch, and $\pm 0.05°/s$ in yaw.

17.11.2   Temperature control

Thermal control is conventional. The bulk of the spacecraft is maintained at approximately ground-ambient temperature by the judicious use of insulation blankets and radiative surfaces. The thermal-band detectors are maintained at a suitable temperature by means of a conventional two-stage radiative cooler.

17.13   SPACECRAFT OPERATION

The mission is controlled from the Tracking and Control Centre, at the Tsuckuba Space Centre, with the receiving station being sited at three tracking and Control Stations at Katsuura near Chiba. Data processing takes place at the Mission Management Facility at the Earth Observation Centre in Hatoyama in Saitama prefecture.

17.14   PRODUCTS

The Earth observation centre produces CCTs and photographic products, at five different levels of rectification, namely:

0 Totally uncorrected, but all necessary calibration and correction data supplied.
1 Level 0, with radiometric correction applied.
2 Level 1, with 'system' (orbital and altitude) corrections applied.
3 Level 2, with precise geometric correction using ground control points.
4 Level 3, with advanced radiometric correction.

Data are supplied by the Remote Sensing Technology Centre (RESTEC), in Tokyo.

REFERENCES

Tsuchiya, K., Arai, K. & Igarishi, T. (1987) 'Marine Observation Satellite' *Remote Sensing Reviews* **3**.
ESA (1987) 'Announcement of Opportunity' for the utilization of MOS-1 data, March 1987.
NASDA (1987) 'MOS-1' Promotional material, NASDA.

18

IRS-1: The Indian Earth-resources satellite

Consultant: **Dr George Joseph**†
Sources: NRSA (1986), Péraldi (1989)

18.1 INTRODUCTION

With the inception of the Indian space programme, the world Space scene could be described as having 'come of age'. Like the INSAT programme, IRS-1 demonstrates that remote sensing satellites do not have to be massively expensive to be effective. The programme is geared strictly towards serving the needs of that relatively impoverished — but rapidly growing — sub-continent, rather than to generating glory or to advancing the art. Two earlier experimental satellites, Bhaskara I and II, have already flown, and the IRS-1 series is regarded as the next logical step towards the introduction of an operational programme. Its task is to provide timely and acceptably accurate information on various renewable and non-renewable resources to assist in the continued development of that country. At the same time it must provide a stimulus for the development of the infrastructure to support a fully operational remote sensing service. A mission life of three years is planned.

The specification for the 'linear imaging self-scanning sensors' (LISS) on IRS 1 was based around the requirements of users already adept at using data from MSS and TM, and who required continuity of characteristics. On the other hand it took full account of the need for a simple and rugged design, with an almost total absence of moving parts‡. For example, frequent revisit capability is important (NRSA, 1986), but there was no question of incorporating angled viewing facilities in order to achieve it. Instead, the viewing angle was made as wide as possible in order to permit a reasonably coarse ground-track pattern. Two different pixel sizes were specified, namely $36\frac{1}{4}$ m and $72\frac{1}{2}$ m, to provide a degree of continuity for existing users of both

---

† Deputy Director, Space Applications Centre, Indian Space Research Organization, Ahmedabad.
‡ The significance of this point will not escape readers who have already studied Chapter 14. To discuss whether Thematic Mapper is 'an exquisite machine', or a nightmare, is left as an exercise for the reader.

MSS and TM data. The wavebands chosen for both, however, were similar to the first four of TM. This was because the TM wavebands are in fact very much better than those selected for the pioneering MSS. A totally non-moving calibration system was devised, based on the use of light-emitting diodes.

Fig. 18.1 — (a) IRS-1. Courtesy: ISRO.

This concept leads almost inevitably to a pushbroom scanning system, and to the use of CCDs. However, the choice of telescope did not fall out so simply. Basically, the choice lay between the traditional use of a single reflecting telescope or, more radically, several refracting instruments (Péraldi, 1989; see also Chapter 3). A reflecting telescope is large, heavy, and difficult to make. It is also ill-suited to providing a wide angle of view. The SPOT system, for example, uses two HRV instruments in parallel partly for this reason. On the other hand, reflective optics lends itself well to multispectral applications because it generates no chromatic aberration. Unfortunately the use of a single telescope demands either beam splitters, with their consequent loss of light, or the dispersal of the detector apertures over the focal plane. These problems can be overcome by splitting the duty among

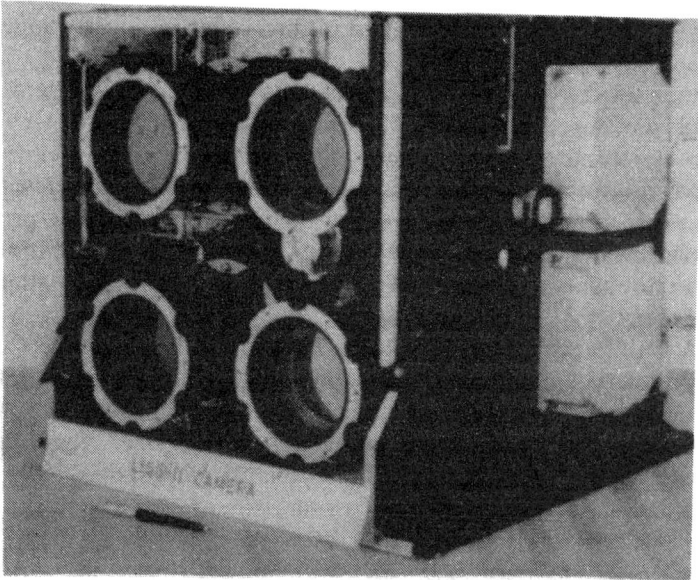

Fig. 18.1 — (b) A LISS II instrument. Courtesy: ISRO.

several telescopes. But such telescopes must, realistically, be fully refracting, and this solution is only applicable where such optics will meet the case.

The solution adopted for IRS-1 was to take the concept to its limit, by providing a separate telescope for each band. However, a single CCD array could not cover an acceptable swath width at the $36\frac{1}{4}$-m pixel size, and so two chips per band were required. The problem of butting up CCD arrays to form a single line of imagery was shown by Matra to be quite difficult (Chapter 16). Therefore the multi-telescope concept was taken even further, by providing a separate telescope for each array. Furthermore, the two resolutions were obtained by the use of two sizes of instrument, LISS I and LISS II, rather than by combining pixels, as is done with SPOT and several other instruments. We are thus led to a total telescope count of 12. To a certain extent the difficult design problems were simply shifted into the laps of the supplier of the telescopes. Stringent performance criteria were imposed which, in general, were all met with comfortable margins. In fact there are two LISS II instruments, each observing one half of the swath, while the single LISS I instrument scans the whole swath. The width of each image is 2048 pixels. Quantization is to 7 bits, giving 128 grey levels.

The orbit is conventional: Sun-synchronous, with a cycle of 307 orbits in 22 days, or 904 km altitude. This leads to a ground track spacing of 130.54 km; which matches nicely the required resolution and the size of the chosen CCD array.

IRS-1 is being flown principally for the benefit of India. No facilities are installed for recording imagery acquired from beyond the reception range of the principle ground station at Hyderabad, and no link-up arrangements have been made with other ground stations. However, a number of neighbouring countries are within the live-data reception range of Hyderabad. Other countries can arrange to receive IRS-1 data if they wish.

In general, the design of the LISS instruments is a remarkably ingenious solution to the problem of developing a new space-borne instrument to a budget. The use of a pushbroom system, the choice of a line length that matches an available CCD chip, that of 7-bit radiometric quantization, the decision to go for a multiplicity of simple refracting telescopes, the obtaining of the high-resolution data by providing two separate instruments, even the choice of a totally non-moving calibration system all testify to a thoroughly realistic set of objectives, and to a determination to get the best out of them.

IRS-1A was launched on 17th March 1988, from the Baikonur Cosmodrome aboard a Vostok rocket. At the time of writing, IRS-1B is due for launch in 1991. It will be identical to IRS-1A. However IRS-1C and -1D will incorporate an additional band in the mid-infra-red, and will feature a 20-m pixel size. There will also be a panchromatic band with a pixel size of better than 10 m (Joseph, 1990). The second generation, IRS-2, is already under discussion.

18.2   THE ORBIT

The IRS-1 orbital cycle is 307 orbits in 22 days, leading to a value of k (the number of orbits in a day, see Chapter 1) of $13\frac{21}{22}$ (13.95). The quoted altitude is 904 km, although the computations in Chapter 1 give 917 km. The descending node is 10:25 am. The altitude is controlled to an accuracy of ± 1 km.

As we have seen, the selection of the orbital cycle was constrained by the choice of a pushbroom scanning system, with but two CCD arrays to cover it at medium resolution. Considering the requirement for frequent revisits, the repeat cycle of 22 days is possibly slightly longer than might have been wished. There were no particular constraints on the ground-track pattern, and a pattern similar to that of MSS emerged from the optimization (Fig. 18.2). Tomorrow's orbits are one track to the west of today's. The defined altitude limits lead to the ground-track pattern being controlled to within ± 14.8 km.

Coverage is from 81°N to 81°S, although that is of little consequence to India.

In choosing the descending node time of about ten o'clock, the following factors were taken into account. Many applications (e.g. agriculture, forest use, land use and water resource applications) are helped by having a high Sun elevation. A descending node gives a slightly higher elevation than an ascending node, for the same local time. Classification accuracy for all crop types is optimum at 11.00 am. Terrain modelling effects are enhanced by a moderate degree of shadowing, i.e. between 09.45 and 10.00. Cloud cover is less extensive in the morning. Maximum scene irradiances occur close to midday. A descending node provides continuity with users experienced in the use of Landsat or SPOT data. Finally, the timings should, ideally, not clash with those of Landsat or SPOT in case some ground stations might wish to acquire both.

A more detailed discussion of second order orbital perturbations, caused by atmospheric drag, terrestrial asphericity etc., is given in NRSA (1986).

Table 18.1

Spacecraft			
Mass — satellite (kg)	950	Launched	3.88
— single LI (kg)	40		
— single LII (kg)	85	Target	Land

Orbit			
Altitude @ Equ.(km)	904	Orbits/da	13.9545 ($13\frac{21}{22}$)
Semi-major axis (km)	9282.34	Orbits/cycle	307
Eccentricity (%)	0.04	Days/cycle	22
Period (min)	103.19	Shift/orb (°)	25.80
Inclination (°)	99.049	(km @ Equ.)	2872
Descending node	10:00	Gnd track spacing (°)	1.173
Ground speed (km/s)	6.45	(km @ Equ.)	130.54

Telescopes			
Type	Multiple refractor		
Focal Length (m)	0.162/0.324	80% blur c. d. (μrad)	27/20
Aperture (m)	0.036/0.072	F.P. dia. (mm)	26.8
f-number	4.5	Field-of-view (°)	± 4.7/ ± 2.35
		IFOV (μrad)	80.2/40.1
		(μm @ F.P.)	13/13
Total obscuration (%)	0	(m on ground)	72.5/36.25
Clear area (m^2)	0.0366		
Exit pupil		Inter-band reg. (pix)	± 0.25

Scanning			
Method	Pushbroom		
Rate	n/a		
Period (ms)	n/a		
Active scan time (ms)	n/a		
Sampling interval (ms)	11.2/5.6		

Channel details (measured characteristics)

Band	Wavebands (μm)	Detectors	Dynamic range (W/m^2/sr/μm)	Noise equiv. Reflectance (% F.S.)	MTF @ pix rate trk/scan	Cooled?
1	0.45–0.52	Si CCD	1–310	—	60/38	No
2	0.53–0.59	Si CCD	1–300	—	60/38	No
3	0.63–0.69	Si CCD	1–250	—	60/35	No
4	0.77–0.86	Si CCD	1–170	—	60/25	No

Electronics	
Consumption (kW)	0.1 (payload only)
Tape recorder?	No
Compression	No

Downlink	
Data rate (Mbits/s)	$5.2/2 \times 10.4$
(GHz)	2.2/8.3 (S-band/X-band)

Data	
Word length (bits)	7
Grey levels	128
Image width (pix)	2048
Image width (km)	148.48

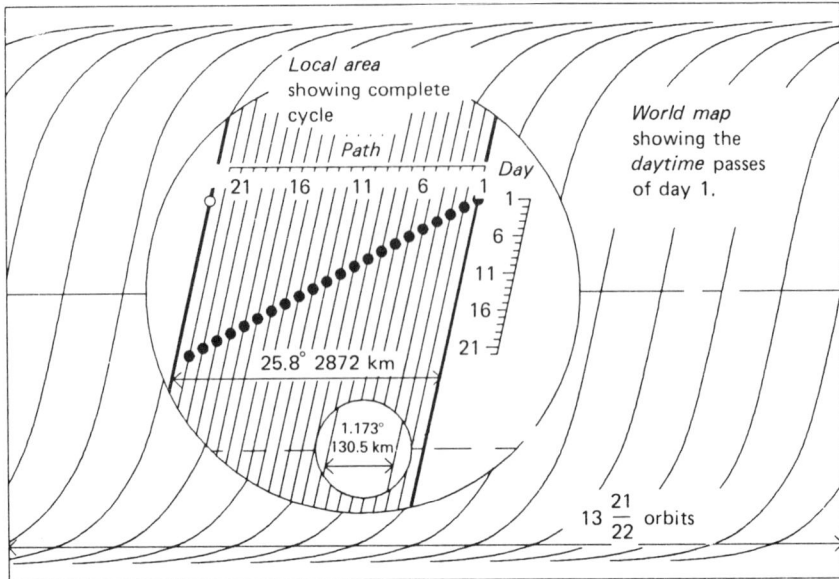

Fig. 18.2 — Ground-track pattern for IRS-1.

18.3 THE TELESCOPES

In accordance with the theme of rugged simplicity, and as has already been discussed, a multiplicity of simple refracting telescopes are used instead of the usual single, complex reflecting instrument. Refracting optics are highly efficient, robust, relatively light and cheap to make — provided always that the aperture required is not too large. The physical aperture requirement is dictated partly by the light-gathering capability needed, and partly by the magnification required (i.e. the focal length, which is equal to detector aperture divided by ground IFOV). Within these limits, image quality can now be made extremely good, and virtually free from distortion, geometric aberrations or polarization effects.

However, chromatic aberration is a serious potential problem and so each spectral band is provided with its own optics. This arrangement also eliminates the problem of beam splitters, a troublesome headache for the designers of the SPOT system. (The loss of light inherent in beam splitting increased the required aperture of the SPOT telescope quite substantially.) On the other hand, however, the policy of simplicity precluded any kind of in-flight adjustment (most of the other instruments described provide such adjustment, although few seem to have used them). The telescopes had to be aligned on the ground to a high degree of accuracy, and mounted sufficiently sturdily to prevent any possibility of movement thereafter.

The ground-track spacing chosen for IRS-1 is 131 km. For a pushbroom instrument, this demands a field of view of $\pm 4.7°$, which, by traditional standards, is remarkably high. It is part of the reason why the SPOT system incorporates a twin telescope system, although the French made a virtue out of necessity by adding an off-nadir viewing capability, and increasing the flexibility of coverage. However, such a viewing angle is by no means unrealistic for modern refracting designs, even when a high resolution is also demanded. We have already seen that later IRS missions will use refracting optics to produce multispectral imagery of a similar performance as that of SPOT (Joseph, 1990). Matra have proposed such optics for other future instruments, with a similar specification (Péraldi, 1989).

The moderate resolution, and the use of CCD chips with their relatively small apertures, makes the focal length much shorter than is usual for such instruments. The focal length of LISS II is 0.324 m, while for LISS I it is down to 0.162 m. Short focal length lenses are easier to make, as well as being stiffer and lighter.

The reader may be surprised to learn that the LISS I telescope is an exact scaled-down version of that of LISS II. This implies that both instruments have the same field of view, whereas we have already said that each LISS II instrument images but one side of the swath. The answer to the paradox is several-fold. First, it leads to economies of design. Second, a degree of over-specification means that future stretch capability is already built in (Joseph, 1990). Third, geometric distortion is reduced (Péraldi, 1989). Pointing the telescopes of both LISS II instruments vertically eliminates the geometric distortion that would result from canting them, and provides a better performance than does reducing the field of view. The feature is analogous to the use of a sliding plate holder in some high-quality tripod cameras.

Thus the IRS-1 payload comprises a total of 12 instruments, conceptually integrated but mechanically separate, as is shown in Fig. 18.3. The four LISS I telescopes are all nominally identical, as are the eight LISS II instruments. However, the slight differences in focal length, that arise from the different wavelengths involved, are adjusted out during the final build of each telescope (Péraldi, 1989). The residual differences in focal length that remained were dealt with by selecting the detector size (Joseph, 1990).

The specification for the LISS lenses presented a considerable challenge. The requirement was to design two lenses with basic parameters as given in Table 18.1. The tolerance on focal length for LISS I was ± 0.15 mm, with a maximum variation within a matched set of four lenses of 10 μm (this was somewhat better than the specified figure of 16 μm, which represents 0.2 pixel). The equivalent figures for LISS II were twice these. Lateral displacement of the optical focal planes of a matched set, referred to a locating feature on the lens mount, had to be maintained to within 1 μm. Variation with temperature was to be within 1 μm between 15 and 25°C.

To accommodate the effects of the residual thermal movement, the depth of focus for LISS I was required to be 25 μm (50 μm for LISS II), which is relatively large considering the tight specification for MTF (spatial frequency response). For LISS I, the MTF had to have fallen to no lower than 70% (or 65% for band 4) at a spatial frequency equivalent to one pixel (see Chapter 5). The figures for LISS II were permitted to be 5% worse than these. For two bands on LISS II, the specification was only just met. For all others, a margin of between 5 and 10% was achieved.

Distortion had to be within 0.1%, to enable adequate registration to be maintained over the whole image. The differences in the distortion between the four

Satellite motion

LISS IIA   LISS I   LISS IIB

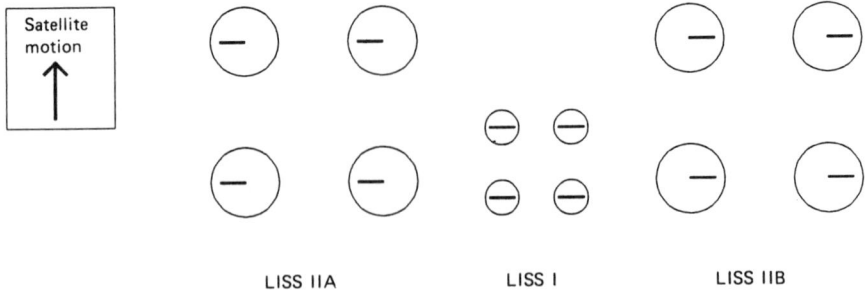

Physical arrangement

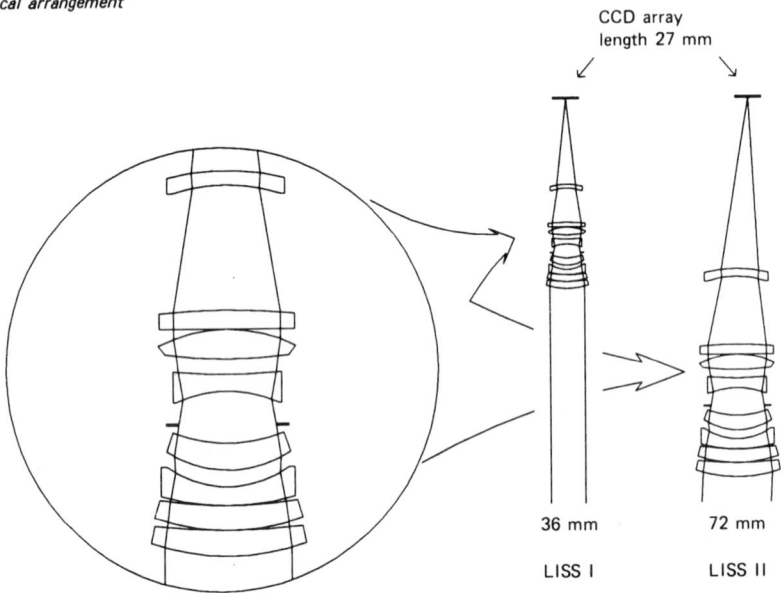

CCD array
length 27 mm

36 mm   72 mm
LISS I   LISS II

Lens configuration

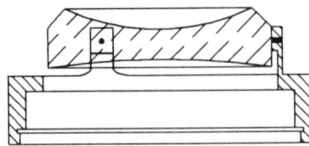

Typical element mount

Fig. 18.3 — The IRS-1 telescopes.

telescopes of a matched set had to be ten times less than this. This is to enable interband comparisons to cancel the residual effect out. Chromatic aberration represented a significant design problem, even over the limited spectral range of a single waveband.

The MTF is, of course, effectively a measure of how well the image is focussed, and it was used to optimize the positioning of the focal plane assembly in relation to the optics, and also to assess compliance with the specification. In general its measurement is not totally straightforward and, in common with other instruments, the question of how well the specification was met posed problems (Jouan *et al.*, 1989). The ISRO view is that the telescope specification was not quite met in certain respects, but that overperformance elsewhere meant that the overall behaviour of the instrument is on target (Joseph, 1990).

The provision of sufficiently stable mountings for the lens elements proved a major challenge. It was necessary to find a method of mounting the eight elements in a fashion whereby they could be positioned to within 5–20 μm, and then fixed sufficiently firmly to hold any further movement within 0.2 μm. There had to be no risk of permanent displacement, or of distortion to even the thinnest of them — under *g*-forces, or vibration levels of up to 20*g* between 20 and 2000 Hz. Dimensional stability had to be within limits over a temperature range from 12 to 28°C. After a trade-off analysis between Invar and aluminium alloy, the latter was chosen for all the major metal components. The glasses used, one crown and one flint, were both chosen for the relative insensitivity of their refractive index to temperature.

Standard lens-mounting techniques involve crimping or the gripping of the edge of the lens with a screwed ring. These did not prove sufficient to prevent the elements moving under *g*-loads or vibration. Instead a separate mount was designed for each element (a typical example is shown in Fig. 18.3) into which the element could be fixed with extreme accuracy, but which still left enough flexibility to cope with temperature variations. Each mount took the form of a ring, from which protruded three lugs equispaced around the ring. The lugs are flexible enough to accommodate residual thermal expansion, but stiff enough to meet the other requirements. Each element was positioned in its mount, to within the 5–20 μm already mentioned, and then glued. The main requirement was in fact to get the lens accurately centred in the mount, because there was no possibility of adjusting out any misalignment that may have crept in. Small longitudinal errors were of less significance, and uncertainty of between 20 and 50 μm could be tolerated (Joseph, 1990). Each lens mount bears against the shoulder of the next one, and so the assembly is built up.

Each lens was optimized individually for its operating wavelength (Péraldi, 1989). Finally, the assembly was inserted into a heated barrel, sized to grip the lens mounts slightly when cool. From then on, no further adjustment was possible. The elements could, however, be removed again if required.

The achievement of these requirements proved quite easy with the LISS I telescope. However, the greater masses of the LISS II lenses meant that they tended to resonate at frequencies around 1.7 kHz, which caused considerable problems to Matra (Jouan *et al.*, 1989). The manufacturing tolerances had to be very carefully controlled to provide adequate grip, and to ensure that they did not slip out of position. With two lenses, random vibration did in fact induce a movement of the optical axis to just outside tolerance.

Each lens assembly was flushed out thoroughly with dry nitrogen, and then sealed. A bursting membrane was incorporated, to allow the nitrogen to escape during launch. Readers may recall reading that, on AVHRR, an inert gas is sealed into the cavity between the infra-red field lenses and the detector elements. However, the reasons for taking the opposite decision in this case were (a) to eliminate pressure stress on the lenses at each end and (b) concern that gas might leak out and alter the optical properties of the lenses. Both these effects would be less for the AVHRR design.

18.4 THE FOCAL PLANE

The focal planes are designed around a single 2048-element Fairchild CCD chip, which was the longest CCD array chip commercially available at the time. Fig. 18.4 shows the arrangement. As in all pushbroom systems, the detector array is oriented across-track, so that a complete line of imagery is obtained at each sampling. The individual detector apertures are the same as those in the SPOT system. However, the shorter focal length of the IRS telescopes leads to substantially greater ground IFOV sizes for the same detector size.

The single LISS I instrument scans a swath 148.5 km wide, which covers the maximum ground-track spacing (of 130.5 km at the Equator) with 18 km of overlap. The two LISS II instruments are aligned to scan either side of the ground track, with an overlap of some 3 km. They provide a combined swath width of 145.5 km.

The level of interband registration specified — and achieved — is 0.25 of a pixel. From the user's point of view, it is of course right that registration should be specified as a function of pixel size. On the other hand, from the point of view of engineering design, the absolute figure in micrometres (at the focal plane) is more indicative of the scale of the problem. On either measure, however, the IRS figures compare favourably with those of, for example, SPOT's 0.3 of a 26-μm pixel (for the multispectral bands only). The achievement is the more remarkable considering the degree of physical separation between the various detector arrays. Clearly, the more mechanical structure there is separating the detectors for the different bands, the more difficult it will be to obtain — and hold — really good registration. For example, the mechanical path between the spectral channels on SPOT's HRV instruments is extremely short (Péraldi, 1989). While the correct alignment of SPOT's various CCD arrays was no trivial task, there is little likelihood of their accidentally becoming misaligned thereafter. On the LISS instruments, however, both the telescopes and the focal plane mountings were candidates for the generation of misalignment. The achievement, and maintenance, of good registration was only achieved by very careful attention to all aspects of the design and construction. We return, yet again, however, to the underlying theme of this instrument, because it would not have been possible to avoid the additional complexity of in-flight adjustment, had not the original specification generated a reasonably short focal length and moderate aperture (Péraldi, 1989).

18.5 THE DETECTORS

A vital set of components, yet to be discussed, is the optical filters. It is difficult to find a consistent yet logical place to cover these. On the IRS instrument they could

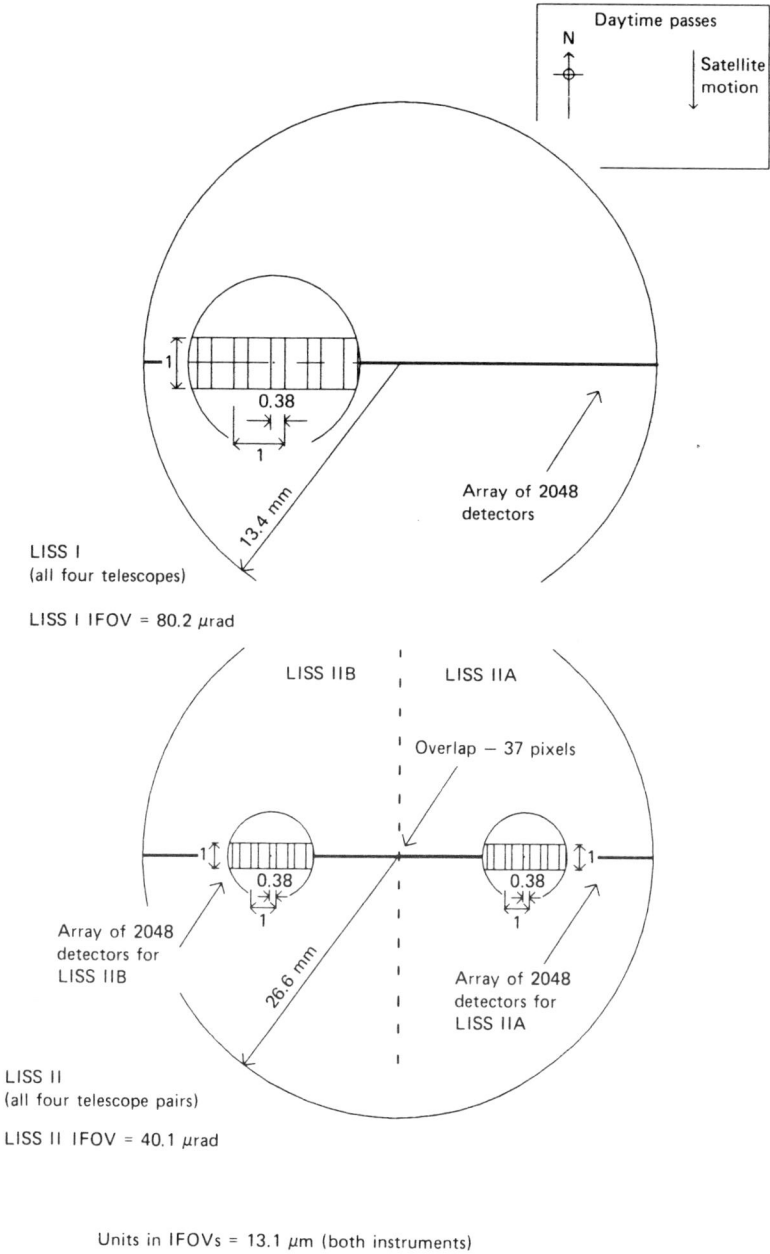

Fig. 18.4 — The logical arrangement of LISS focal planes.

reasonably be considered to be part of the telescope, and yet on Thematic Mapper, and several other instruments, the filters are built into the detector housing. On AVHRR, and related radiometers, they are effectively part of the beam-splitting arrangements. For better or for worse, this section is where they are given a home.

The requirements for the spectral filters was fairly stringent, demanding a pass band which had a flat top, to within ±5%, and steep sides. That this is not always achieved can be seen from Fig. 16.6, or the equivalent for many other instruments. As Fig. 18.6 (Section 18.10) shows, the requirements were met; although, in the case of Band 2, only just.

The filters are mounted on the upstream end of the lens assembly. Mounted outside the main filter is a neutral density filter. Both are removable, and the neutral density filter is intended to be re-coated to compensate for variations in lens transmission and in CCD sensitivity. The limited dynamic range of CCDs makes it necessary to match the incoming flux as closely as possible to that which they can handle. The maximum expected flux is adjusted to represent 80% of the CCD saturation value (Joseph, 1990).

To meet the specification required a four-element interference filter. As is discussed in Chapter 4, interference filters are sensitive to incidence angle. The IRS filters are therefore sited in front of the telescope, where incidence angle differences are minimized, in spite of the fact that they must then be very much larger than those of most other instruments. In fact producing filters of less than 4 in (10 cm) diameter poses no particular problems (Joseph, 1990).

As with electronic filters, sharp cut-off can only be obtained at the expense of in-band performance, unless a large number of elements is used which brings its own penalties. A degree of ringing in the pass band is therefore a natural accompaniment to a sharp cut-off.

An interesting problem in the design of the filters concerned the use of glue. Two interference elements were formed on each of two glass discs, which then had to be bonded together. The glue tended to diffuse into the coating and affect its optical properties, causing scattering of the light. This led to a level of veiling glare outside specification. The 'correct' solution would have been to redesign the filters, by mounting all the coatings onto a single disc. However there was insufficient time for this, and the effect was successfully controlled by reducing to a minimum the amount of glue used.

The detector array for each band of each instrument is a single 2048-element chip of CCDs. The operation of a CCD is described in more detail in Chapters 5 and 16. Fairchild chips are used, similar to those used on the first two SPOT systems (except that those used for SPOT incorporated only 1728 elements). By limiting the image width to the capacity of a single chip, one of the major headaches faced by the designers of the SPOT instrument (Chapter 16) was eliminated. The light loss associated with the SPOT solution was also avoided, with knock-on benefits to other parts of the instrument.

18.6   THE DETECTOR ELECTRONICS

Each LISS instrument has totally independent electronics, right through to the transmitter. LISS I and LISS II have identical circuitry, except for the differences in sampling rate. The electronics for a single LISS instrument are shown in Fig. 18.5.

Much of the circuit is broadly similar to that of the SPOT panchromatic channel.

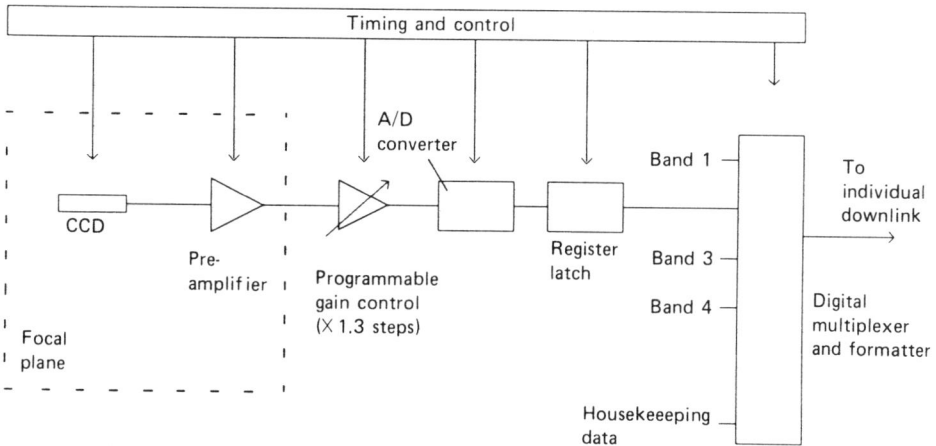

A typical LISS instrument

Note the source material does not explain the function
of the register latch or the formatter

Fig. 18.5 — IRS-1 electronics.

As discussed in Chapter 16, a CCD system requires no explicit anti-aliasing filter. The analogue signal bursts, obtained from sampling the elements of each CCD, are passed to a fixed gain pre-amplifier, and hence to a post-amplifier whose gain is adjustable, under ground control, in steps of times 1.3. By this means the expected signal level can be manipulated to fill the dynamic range of each digitizer. (Sensitivity differences between the CCDs have already been taken out, as previously discussed, by adjusting the transmittance of neutral density filters fitted to the telescopes.) Each channel is provided with its own 'successive approximation'-type A/D converter. The data are passed via a 'register latch' to a common digital multiplexer and 'formatter' (the source material does not explain the function of the latch or the formatter).

Scanning is repeated, under the control of the timing clock, at a rate that allows the spacecraft to advance by $36\frac{1}{4}$, or $72\frac{1}{2}$ m per scan.

The area of interest is confined to the Indian sub-continent, which is all within the reception area of the Hyderabad ground station. Therefore all the imagery of interest to the NRSA can be captured in real time. Thus no overwhelming need was felt to depart from the principle of rugged simplicity by installing temperamental and potentially unreliable tape recording facilities for the imagery. However, a bank of solid-state memory was installed for recording housekeeping data.

18.7 THE DOWNLINK

The only official receiving station for IRS data is the Indian Receiving Station at Hyderabad, strategically situated close to the centre of the sub-continent. In the absence of recording facilities, the reception area of this station also represents the official area of availability of IRS data.

Two downlinks are provided, of different capacities, to cover the different requirements of the two instruments. LISS I data are transmitted at 5.2 Mbit/s via a 2.2 GHz S band link, while combined data from the two LISS II instruments are transmitted at 20.8 Mbit/s via an 8.3-GHz X-band link. Transmitting the data from LISS I and LISS II separately increases flexibility and also the ultimate reliability.

Data are recorded on two high-density data recorders, working in parallel, thus providing full duplication. An online quick-look facility can display, in real time, any one band from any one camera. It can also be recorded on 70-mm film.

A telemetry system operates in the S-band, and also provides a command capability. While the main uplink is also in the S-band, a VHF system is also available.

The S-band tracking is expected to provide accuracies of 30 m in range and 0.1 m/s in range rate (both at 1 standard deviation). However an X-band beacon has also been included to facilitate auto-tracking.

18.8 CALIBRATION AND IN-FLIGHT CHECKS

The system was thoroughly checked out and calibrated before launch. Such features as light transfer characteristics, spectral responsivity, dark current, dynamic range, and shading characteristics were all evaluated.

The 'no moving parts' principle has been followed for the internal calibration system, although it means that the optics cannot be included in the calibration. Light-emitting diodes (LEDs) are used as the calibration source instead of the traditional tungsten filament lamps. LEDs offer high efficiency, and can be placed close to the detectors without risk of affecting their temperature. They are reliable, and their output does not fall off with use. Their response is extremely rapid, and using them in pulsed mode gives an additional mechanism for controlling excitation levels. Used in pulsed mode, with reasonably low-duty cycles, high intensities can be obtained without shortening their life.

Two LEDs are used per CCD array, mounted at an angle of 30° either side of the telescope axis. The optical arrangement ensures that the illumination is more or less uniform along each array (Joseph, 1990). However, intensity profiles were measured before launch and are used to take account of residual differences. The two lamps generate different intensities at the CCD. They are used in pulsed mode, with four different pulse widths. Excitation current to the two LEDs is switchable independently between two different levels. In practice, this results in 12 different calibration levels and zero.

Calibration is carried out during a night-time pass, when the amount of stray light reaching the telescope is much smaller than the lowest count (Joseph, 1990). Each complete calibration cycle occupies 4096 scan lines and takes about 46 s for LISS I, and 23 s for LISS II. During the initial period, calibrations were carried out once a week. For the remainder of the mission they are carried out monthly, to study the degradation in system performance.

18.9 DATA QUALITY (AND QUALITY CHECKING)

The two swaths from the LISS II cameras will have a sidelap of 3 km, so the combined swath width will be 145.5 km, as against 148.5 km for LISS I. Apart from that, they will be synchronized.

Interband registration has been nominally set at 0.25 pixel.

18.9.1 Radiometric performance

A noise-equivalent (NE$\Delta\rho$) of 0.5%, of full scale, was taken as the design goal. Thus 128 grey levels are sufficient, permitting a certain economy in data mass.

Signal/noise ratios have been evaluated post-launch and, at better than 255 for all bands, are significantly better than expected.

18.9.2 Spectral performance

The specification for the filter profile is given in Jouan *et al.* (1989), as discussed in Chapter 7. It is clearly a tight specification, both as to the steepness of the sides and the flatness of the top. The requirements appear to have been met, though with little margin in the case of Band 2. The measured profiles, derived by Matra, are given in Fig. 18.6. They do not include the response of the CCD detectors. The plots are adapted to give relative profiles to the book standard scaling.

18.9.3 Spatial frequency response

Jouan (1989) gives detailed predictions of how the MTF should vary over the focal plane, and along the optical axis in the region of the focal plane. The curves are used to find the depth of acceptable focus, and the optimum distance at which to site the focal plane assembly. The curves take account of predicted manufacturing degradations, which turned out to be somewhat pessimistic. The lenses are claimed to meet the specification with a slightly higher margin than anticipated.

The worst MTF (which is limited by the detector in the near infra (red) was expected to be around 20%.

The available data are shown, in book standard format, in Fig. 18.7.

18.9.4 Interband registration

Band 2 of LISS I has proved to be '1 pixel off with respect to Band 1'. All other relationships are within the specified $\pm\frac{1}{4}$ pixel.

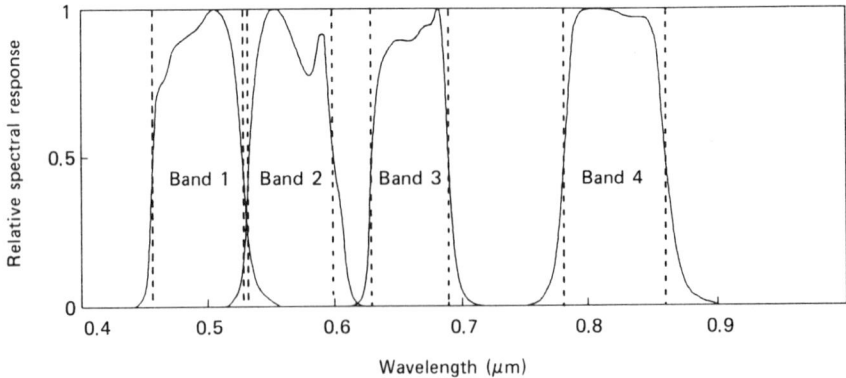

Dotted lines show nominal band limits

Fig. 18.6 — IRS-1 spectral responses.

18.10   THE HOST SPACECRAFT

The main structure is a central load bearing cylinder with four vertical, and two horizontal, honeycomb panels upon which the various systems are mounted. It is made of aluminium throughout, and weighs approximately 950 kg.

Solar arrays generate a power of 709 W (at end of life). Two nickel cadmium rechargeable chemical batteries even out the power to the electrical subsystems.

18.10.1   Attitude control

Attitude control is conventional; using three orthogonal reaction wheels, with a magnetic torquer braking system dumping excess momentum to the Earth's magnetic field (see Chapter 12). A fourth, skewed, wheel is installed as a spare. The skewed wheel induces rotation about all three axes. It can therefore substitute for the failure of any one reaction wheel, using the remaining good wheels to correct for the rotation induced about the other two axes. Sun sensors are used for initial acquisition. Infra-red horizon sensors drive the roll and pitch controls, while the yaw sensing is done through a 'precision yaw sensor' (presumably a rate integrating gyro). A star sensor is used for accurate determination of the attitude delivered by the control system.

Required pointing accuracies are $0.4°$, with a required maximum drift rate of 3.10^{-4}degrees. and a maximum jitter of 3.10^{-4}degrees.

18.10.2   Temperature control

In the absence of a thermal band, there is no requirement for a strongly cooled zone.

The thermal control system is based on insulation and controlled heaters. The electro-optic modules, including the lenses and the detectors, are controlled at $20°C \pm 5$. The batteries are controlled at $5°C \pm 5$. The electronics are allowed to range between 0 and $40°C$.

- Quoted MTF figure

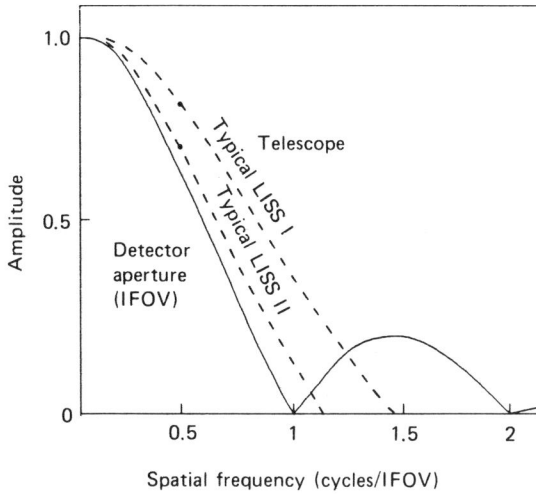

Spatial frequency (cycles/IFOV)

Component MTFs — deduced

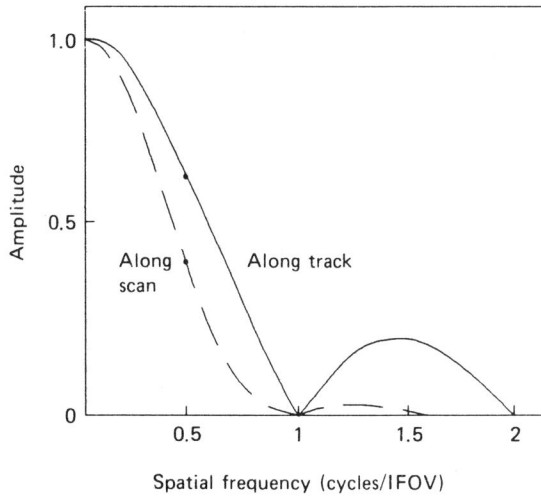

Spatial frequency (cycles/IFOV)

Net MTFs — deduced

Fig. 18.7 — IRS-1 spatial responses.

18.11 PRODUCTS

In the absence of any imagery recording capability on IRS-1, or any formal links with foreign ground stations, interest in the cataloguing and sale of IRS data naturally concentrates on the local area. Thus, although the paths and rows of the IRS WRS are defined for the worldwide capture area of the instruments, they are defined in local terms, and the world map indicates only every fifth path and row. A detailed map has, of course, been prepared covering the visibility region of the Hyderabad ground station. The WRS is defined in terms of the size of a LISS I scene. A LISS II scene effectively covers one of the four quadrants of a LISS I scene, and is defined as discussed below.

Paths (ground tracks) are numbered westwards sequentially from the approximate eastern limit of reception of the Hyderabad ground station. Their exact position is, however, dictated by the need to avoid having the satellite pass directly over the ground station. This is because of a limitation in the tracking speed of the antenna, which would make it lose lock as the satellite passed directly overhead. The orbital tracks are aligned so that the ground station lies half way between Paths 25 and 26.

The splitting of a swath into scenes is, of course, arbitrary. Unlike many other systems however, IRS-1 images are not square. Instead, the un-overlapped portion of each scene is square, at 2048 lines of LISS I imagery, and then additional lines are added, for overlap, to make the total height of an image 2400 lines (this is the number of lines that will fit conveniently, at a scale of 1:1 M, onto a roll of 240-mm film). This being so, the Rows of the IRS WRS are 2048 LISS I lines apart. They are defined so that the Equator lies on the centreline of Row 69. This ensures that Row 1 represents the northernmost extremity of coverage, at approximately 81°N.

A LISS II scene is defined by the path and row of the LISS I scene of which it represents a quadrant. 'A' or 'B' is added to the path number, according to whether it is situated to the left of the right side of the LISS I image. To specify whether it is an upper or a lower quadrant, '1' or '2' is added. Thus LISS I scene P:R spawns the LISS II scenes PA1:RA1, PB1:RB1, PA2:RA2 and PB2:RB2.

A full range of standard products is supplied, ranging from 'quick look', which is supplied in 'real time' on 70-mm film, through various degrees of radiometric and geometric correction to 'special' with a turnround time of 15 days. Special products include extracts covering specified areas only, and/or imagery with band ratioing or principal components analyses already carried out. Data are available both as photographic products or on computer-compatible tape. The take-up rate has been extremely healthy. Indeed, a graph given in ISRO (1989) indicates that the number of products disseminated during the summer of 1989 were three times the total for Landsat and SPOT combined.

18.12 APPLICATIONS

NRSA (1986) gives Table 18.2, outlining the principle applications of the various bands. Its gist will already be familiar to many readers.

IRS-1 data are used intensively over the full range of land and ocean use applications. Preliminary evaluations showed that the imagery was adequate for the updating of maps up to 1:250 000 scale. Late-arriving material indicates that, for land use, the usefulness of the data are somewhere between those of MSS and Thematic Mapper. The deficiency, compared with Thematic Mapper is presumably mainly due to the absence of the longer wavebands. However, the later passage of IRS, and the consequent higher Sun angle, means that there is less visual modelling of ground

Table 18.2

Band	Waveband (μm)	Applications
1	0.45–0.52	Coastal environmental studies (coastal morphology and sedimentation). Soil/vegetation differentiation. Deciduous/coniferous forest cover discrimination.
2	0.52–0.59	Vegetation vigour. Rock/soil discrimination. Turbidity and bathymetry in shallow waters.
3	0.62–0.68	Sensitivity to chlorophyll absorption by vegetation. Differentiation of soil and geological boundaries.
4	0.77–0.86	Delineation of water features. Landform/geomorphic studies. Sensitivity to green biomass and moisture in vegetation.

features than in Thematic Mapper imagery. Imagery acquired during the winter is, naturally, better in this respect.

REFERENCES

ISRO (1989) *IRS Newsletter*, **2**, 1, October 1989.

Joseph, C. (1990) Corrections in draft.

Jouan, J., Martinuzzi, M. M., Mallinge, N. & Coussot, M. (1989) 'Wide field high performance lenses' Matra Space Branch.

NRSA (1986) *IRS Data Users' Handbook*.

Péraldi, A. (1989) A. Péraldi, Scientific Director, Matra, Toulouse; corrections in draft.

Index

aberrations, *see* optical
active/passive components, 99
AD conversion, 167, 189, 192, 209, 228, 231, 262, 288, 323
aiming path, 253, 257, 268
airborne remote sensing, 12
Airy diffraction pattern, 43, 44
Airy disc, 44, 82, 89, 257, 268, 289
albedo, 77, 79, 80, 113, 191, 264
aliasing, 84, 118, 195, 231, 289
  noise, 76, 86, 231
altitude boosting, 21, 220, 278
amplifier, 103–106
  chopper stabilized, 203
  compression, 229
  current, 71, 104, 189
  differential, 105
  gain of, 103
  loading of, 105
  logarithmic, 132
  operational, 7, 103, 189, 260
  post-, 189, 209, 229, 323
  pre-, 75, 81, 82, 95, 102, 105–106, 164, 187, 192, 209, 228, 260, 289, 296, 323
    long tailed pair, 105, 106
    parallelled, 164
    source follower, 105, 260, 260
    transimpedance, 104, 228
  track and hold, 231, 262, 287
  voltage, 104
analogue shift register, 95
angled viewing, 24, 253, 275, 280, 281, 293
  pixel elongation, 181, 281
  stereo effect, 276, 281
angular frequency, 37
antenna despinning
  electronic, 169
  mechanical, 143–144
anti-reflection coating, 61
  MgFl2, 185
  optical black, 254
aperture, telescopedetector, 42, 83
APT (analogue picture transmission), 176, 190
astigmatism, 46

atmosphere, the, 30–40
  absorption, molecular, 33
  active constituents, 30–36
  aerosol, 31
    continental, 31
    maritime, 31
    rural, 32
    urban, 31
    volcanic, 33
  attenuation, 31, 39
    extinction, 33
  models, 37
    LOWTRAN, 37, 79
    predictions, 34–35, 39
  particulates, 31
  path radiance, 38
  properties (Fig.), 6
  scattering
    Mie, 31
    molecular, 33
    Rayleigh, 33
  US Standard, 1962, 36, 37
atmospheric sounding, 135, 141, 173, 195
ATS programme, 135
attitude control, 19, 133, 143–145, 244, 269, 295, 310, 326
  during orbital transfer, 144
  gyro
    rate (integrating), 196, 269, 295, 326
    control moment, 197
  momentum wheel, 197
    biassed, 197, 213, 310
  nutation, 145, 169
  three-axis, 156, 169, 195–197, 201, 213
  reaction wheel, 197, 270, 295, 326
  sensing, 144, 170, 196, 213, 258, 264, 269, 295, 326
  spin, 143–144, 148, 169–170, 197
  yaw, 196, 295
autocorrelation function, 43
AVHRR, 27, 49, 67, 73, 79–82, 94, 117, 176–200

background irradiance, 73, 81, 185, 191, 254, 264
band-merged sources, 272

batteries, 142, 268, 271, 295, 326
beam splitting, 66–69, 115, 306, 316
　　dichroic, 66–69, 184, 185, 192, 206, 284, 306
　　gold, 206
　　neutral density, 66, 184, 284
　　inconel, 184
Bessel function, 43
Bhaskara missions, 311
black body, 32, 38, 166, 171, 173, 192, 248, 267,
　　see also calibration
blur circle, 44, 89, 283
bolometer, 95
Boltzmann's constant, 75
Boltzmann's factor, 92
boundary layer, Earth's, 31, 35, 39
Brownian motion, 33
bus, *see* SPOT

calibration, 109, 133, 233–235, 263–265, 289,
　　292–293, 309, 324
　　absolute, 136, 166, 190, 211, 235, 263, 292, 293
　　interband, 293
　　LED system, 167, 324
　　multidate, 293
　　non-reliance on, 236
　　relative, 263, 292
　　right through, 37, 235
　　sources
　　　black body, 166, 191, 211, 233, 305
　　　oscillating flag, 233
　　　stellar objects, 263
　　　sunlight, 292
　　　tungsten lamps, 233, 292
CCDs, 51, 95–98, 228, 277, 286–287, 289, 292,
　　296, 302, 306, 309, 322
　　anti-aliasing provisions, 96, 289, 323
　　2D arrays, 95
　　butting, 121, 313
　　calibration, 277, 289, 293
　　dynamic range, 286, 322
　　integrating detector, as, 286
　　matching, 289
　　noise, 76
　　potential well, 96, 286
　　response, 287, 289
　　transfer register, 95
charge-coupled devices, *see* CCD
cloud analysis, 173
CMV (cloud motion vectors), 172, 215
cloud top height, *see* CTH
coherence length, 61
columns, *see* SPOT
computers, on board, 262
cooling
　　cold trap, 214
　　cool/cold patch, 160, 170, 185, 197, 213, 245
　　　heat loads (AVHRR), 198
　　liquid helium, 249, 270, 271
　　　supercritical, 271
　　　superfluid, 270
　　need, the, 75, 81, 92

radiative cooler, 146, 170–178, 198, 213, 245,
　　310
　　three-stage cooler, 270
copyright, 278, 297
cosmic rays, 262
cross correlation, 121
CTH (cloud top height), 30, 173

damping, 57
DCP (data collection platform), 166
data compression, 190, 199, 262–263, 290
data quality, 111–122, 167, 191–195, 211,
　　235–244, 265–268, 293–295
data tables, 127, 138, 150, 158, 179, 204, 219,
　　252, 279, 303, 315
dB (decibel), 115
deformation model, 172
design life, 143, 156, 169, 202, 295, 302, 311
detectivity, 79
detectors, 70–98, 132, 141, 149, 162–164,
　　185–187, 209, 228, 258, 286–287, 306, 322
　　bias, 93, 94, 186, 187, 209, 260
　　dark current, 79, 90, 92
　　differential detector, 187
　　exposure, 73, 132, 221
　　failed, substitution for, 235
　　noise, 71
　　quantum efficiency, 81, 90
　　semiconductor devices, 92–94
　　　band gap, 92
　　　depleted region, 93
　　　intermediate energy level, 92
　　　n- and p- type material, 93
　　　solid-state diode, 92–94
　　　photoconductive device, 80, 92, 141, 162, 186,
　　　　209
　　　　extrinsic, 92
　　　　intrinsic, 92
　　　photoconductive mode, 94, 162
　　　photovoltaic mode, 81, 93, 186
　　　PIN photodoide, 94
　　　pyroelectric device, 95
　　　thermistor, 95
　　　thermocouple, 94
　　　thermopile, 94
　　semiconductor materials
　　　GeGa, 252
　　　HgCdTe, 57, 80, 92, 141, 149, 162, 186, 189,
　　　　209, 228, 239, 306
　　　InSb, 81, 92, 141, 186, 228
　　　SiAs, 252
　　　silicon, 57, 79, 115, 132, 162, 168, 185, 208,
　　　　228, 233, 239, 306
　　　SiSb, 252
　　vacuum tube devices, 89–92
　　　anode and cathode, 90
　　　dynode, 90
　　　PM tube, 90, 132, 141
　　　secondary emission, 90
　　　work function, 90
dielectric constant, 58
differentiating circuits, 106

diffraction limitation, 268, 257
downlink, 133, 143, 151, 166, 190, 211, 232, 290, 309, 324
　limitations, 302, 305, 309
duplication, 143, 149, 159, 162, 168, 206, 213, 257, 269, 296, 306, 324
dynamic range, 11, 96, 238

e (charge on electron), 71
ear, the, 11, 84
Earth's field magnetometer, 137
eclipse, 19, 142, 146, 170, 197, 269, 271, 292, 295
EIFOV (effective IFOV), 84, 240
electric field, 90
electron volt, 90
electronics, detector, 99–110, 132, 141, 149, 164, 187–190, 209, 228–232, 260–262, 287–290, 309, 323
　active components, 99
　drift correction, 108–110, 189, 209, 231
　pick-up, 105, 228, 229, 231, 260
　potentiometer chain, 99
　resistors, effect of cold, 260
　virtual earth, 103, 262
emissivity control, 69
energetic particle sensor, 137
Eosat Inc., 217,
ESM (equipment support module — AVHRR), 187, 190
EROS data Center, 216,
eye, the, 13

Fabry–Perot Etalon, 63
fast Fourier transform, *see* FFT
feedback, 103
　resistor, 103
　theory, 189
FFT, 120
fibre optics, 130, 141
filters, electronic, 84, 88, 106–108
　anti-aliasing, 133, 164, 189, 194, 231, 241, 262
　　with CCDs, 96, 290, 323
　AVHRR, 189
　Bessel, 164
　　as delay line, 262
　Butterworth, 108, 133, 209
　Goldberg, 229
　linear phase, 189
　order, 108
　RC, 106
　roll off, 108
　spatial implications, 130, 195, 211, 231, 290, 291, 305
　tuned-circuit, 108
filters, optical spatial, 44
filters, optical spectral, 54–69, 114–117, 141, 186, 214, 228, 239, 306
　band stop, 56
　bulk absorption, 54, 55–57, 117, 187, 258, 287
　materials
　　black polythene, 54, 56
　　HgCdTe, 57

PTFE, 56
　molecular resonance, 56
　quartz, 56
　Schott glass, 187
　semiconductor, 56
　reststrahlen, 56
channel flatness, 115, 187
flatness, 115
in-orbit checks, 240, 267
interference/dichroic, 55, 41–66, 164, 183, 187, 258, 287, 322
　band-pass, 56, 46–47, 185
　edge, 62, 63, 187
　construction (IRAS), 255
　Fabry–Perot (Etalon), 63–64
　half-wave cell, 63
　quarter-wave cellcavity, 58, 60
　ringing, 117
　transmission, 55
　wheel, 141
　thermal barrier, as, 198
focal length, 41
focal plane, 41, 148, 162, 183–185, 206–208, 226–228, 255–258, 284–286, 306, 320
　curvature, 46, 50, 258, 283
focal point, 41
frequency domain, 120
Fourier operator, 106
Fourier techiques for fault finding, 159
Fourier transform, 120
frequency response
　spatial, 84, 117–120, 133, 167, 194, 211, 241, 268, 325
　measurement, 86, 87, 260, 295
　see also MTF
　spectral, *see* spectral response
　temporal, 106, 118
frequency/wavelength, 37, 55, 58
Fresnel formula, 58

gamma ray shielding, 260
geometric correction, 171
Global area coverage (AVHRR), 200
Global network of geosynchronous satellites, 134
GMS-4, 147–152
GOES, 134–146
gravitational constant, 22
gravity potential minima, 137
ground tracks, 23, 180, 320, 328
　accuracy, 280
　pattern, 18, 23, 128, 178, 221, 253, 280, 304, 316
GVHRR (ATS-6), 202

helium, liquid, *see* cooling
hertz, 37
HIRS/2, 195
holes, 93
horizon sensors, 144, 170, 196, 213, 269, 295, 326
hours confirming, 257, 273
HRPT (AVHRR), 199
APT (AVHRR), 199

HRV (SPOT), 281–290

IFOV, 8, 42, 83, 218, 231, 238, 241, 320
  detectors, 140, 162, 208, 226, 228, 258, 284
  exposure, 185
  pixel size, 82, 129
  registration, 241
  resolution, 268, 284, 289
  scanning, 84, 139, 208, 231, 305
imaging spectrometry, 13, 54
impedance, 106
impulse response, 87
inclination correction, 278
input impedance, 104
input resistor, 103
INSAT, 201–215
instantaneous field of view, see IFOV
integrating circuits, 106
introudction, 11,
IRAS, 12, 66, 92, 248–274
  location accuracy, 257
IRS-1, 50, 55, 94, 117, 121, 311–329
ISO (Infrared Space Observatory), 249

Laplace operator, 107
lattice vibrational absorption, 56
Le Chatelier's principle, 108
LED, 94
lens, 317–320
  aspheric, 51
  double gauss, 51
  field, 162, 186, 258
    aplanat, 186, 208
    corrector, 283
    effect (TM), 226
  gauss, 306
  materials, see optics
  Petzval, 51
  selfoc, 292
  see also optics
Lenz's law, 108
light, speed of, 37
line spread function, 120
LISS instruments (IRS-1), 316–324
Local area coverage (AVHRR), 200
logical focal plane, 183, 226

magnetic torquer, 197, 213, 295, 310, 326
materials
  coating
    assorted, 245
    silver teflon, 213
  other, see detectors, optics, etc.
  structural
    aluminium, 319, 326
    beryllium, 139, 254
    carbon fibre, 160, 296
    filament wound fibreglass, 245
    graphite epoxy, 225, 284
    invar, 160, 185, 206, 319
    negative thermal expansion, 296
MESSR (MOS-1), 305–309

METEOSAT, 153–174
MIRP (AVHRR), 189
mirrors, 64–66
  coatings
    AGThF4, 160
    aluminium, 183, 254
    nickel, 180, 222
    silver, 183,
    thorium, fluoride, 160
  materials
    beryllium, 66, 180, 222
    ULE glass, 160, 206
    zerodur, 160
  overcoating, 66
  swing, effect of, 222
  constant speed strategy, 224
mixel, 291
modulation transfer function, see MTF
MOI (moment of inertia), 143
MOMS telescope, 53
MOS-1, 299–310
MSS, 83, 126–133
MTF, 87, 118–120, 195, 241, 319, 325, see also
  frequency response

NGC 6543 (planetary nebula), 263
NIVR (Netherlands Aerospace Agency), 248
NOAA weather satellite series, 175
noise, 70–82, 111
  $1/f$, 76, 192
  aliasing, 76, 86
  background, 73, 78, 81, 82, 192, 264
  bandwidth, 76
  CCD, 76
  D, 78
  D^*, 78
  D^{**}, 78
  D^*_{blip}, 78
  detector, 71
  dielectric loss, 76
  equivalent
    flux density, see NEFD
    ground reflectance, see NEΔρ
    irradiance, see NEI
    power, see NEP
    radiance, see NER
    temperature difference, see NEΔT
  excess, see noise, $1/f$
  flicker, see noise, $1/f$
  generation/recombination, 75
  Johnson, 75
  NEΔρ, 78, 112
  NEΔT, 78, 113
  NEFD, 78, 264, 265
  NEI, 78
  NEP, 78, 113, 267
  NER, 81, 113
  Nyquist, 75
  pattern, 76
  performance measures, 77–79
  photon, 71, 114
  quantum, 71

radiation noise, 71
  shot, 75
  Shottky, 75
  SNR, 77, 80, 82, 111, 113, 167, 237, 294, 325
  system, 74
  temperature, 75
  thermal, 75
  worked examples, 79–82
normalised detectivity, *see* noise, D*
Nyquist frequency, 85, 120

one bar per IFOV, 119, 120
one over *f* knee frequency, 80
operation of spacecraft, 214, 271, 291, 310
optical
  aberrations, 45–46
    chromatic, 45, 50, 51, 184, 218, 224, 283, 302, 306, 319
    coma, 46
    distortion, 45, 317
    geometric, 45, 45, 50, 316
  bench design, 177, 181, 183
  cross talk, 265
  dispersivity, 51
  field stop, 89, 182, 186, 282, 318
  line divider, 294
  materials
    crown glass, 319
    exotic glasses, 283
    flint glass, 61, 319
    germanium, 140, 185, 208, 258
    halogen salts, 224
    sapphire, 185
    Schott glass, 187
    semiconductors, 61
    silicon, 258
    soda glass, 61
    ZnSe, 185
  polarization, 67, 184, 316
  transmission loss, 191
optics
  basic, 41–46
  geometric, 41–43
  physical, 43
  real and virtual image, 42
  relay, 225
orbital
  accuracy, 24
  cycle, 18, 128, 178, 220, 249, 278, 302, 314
  lives, 21
  node, 19, 178, 280, 314
  parameters (Table), 28
orbits, 17–29
  atmospheric drag, 21, 196, 278
  AVHRR, 27
  geostationary, *see* orbits, geosynchronous
  geosynchronous, 17–18, 23, 137, 147, 157, 205
  inclination of, 21
  Molniya, 18
  near polar, 18
  sun-synchronous, 8, 18–27, 126–129, 178, 220, 249–251, 278–281, 302–305, 314–316

outgassing, effects of, 167, 213, 238, 265
output impedance, 105

PA band, *see* SPOT
PDUS (Primary Data User Station), 166
photosynthesis, 13
PI (precipitation index), 173
piezoceramic inchworms, 225
pixel, 43, 199, 291
  size, 129, 159, 181, 218, 278, 314
  registration, 120–122, 278, 320
  resolution, 82
  IFOV, *see* IFOV
  area imaged, 83, 130, 305
  elongation, 181, 281
point spread function, 43, 89
position sensing, 264
products, 171–173, 199, 215, 271–273, 296–298, 310, 328

radiation pressure, solar, 196, 197, 201, 278
radiometer, 108, 136, 156, 175, 203, 302
radiometric
  performance, 111–114, 129, 132, 167, 176, 191, 211, 236–239, 265, 294, 325
  quantization, 132, 244, 262, 290
RAL (Rutherford Appleton Laboratory), 248
Rayleigh atmosphere, 293
Rayleigh criterion, the, 82
red edge, 302
redundancy, *see* duplication
reflecting integrating cavity, 260
registration, 120–122, 130, 162, 171, 185, 206, 225, 228, 241, 286, 320, 325
  statistical evaluation, 121
relative information content, 244
relative spectral response, *see* RSR
remote sensing, 11–14, 30, 31, 37, 43, 54, 111, 236
resampling, 130, 226, 273
resolution, 44, 82–84, 257
RESTEC (Tokyo), 310
RSR, 115

sampling, 84, 96, 118, 129, 131, 181, 189, 209, 231, 262, 289
  rate/IFOV, 231
  integrate and dump, 231
  track and hold, 231
SAR (search and rescue), 195
satellite time (AVHRR), 200
scan line corrector, 222
scanning, 148
  2D step scan, 205
  per IFOV, 83
  pushbroom, 276, 284, 302, 305, 312
  raster, 129
  spin scan, 137, 157–159
  spun mirror, 178, 305
  step scan, 139, 148, 157
  swing entire satellite, 251
  swing main mirror, 157

symmetrical zigzag, 221
timing problems, 121
whiskbroom, 220
scenes, 18, 236, 225, 275, 297, 305, 316
SDUS (Secondary Data User Station), 166
sea surface temperature, *see* SST
seconds confirming, 257, 272
Seebeck effect, 94
shuttle, servicing by, 220
sidereal day, 22
signal/noise ratio, *see* SNR
sky tracks, 249
slitless spectrograph, 272
SMS program, 135
solar arrays, 326
solar cell, 94,
solar flux, 164
solar irradiance (Fig.), 36
solar panels, 197, 310
solar sail, 197, 202
solar X-ray sensor, 137
SOP (satellite observation period), 271
South Atlantic Anomaly, 254
spatial domain, 96, 120, 289
spatial frequency response, *see* frequency
    response and MTF
spatial frequency, 84–89
specific detectivity, *see* noise, D*
spectral response, 36, 114–117, 168, 193, 239,
    266, 294, 325
spherical aberration, 46
SPOT, 55, 88, 121, 275–298
    bus, 277, 295
    PA and XS bands, 43, 275, 278, 286
square wave response, *see* SWR
SST (sea surface temperature), 173
SSU, 195
stabilization, *see* attitude
star sensor, 257, 326
step response, 87
steradian, 113
stray radiation, *see* background
striping, 111
sun sensor, 144, 170, 269, 295, 326
swath, 17, 23, 53, 96, 220, 257, 275, 280, 281, 306
    split, 305
SWR, 88, *see also* MTF and frequency response

tape recorder, 190, 200, 262, 271, 309, 324, 290,
    291
TDRSS, 232
telescope, 41–53, 130, 139–140, 148, 160–162,
    206, 224–226, 254–255, 282–284, 305
    afocal, 206
    aplanatic, 160
    designs (Table), 48
    focus checks in orbit, 88, 162, 225, 295, 316
    reflective, 47–50
        afocal/confocal, 49, 181, 206
        Baker and Paul, 283
        cassegrain, 47
        central blockage, 44, 283

corrector plate, 49, 283
        multiple mirror, 49
        newtonian, 47
        Ritchey Chretien, 8, 49, 130, 139, 160, 224,
            254, 258, 283, 305
        Schmidt, 49, 283
    refractive, 50–53
        Galileo's, 46
        multiple, 51, 121, 305, 306, 316–320
temperature control, 145, 170–171, 197–198, 213,
    245, 270, 296, 310, 326
Thematic Mapper, 112, 115, 116, 216–247
thermal inertia, 14
TIROS-N weather satellite series, 175
toroidal mirror, 235
TOVS, 195
transfer function, 106
transistor, the, 99–103
    bipolar, 100–102
    IGFET, 102
    JFET, 102, 164, 187, 209, 228, 260
    gate, 102
    n- and p-channel, 102
    MOSFET, 102, 262
    MOST, 102
twin vertical viewing mode, 281

UV, 13
units
    noise power, 70, 77–78
    spectral, 55, 77
UTH (upper tropospheric humidity), 173

VAS (GOES), 136, 139 142
vernier thrusters, 144
VHRR
    INSAT, 205–211
    NOAA, 176
viewing angle, 32
VISSR
    GMS, 148–149
    GOES, 135
VTIR (MOS-1), 305–309

wave reflection, 58
wavebands, 36, 79–82
    far IR (25–120 fm), 252
    mid IR (3.7 fm), 73, 81–82, 114
    near (or SW) IR (1.6 fm), 177, 219, 278
    PA and XS, *see* SPOT
    reflective, 12, 33, 36, 79, 112, 126, 141, 149,
        156, 179, 208, 219, 279, 303, 315
    thermal, 12, 13, 33, 36, 38, 80–81, 113, 141,
        156, 179, 209, 219, 303
    water vapour (6–7 fm), 156, 167, 156, 303
wavelength/frequency, 37, 55, 58
wavenumber, 37
weeks confirmed test, 273
WEFAX service, 166

world weather watch, 147
WRS (World Reference System)
   IRS, 327
   MOS-1, 304
   SPOT, 281, 296

XS bands, *see* SPOT

zenith angle, 32
zenith point, 249
zenithal halo, 168